Nonlinear Dynamics in Economics, Finance and the Social Sciences

"Small text"
dynamic
agent-based
modeling
("complexity"
"big text"
dynamic compl
4 C's

Gian Italo Bischi • Carl Chiarella •
Laura Gardini
Editors

Nonlinear Dynamics in Economics, Finance and the Social Sciences

Essays in Honour of
John Barkley Rosser Jr

Springer

Editors
Professor Gian Italo Bischi
Department of Economics and
Quantitative Methods – DEMQ
Faculty of Economics
University of Urbino "Carlo Bo"
Via Saffi, 42
61029 Urbino
Italy
gian.bischi@uniurb.it

Professor Carl Chiarella
School of Finance and Economics
The University of Technology,
Sydney
Haymarket, NSW 2007
Australia
carl.chiarella@uts.edu.au

Professor Laura Gardini
Department of Economics and
Quantitative Methods – DEMQ
Faculty of Economics
University of Urbino "Carlo Bo"
Via Saffi, 42
61029 Urbino
Italy
gardini@econ.uniurb.it

ISBN 978-3-642-42453-3 ISBN 978-3-642-04023-8 (eBook)
DOI: 10.1007/978-3-642-04023-8
Springer Heidelberg Dordrecht London New York

Cover design: WMXDesign GmbH, Heidelberg, Germany

Printed on acid-free paper

Springer is part of Springer Science+Business Media (www.springer.com)

Foreword

This edited volume contains a collection of selected and refereed papers presented at the Fifth MDEF Modelli Dinamici in Economia e Finanza (Dynamic Models in Economics and Finance) International Workshop held at the Department of Economics and Quantitative Methods of the University of Urbino, Italy, on the 25th–27th of September, 2008.

It is true without doubt that scientific meetings derive their value not only from the scientific results presented during the formal sessions, but also and perhaps more importantly from the atmosphere created amongst the participants during the breaks and dinners that afford them the opportunity to meet again with old friends and make new ones. In that respect we are happy that the fifth MDEF meeting gave us the opportunity to pay tribute to **John Barkley Rosser, Jr**, an outstanding scientist and good friend of many of the participants, who celebrated his sixtieth birthday in 2008.

We shall not attempt to give a full list of Barkley's numerous scientific achievements here; we shall only mention his important contributions to various areas that fall within the focus of the Workshop, in particular, his work on the importance of nonlinearity and complexity in the economic sciences. With some of his earliest work on these topics appearing in the 1980s, it would be fair to say that Barkley has been one of the pioneers in the area. The first edition in 1991 of his book *From Catastrophe to Chaos: A General Theory of Economic Discontinuities* brought together and gave a perspective on many of the early contributions. This book, and its second edition in 2000, had a profound influence on many researchers who have contributed to nonlinearity and complexity in the economic and social sciences over the intervening two decades. As one would expect of a scholar who has very broad and eclectic interests, Barkley's influence extends beyond the area in which he originally made his name, and he has made contributions to areas such as the new traditional economy, econochemistry, the megacorpstate, economic inequality and the underground economy. No doubt because of his contributions across so many areas, Barkley assumed the role of Editor of the *Journal of Economic Behavior and Organization* in 2001. In an era when many top journals seem to be more and more closed to new or different ways of thinking, this journal remains open to new and innovative ideas in the economic and social sciences, and as such it provides the perfect foil for Barkley's very broad range of interests.

After completing a bachelor's degree with a major in economics and a minor in mathematics, Barkley completed his master's and doctoral degrees in economics, all at the University of Wisconsin-Madison. He has built his entire career with the Department of Economics at the James Madison University, which he joined in 1977 and where he has occupied various positions. In 1996, Barkley was appointed as the Kirby L. Cramer Jr. Professor of Business Administration, a position that he still occupies.

The fifth MDEF workshop was held in a period of severe global economic crisis, probably the worst since the 1930s. The crisis has not only brought about widespread influences on social and economic life in many countries, but also poses a challenge to what had developed as the mainstream economic consensus from about the mid-1970s. This challenge is highlighted by the fact that in order to manage the crisis, policy makers in the major economies have had to adopt policies that run counter to the principal tenets of that mainstream orthodoxy. It therefore seems very apposite against this backdrop to hold a workshop on issues devoted to complexity, nonlinearity and heterogeneity in economic science, and that the workshop should be dedicated to Barkley who has been at the forefront in criticising the mainstream orthodoxy and developing new ways of thinking about economic science.

The papers that appear in this volume deal with a number of different topical areas involving the application of concepts from the theory of nonlinear dynamical systems, from dynamic models that describe the interactions between economic activities and the environment, a topic that has been repeatedly stressed in many papers and books by Barkley, to the description of the wild dynamics of financial markets, both through deterministic as well as stochastic models. There is also a set of papers dealing with strategic interaction in economics and the social sciences by the use of the methods of game theory, as well as some contributions on markets with heterogeneous agents or models dealing with expectations and learning in economic systems, an issue that is currently topical in economics, finance and the social sciences. Some applications of deterministic dynamical systems to business cycles and labour markets are also presented, as well as dynamic oligopoly games and nonlinear evolutionary games for the description of social systems and the sustainable exploitation of natural resources. Such a broad spectrum of applications, as well as the various mathematical methods used to analyse the corresponding models, are intended to give some perspective on the different streams of the growing literature in this field. It is thereby our hope that this special volume will stimulate further collaborations amongst researchers from different fields, through a fruitful trade-off between theoretical issues and applications. We hope, furthermore, that this special volume will help the reader to gain an entrée into the main topics in nonlinear dynamics applied to economics, finance and the social sciences, as well as their recent advances.

We now give, for the convenience of the reader, a brief review of the twenty contributions, chosen after a careful selection and revision, in order to give a broad idea of the kind of dynamical models proposed, the mathematical methods used, and to show how they reflect some common themes and features.

In the first paper, Antoci and Borghesi propose a two dimensional evolutionary game in continuous time, to describe a perverse effect by which environmental degradation may induce agents to adopt self-protection strategies that generate negative externalities by further increasing environmental degradation. The second paper, by Antoci, Russu and Ticci, considers a model of a small open economy in order to mimic the situation of developing countries where economic agents differ not only with respect to income, but also with respect to their vulnerability to environmental depletion. Their model takes into account two main factors that have been partially neglected by the economic development literature: the environmental externalities of human activity and agent heterogeneity in terms of asset endowment and, consequently, in terms of income source and vulnerability to depletion of natural resources. In a similar vein, the paper by Antoci, Naimzada and Sodini proposes an overlapping generations model, a quite natural framework in which to represent problems of sustainable development (a typical intergenerational issue), to analyse possible feedback effects on environmental degradation, consumption and economic growth. Another important topic in environmental dynamics (on which Barkley Rosser has written many interesting papers) is the commercial exploitation of natural renewable resources, such as fisheries. This is also the topic of the paper by Gu and Lamantia, where a discrete-time dynamic model is proposed to model different harvesting policies of a single species with age structure, where the exploiters can compete or cooperate, so that they can try to maximize the profit of a coalition instead of the individual profit.

The paper by Marta Biancardi investigates the stability of international agreements for environmental protection in a dynamic model of emissions reduction where the countries involved in the agreement determine their abatement levels in a dynamic setting, given the dynamics of pollution stock and the strategies of the other countries. The problem is studied in a differential game setting. Also in a dynamic game framework, Giovanni Villani analyzes a model of R&D cooperation where strategic alliances that create synergies are considered, and additional information increases the probabilities of success of R&D projects, where firms are divided into leaders and followers and R&D investments are assumed to be characterized by positive network externalities that induce more benefits in case of reciprocal R&D success. Within the theme of leaders and followers and in the framework of dynamic oligopoly models, Tönu Puu makes an important contribution with his attempt to unify the Cournot and Stackelberg approach, where a Cournot duopolist can shift to Stackelberg leadership if too disappointed by current profits. Issues on strategy-switching dynamics are also analyzed from a general point of view, by Weihong Huang, who applies his results to explore the significance of adopting price-taking strategy in a quantity-competed oligopoly. R&D public expenditure and knowledge spillovers are considered in the paper by Commendatore, Kubin and Petraglia, who propose a dynamic capital model with publicly financed R&D activities under alternative assumptions on the intensity of knowledge spillovers, and obtain new results about global stability properties of boundary equilibria. The dynamics in non-binding procurement auctions, with boundedly rational bidders, are considered by Colucci, Doni and Valori, who study a procurement auction game where buyers rank different bids on the basis of both the prices submitted and the quality of each bidder, which is their private information. These authors assume that bidders have bounded rationality because they form expectations on market price rather than on the best price of competitors and also because they update expectations adaptively.

A general analysis of delay differential nonlinear economic models is provided by Matsumoto and Szidarovszky, who compare fixed and continuously distributed information lags and show that the two types of models generate identical local asymptotic behaviour when small delays with exponentially decreasing kernel functions are considered, whereas for large delays the asymptotic properties become quite different. They apply these general results to the business cycle models of Goodwin and Kaldor, augmented with a Kaleckian investment lag, and to a Cournot oligopoly model. The paper by Ferri and Variato studies the relationship between imperfect competition and economic fluctuations in a macro model with uncertainty. In such a model, imperfect knowledge economics suggests that the relationships between the agents and the environment become complex, while a learning process capable of generating endogenous dynamics takes place. Colombo and Weinrich analyse persistent disequilibrium dynamics in a theoretical dynamical model involving temporary equilibria with quantity rationing in each period and price adjustment between periods. The resulting dynamic system may present a variety of dynamic behaviours, ranging from the convergence to stationary or quasi-stationary states, to complex or even chaotic dynamics. Fabio Privileggi analyzes the transition dynamics in a continuous time endogenous growth framework in which knowledge evolves according to the Weitzman recombinant process, and finds a suitable transformation for the state and control variables in the dynamical system diverging to asymptotic constant growth, so that an equivalent 'detrended' system converging to a steady state in the long run can be tackled. An interesting application of continuous-time stochastic dynamic modelling with optimal control is provided by Longo and Mainini, who study a model of electoral competition, where elections serve as a device for selecting talented politicians, by using dynamic programming techniques. Saltari and Travaglini propose a behavioral approach to portfolio choice by adopting the theory of disappointment aversion to show how disappointment aversion affects the optimal portfolio choice when risk is small. Indeed, the standard portfolio model predicts a large equity position for most households, whereas empirical evidence shows however that household wealth is characterized by a small proportion of risky assets. To solve this paradox, the authors employ the axiomatic theory of disappointment aversion.

Finally, some applications of dynamic modelling to the description of financial markets complete the spectrum of new trends of nonlinear dynamics contained in this book. Of course, the development of new mathematical approaches to simulate and control the dynamic behaviour of financial markets is particularly apt at a time characterised by a general financial crisis. For example, the paper by Frank Westerhoff proposes an agent-based financial market model where agents following technical and fundamental trading rules to determine their speculative investment positions interact and may decide to change their trading behaviour. Despite its simple mathematical structure, this model is able to replicate some salient features of asset price dynamics. Along the same line, Tramontana, Gardini, Dieci and Westerhoff consider a three-dimensional nonlinear dynamic model of interacting stock and foreign exchange markets jointly driven by the speculative activity of heterogeneous investors, and give a global dynamic analysis by using analytical and numerical tools. A heterogeneous Capital Asset Pricing Model (CAPM) is proposed by Chiarella, Dieci and He, where agents are assumed to form optimal portfolios based upon their heterogeneous beliefs about conditional means and covariances of the risky asset returns, and in this framework they are able to obtain the exact relation between heterogeneous beliefs and the market equilibrium returns. The impact of the dynamics of interest rates of a Central Bank on the behaviour of commercial banks is analyzed in the paper by Casellina and Uberti, which takes into account the expectations of economic agents. Their model is calibrated by the VAR approach on Italian quarterly data from 1990 to 2007.

Before ending this foreword, we wish to thank the various academic colleagues around the world who have provided prompt and insightful referee reports on all the papers that were submitted to this special volume. Thanks are particularly due to Fabio Tramontana who greatly assisted the editors in putting the papers into the required Springer format. We would also like to express special thanks to Mrs Dr. Martina Bihn, the Springer Editorial Director, who facilitated the book's publication and carefully guided the entire editorial process. Finally, we thank all the participants at MDEF, whose efforts initiated a very interesting series of fruitful seminars, and who submitted so many interesting papers to us.

Urbino, Italy — Gian Italo Bischi
Sydney, Australia — Carl Chiarella
Urbino, Italy — Laura Gardini

Contents

Contributors

Angelo Antoci Dipartimento di Economia Impresa e Regolamentazione, Università di Sassari, Via Sardegna 58, 07100 Sassari, Italy, antoci@uniss.it

Marta Elena Biancardi Facoltà di Economia, Università di Foggia, Via IV Novembre 1, 71100 Foggia, Italy, m.biancardi@unifg.it

Simone Borghesi Faculty of Political Science, University of Siena, Via Mattioli 10, 53100 Siena, Italy, borghesi@unisi.it

Simone Casellina Banca d'Italia, Via Nazionale 90, Rome, Italy, simoneenrico.casellina@bancaditalia.it

Carl Chiarella School of Finance and Economics, University of Technology, Sydney, P.O Box 123, Broadway, NSW 2007, Australia, Carl.Chiarella@uts.edu.au

Luca Colombo Università Cattolica di Milano, Istituto di Economia e Finanza, Via Necchi 5, 20123 Milano, Italy, lcolombo@mi.unicatt.it

Domenico Colucci Facoltà di Economia, Università di Firenze, Via Lombroso 6/17, 50134 Firenze, Italy, domenico.colucci@dmd.unifi.it

Pasquale Commendatore Dipartimento di Teoria Economica e Applicazioni, Università di Napoli, Via Rodinò 22, 80138 Napoli, Italy, commenda@unina.it

Roberto Dieci Facoltà di Economia del Polo di Rimini, Università degli Studi di Bologna, Via D. Angherà 22, 47900 Rimini, Italy, roberto.dieci@unibo.it

Nicola Doni Dipartimento di Scienze Economiche, Via delle Pandette 9, 50127 Firenze, Italy, nicola.doni@unifi.it

Piero Ferri Dipartimento di Economia, Università di Bergamo, Via dei Caniana 2, 24127 Bergamo, Italy, pietro-enrico.ferri@unibg.it

Laura Gardini Dipartimento di Economia e Metodi Quantitativi, Università di Urbino "Carlo Bo", Via Saffi 42, 61029 Urbino, Italy, laura.gardini@uniurb.it

En-Guo Gu Department of Mathematics, South-Central University for Nationalities, Wuhan 430074, P.R. China, guenguo@163.com

Xue-Zhong (Tony) He School of Finance and Economics, University of Technology, Sydney, P.O Box 123 Broadway, NSW 2007, Australia, Tony.He-1@uts.edu.au

Weihong Huang Nanyang Technological University, Nanyang Avenue, Singapore 639798, Singapore, Awhhuang@ntu.edu.sg

Ingrid Kubin Vienna University of Economics and Business Administration, Augasse 2-6, 1090 Vienna, Austria, ingrid.kubin@wu.ac.at

Fabio Lamantia Department of Business Sciences, University of Calabria, Via P. Bucci 3C, Rende (CS), Italy, lamantia@unical.it

Michele Longo Università Cattolica di Milano, Largo Gemelli, 1, 20123 Milano, Italy, michele.longo@unicatt.it

Alessandra Mainini Università Cattolica di Milano, Largo Gemelli, 1, 20123 Milano, Italy, alessandra.mainini@unicatt.it

Akio Matsumoto Department of Economics, Chuo University, 742-1, Higashi-Nakano, Hachioji, Tokyo 192-0393, Japan, akiom@tamacc.chuo-u.ac.jp

Ahmad Naimzada Dipartimento di Metodi Quantitativi, Università di Milano–Bicocca, Piazza Ateneo Nuovo 1, 20126 Milano, Italy, ahmad.naimzada@unimib.it

Carmelo Petraglia Dipartimento di Economia, Via Cintia, Monte Sant'Angelo 80126, Università "Federico II" di Napoli, Napoli, Italy, carmelo.petraglia@unina.it

Fabio Privileggi Dip. di Politiche Pubbliche e Scelte Collettive Polis, Università del Piemonte Orientale, Via Cavour 84, 15100 Alessandria, Italy
and
IEL - Collegio Carlo Alberto, Via Real Collegio, 30 10024 Moncalieri, Torino, Italy, fabio.privileggi@sp.unipimn.it

Tonu Puu Department of Economics, Umea University, 90187 Umea, Sweden, Tonu.Puu@econ.umu.se

Paolo Russu Dipartimento di Economia Impresa e Regolamentazione, Università di Sassari, Via Sardegna 58, 07100 Sassari, Italy, russu@uniss.it

Enrico Saltari Department of Public Economics, University of Rome La Sapienza, Via del Castro Laurenziano 9, 00161 Roma, Italy, Enrico.Saltari@uniroma1.it

Mauro Sodini Dipartimento di Statistica e Matematica Applicata all'Economia, Università di Pisa, Via Ridolfi 10, 56124 Pisa, Italy, m.sodini@ec.unipi.it

Ferenc Szidarovszky Systems and Industrial Engineering Department, University of Arizona, Tucson, AZ 85721-0020, USA, szidar@sie.arizona.edu

Elisa Ticci Department of Economics, University of Florence, Via delle Pandette 9, 50127 Firenze, Italy, elisa.ticci@unifi.it

Fabio Tramontana Dipartimento di Economia, Università Politecnica delle Marche, Piazzale Martelli 8, 60121 Ancona, Italy, f.tramontana@univpm.it

Giuseppe Travaglini Dipartimento di Economia e Metodi Quantitativi, Università di Urbino "Carlo Bo", Via Saffi 42, 61029 Urbino, Italy, giuseppe.travaglini@uniurb.it

Mariacristina Uberti Dipartimento di Statistica e Matematica Applicata, Università degli Studi di Torino, Corso Unione Sovietica 218/bis, 10134 Torino, Italy, uberti@econ.unito.it

Vincenzo Valori Facoltà di Economia, Università di Firenze, Via Lombroso 6/17, 50134 Firenze, Italy, vincenzo.valori@dmd.unifi.it

Anna Maria Variato Dipartimento di Economia, Università di Bergamo, Via dei Caniana 2, 24127 Bergamo, Italy, anna.variato@unibg.it

Giovanni Villani Facoltà di Economia, Università di Foggia, Via IV Novembre 1, 71100 Foggia, Italy, g.villani@unifg.it

Gerd Weinrich Facoltà di Economia, Università Cattolica di Milano, Largo Gemelli 1, 20123 Milano, Italy, gerd.weinrich@unicatt.it

Frank Westerhoff Department of Economics, University of Bamberg, Feldkirchenstrasse 21, 96045 Bamberg, Germany, frank.westerhoff@uni-bamberg.de

Transferring Negative Externalities: Feedback Effects of Self-Protection Choices in a Two Hemispheres Model

Angelo Antoci and Simone Borghesi

1 Introduction

In this paper we analyze an economy in which self-protection choices made by economic agents to face environmental degradation generate environmental negative externalities on other agents. By self-protection choices we mean choices that agents may do to protect themselves against some form of social degradation (e.g. crimes, lack of leisure, depletion of social capital) or environmental degradation (air and water pollution, loss of biodiversity, growing scarcity of green areas, etc.). The notion of self-protection choices is not new in the literature. Hirsch (1976) was the first to introduce the concept of defensive consumption, that is, consumption induced by a growth in negative externalities. The notion originally proposed by Hirsch concerned a wider set of choices than those induced by environmental deterioration. The concept, however, has become particularly popular in the environmental literature where there is a major debate on how the Gross National Product as a measure of welfare should be corrected to take into account defensive expenditures and environmental depletion.[1] There exist many alternative classifications of environmental defensive expenditures generated by self-protection choices (e.g. Hueting, 1980; Leipert, 1989; Leipert and Simonis, 1989). Among them, a particularly interesting taxonomy is the one proposed by Bird (1987) and Shogren and Crocker (1991) who distinguish between environmental self-protection choices that generate negative externalities and those that produce positive externalities. The former choices *transfer* the environmental damage to other agents, while the latter *filter* it, transferring the *reduction* of the environmental damage to other agents. In what

[1] See, for instance, Aronsson et al. (1999), Kadekodi and Agarwal (2000), Vincent (2000) and the special issue (n. 5, 2000) devoted by the journal "Environment and Development Economics" to the research on this topic.

S. Borghesi (✉)
University of Siena, Faculty of Political Science via Mattioli 10 53100 Siena, Italy,
e-mail: borghesi@unisi.it

G.I. Bischi et al. (eds.), *Nonlinear Dynamics in Economics, Finance and the Social Sciences*, DOI 10.1007/978-3-642-04023-8_1,

follows we will focus on the former category since it appears to be the larger one (cf. Shogren and Crocker, 1991).

Everyday life provides many examples of environmental self-protection choices that may cause negative externalities. In most industrialized countries, for instance, people spend increasingly more on mineral water since tap water is often undrinkable. This increasingly frequent consumption habit, however, ends up raising production of plastic bottles and the correspondent recycling costs for the community as a whole. Similarly, many beaches have become progressively more dirty in many large urban centers, so that individuals may prefer to buy an expensive holiday in some uncontaminated resort rather than go to the open access, polluted beach near home. Also this self-protection choice, however, may generate negative externalities as it tends to raise the use of cars, ships and airplanes to reach the holiday resort (depending on its distance from home) and thus also the atmospheric and water pollutants released by these means of transport. Air conditioners provide another example of environmental self-protection choice that tend to damage the other agents. Keeping the inner temperature stable inside buildings and cars, air conditioners manage to protect the person who buys them from the observed climate change, but cause the others to suffer from the emission of hot air produced by air conditioners themselves.

The "textbook" examples of environmental self-protection choices provided so far are far from exhaustive. As argued by the literature on this issue (see, e.g. Antoci, 2009; Antoci and Bartolini, 1999, 2004; Antoci et al., 2005, 2008; Bartolini and Bonatti, 2002, 2003; Hueting, 1980), individuals may react to environmental deterioration in many different ways that generate many kinds of self-protection choices. What all these different kinds of choices have in common is that, generally speaking, when the environment deteriorates, individuals are more incentivated to adopt consumption patterns based on the use of private goods rather than of free access environmental goods. Thus, for instance, spending a day on an uncontaminated beach close to home (the free access environmental good) can be more rewarding (and generally requires the consumption of a lower quantity of private goods) than spending a day in town – where the opportunities for spending one's free time without sustaining expenses are rare; nevertheless, the latter option becomes relatively more remunerative if the quality of the beach is compromised.

In order to analyse the economic consequences of the self-protection choices, Shogren and Crocker (1991) have proposed a static model that focuses on symmetric Nash equilibria, where all agents make the same choice. Differently from their approach, in this paper we analyze a dynamic evolutionary model that allows for heterogeneity of choices; furthermore, our approach underlines the strong path-dependence nature of the diffusion process of the choices. Other works in the literature have studied the diffusion of self-protection choices in a dynamic context (see, e.g. Antoci, 2009; Antoci et al., 2005; Bartolini and Bonatti, 2002, 2003). These works, however, focus on the implications of self-protection choices for economic growth and capital accumulation. In our work we study the diffusion process in a simpler analytical context which allows us to address the problem of the possible feedback effects due to the interaction between economic agents belonging to two hemispheres (the North and the South). Think, for instance, of sea pollution in the North. The increasing degradation of many Northern beaches may induce agents in the North to self-protect by purchasing an expensive holiday in a Southern country where beaches are still relatively clean. However, if the number of Northern agents that go on holiday to the South is relatively high, this may cause the exploitation of natural resources in the South, generating an increasing interdependence between the environmental quality of the two hemispheres. Similarly, the increasing production of goods for self-protection purposes in one hemisphere tends to enhance global pollutants like carbon dioxide that end up damaging the environmental quality in the other hemisphere as well. In this context we show that there may exist multiple Nash equilibria (corresponding to attracting fixed points of the dynamics) which can be ordered in the sense of Pareto. Furthermore, we show that if the North transfer negative externalities to the South, this can give rise to a feedback mechanism that may end up decreasing the Northern individuals' welfare.

The paper has the following structure. Section 2 introduces the model. Section 3 provides the basic mathematical results. Section 4 examines the well-being in the two hemispheres. Section 5 investigates the effects on well-being of transferring the environmental impact of Northern production to the South. Finally, Sect. 6 concludes.

2 The Model

There are two hemispheres: North (N) and South (S). There are two populations of economic agents: the population of the North (N-agents) and the population of the South (S-agents).

Time is continuous. At every moment of time t, j-agents ($j = N, S$) have to choose between two strategies:

1. **Strategy P:** Agents adopting this strategy choose to self-protect against environmental deterioration.
2. **Strategy NP:** Agents adopting this strategy choose no self-protection device against environmental deterioration.

Let us indicate with $x \in [0, 1]$ the share of agents that choose strategy P in the North at time t (consequently, $1 - x$ is the share of agents that choose strategy NP).

Similarly, we indicate with $z \in [0, 1]$ the share of agents that choose P in the South at time t

Following Bird (1987) and Shogren and Crocker (1991) we assume that, in both hemispheres, agents' payoffs at time t depend negatively on the share x_t and z_t of agents choosing P at time t. Under this assumption, self-protection choices generate negative externalities that reduce the payoffs of strategies P and NP, both in the South and in the North. For the sake of simplicity, we assume the following linear payoff functions:

$$\Pi_P^N(x,z) = a_1 - b_1x - c_1z, \qquad \Pi_{NP}^N(x,z) = a_2 - b_2x - c_2z,$$
$$\Pi_P^S(x,z) = d_1 - e_1x - f_1z, \qquad \Pi_{NP}^S(x,z) = d_2 - e_2x - f_2z,$$

where Π_P^N and Π_P^S (Π_{NP}^N and Π_{NP}^S) are the payoffs of strategy P (strategy NP) for N-agents and S-agents, respectively, and b_k, c_k, e_k, f_k, $k = 1,2$, are strictly positive parameters.

For the sake of simplicity, we assume that the dynamics of x and z are given by the so-called "replicator dynamics" (see, e.g. Weibull, 1995):

$$\begin{cases} \dot{x} = x(1-x)\left(\Pi_P^N - \Pi_{NP}^N\right) = x(1-x)\left(a - bx - cz\right), \\ \dot{z} = z(1-z)\left(\Pi_P^S - \Pi_{NP}^S\right) = z(1-z)\left(d - ex - fz\right), \end{cases} \tag{1}$$

where $\dot{x}$ and $\dot{z}$ indicate the time derivative of the variables x and z, and $a{:=}a_1{-}a_2$, $b := b_1 - b_2$ and so on. We assume that $b, c, e, f < 0$; that is, the negative effect of an increase in x and z on the agents choosing P is lower than the negative effect on the agents adopting NP. In other words, individuals choosing strategy P are more protected against the negative externalities produced by the diffusion of self-protective behaviour in the economy.

Dynamics (1) describes an adaptive process based on an imitation mechanism: each period, part of the population changes its strategy, adopting the more remunerative one. Differently from the typical context in which replicator dynamics are introduced, that is, random pairwise matching between economic agents, in the present model the payoff of an individual adopting a given strategy at time t depends on the strategies that *all* individuals are choosing at that same moment. Replicator dynamics may be generated by several learning mechanisms in a random matching context (see, e.g. Borgers and Sarin, 1997 and Schlag, 1998); however, rationales for such dynamics can also be found in our context (see, for instance, Sethi and Somanathan, 1996, for an application of replicator equations in a context similar to ours).

3 Basic Mathematical Results

3.1 Fixed Points

The dynamic system (1) is defined in the square Q:

$$Q = \{(x,z) : 0 \le x \le 1, 0 \le z \le 1\}.$$

In what follows we will denote with $Q_{x=0}$ the side of Q where $x = 0$, with $Q_{x=1}$ the side where $x = 1$. Similar interpretations apply to $Q_{z=0}$ and $Q_{z=1}$. All sides of this square are invariant; namely, if the pair (x,z) initially lies on one of the sides, then the whole correspondent trajectory also lies on that side.

Note that the states $\{(x, z) = (0, 0), (0, 1), (1, 0), (1, 1)\}$ are always fixed points of the dynamic system (1). In such states, only one strategy is played in each hemisphere.

Other fixed points are the points of intersection between the interior of the sides $Q_{x=0}$, $Q_{x=1}$(where it holds $\dot{x} = 0$) and the straight line $d - ex - fz = 0$ (where $\dot{z} = 0$):

$$(x, z) = \left(0, \frac{d}{f}\right), \left(1, \frac{d-e}{f}\right), \text{ that exist if } \frac{d}{f} \in (0, 1) \text{ and } \frac{d-e}{f} \in (0, 1)$$

and the points of intersection between the interior of sides $Q_{z=0}$, $Q_{z=1}$(where it holds $\dot{z} = 0$) and the straight line $a - bx - cz = 0$ (where $\dot{x} = 0$):

$$(x, z) = \left(\frac{a}{b}, 0\right), \left(\frac{a-c}{b}, 1\right), \text{ that exist if } \frac{a}{b} \in (0, 1) \text{ and } \frac{a-c}{b} \in (0, 1).$$

In such fixed points, there is an hemisphere where both available strategies are played while in the other all agents choose the same strategy. The remaining possible fixed points are those in the interior of Q at the intersection between the two lines $a - bx - cz = 0$ and $d - ex - fz = 0$; so, generically, the interior fixed point is at most one and has the coordinates:[2]

$$\begin{aligned}(x, z) &= \left(\frac{af - cd}{bf - ce}, \frac{bd - ae}{bf - ce}\right), \text{ that exists if } \frac{af - cd}{bf - ce} \in (0, 1) \\ &\text{and } \frac{bd - ae}{bf - ce} \in (0, 1).\end{aligned}$$

In such state, all strategies coexist. Notice that the system (1) generically admits at most nine fixed points (one in the interior of Q, one in the interior of each side of Q and the four vertices of Q).

3.2 *Stability of the Fixed Points*

The Jacobian matrix $J(x, z)$ of system (1) is

$$\begin{bmatrix} (1 - 2x)(a - bx - cz) - bx(1 - x) & -cx(1 - x) \\ -ez(1 - z) & (1 - 2z)(d - ex - fz) - fz(1 - z) \end{bmatrix}.$$

So, as it can be easily verified, the following proposition applies.

[2] This result does not hold only if the two lines $\dot{x} = 0$ and $\dot{z} = 0$ are coincident, in which case there exists an infinite number of fixed points.

Proposition 1. *The eigenvalue of* $(0,0)$ *in direction of* $Q_{z=0}$ *is equal to* a *and the eigenvalue in direction of* $Q_{x=0}$ *is equal to* d.

The eigenvalue of $(0,1)$ *in direction of* $Q_{z=1}$ *is equal to* $a-c$ *and the eigenvalue in direction of* $Q_{x=0}$ *is equal to* $f-d$.

The eigenvalue of $(1,0)$ *in direction of* $Q_{z=0}$ *is equal to* $b-a$ *and the eigenvalue in direction of* $Q_{x=1}$ *is equal to* $d-e$.

The eigenvalue of $(1,1)$ *in direction of* $Q_{z=1}$ *is equal to* $b+c-a$ *and the eigenvalue in direction of* $Q_{x=1}$ *is equal to* $e+f-d$.

Proposition 2. *The eigenvalues of the fixed points* $(i,\overline{z})$ *in the interior of the edges* $Q_{x=i}$ $(i=0,1)$ *are equal to* $-f\overline{z}(1-\overline{z})>0$ *in direction of* $Q_{x=i}$ *and* $(1-2i)(a-bi-c\overline{z})$ *in direction of the interior of* Q.

The eigenvalues of the fixed points $(\overline{x},i)$ *in the interior of the edges* $Q_{z=i}$ $(i=0,1)$ *are equal to* $-b\overline{x}(1-\overline{x})>0$ *in direction of* $Q_{z=i}$ *and* $(1-2i)(d-e\overline{x}-fi)$ *in direction of the interior of* Q.

Note that, given a fixed point in the edge $Q_{j=i}$, $j=x,z$ and $i=0,1$, the sign of its eigenvalue in direction of $Q_{j=i}$ is always strictly positive; therefore, all the (hyperbolic) fixed points in the interior of the edges are saddles or sources.

The following proposition concerns the stability of the fixed point $(\overline{x},\overline{z})$ in the interior of the square Q, where both choices coexist in the North and in the South.

Proposition 3. *The fixed point* $(\overline{x},\overline{z})$ *is a saddle if* $bf-ce<0$ *and a source if* $bf-ce>0$.

The following subsection highlights the basic features characterizing the dynamics of system (1).

3.3 Analysis of the Dynamics

Let us first observe that both $a-bx-cz=0$ (where $\dot{x}=0$) and $d-ex-fz=0$ (where $\dot{z}=0$) have negative slope; furthermore, above (below) $a-bx-cz=0$ it holds $\dot{x}>0$ (respectively, $\dot{x}<0$) and above (below) $d-ex-fz=0$ it holds $\dot{z}>0$ (respectively, $\dot{z}<0$). The signs of $\dot{x}$ and $\dot{z}$ derive from the fact that the adoption process of P has a self-enforcing character: the higher is the share of individuals choosing to adopt P in both hemispheres, the higher is the incentive to adopt P. This is due to the fact that self-protection choices damage relatively more the agents who do not protect, therefore they will be induced to change their strategy and "imitate" what others do in the population. Consider, for instance, air conditioners as a self-protection device against global warming. As pointed out above, their use helps cooling the interior of homes and offices, but give off heat to the exterior, further contributing to raise the external temperature. Therefore, if an increasing number of people use air conditioners, then those who do not use them end up suffering even more from the consequent increase in the temperature, forcing them to modify their choice while "reinforcing" the others' decision to use air conditioners.

The following proposition characterizes dynamics (1).

Proposition 4. *The system (1) has the following features:*
(a) Every trajectory of the system approaches a fixed point.
(b) Only the fixed points $(0,0)$, $(0,1)$, $(1,0)$, $(1,1)$ *can be attractive.*

Proof. The proof of point (b) is straightforward and follows immediately from local stability analysis in the preceding section. To prove point (a) we have to show that limit cycles cannot exist. This is obviously the case when the interior fixed point $(\overline{x},\overline{z})$, $0 < \overline{x},\overline{z} < 1$, does not exist or, if existing, it is a saddle point. If $(\overline{x},\overline{z})$ is a source, then it is easy to see that the regions of Q where $\dot{x}$ and $\dot{z}$ have the same sign are positively invariant; this implies that no oscillatory behavior of trajectories may occur.

Remark. Notice that the vertices of Q can be simultaneously attractive; in particular, this is the case when the following conditions hold:

$$b < a < c, \tag{2}$$

$$f < d < e. \tag{3}$$

When conditions (2) and (3) hold, the fixed point in the interior of Q is a source and those in the interior of the edges of Q are saddles. This case is shown in Fig. 1 in which attractive fixed points are represented by full dots (•), repulsive fixed points by open dots (∘). Notice that almost every trajectory approaches a vertex of Q, where each hemisphere ends up choosing a unique strategy (either adopting P or NP).[3] The basins of attraction of the vertices are delimited by the stable manifolds of the saddle points in the interior of the sides of Q.

4 Well-Being Analysis

In this section we will examine the average well-being level in the two hemispheres. The average well-being level in the North and in the South is equal to, respectively:

$$\overline{\Pi}^N(x,z) := x \cdot \Pi_P^N(x,z) + (1-x) \cdot \Pi_{NP}^N(x,z),$$

$$\overline{\Pi}^S(x,z) := z \cdot \Pi_P^S(x,z) + (1-z) \cdot \Pi_{NP}^S(x,z).$$

The following proposition applies.

[3] We do not converge to a vertex in a zero measure set, given by the fixed point inside Q and the stable manifolds of the saddle points along the sides of Q.

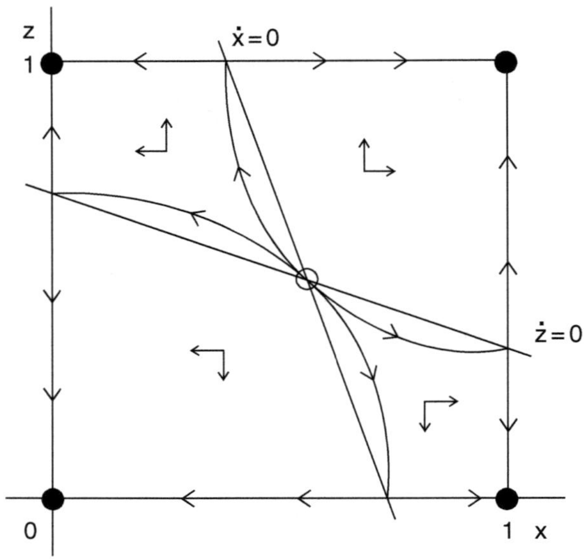

Fig. 1 Dynamic regime with simultaneously attracting vertices

Proposition 5. *If the fixed point* $(0,0)$ *is a sink, then it Pareto-dominates all the other attracting fixed points of system (1), that is* $\overline{\Pi}^N(0,0) > \overline{\Pi}^N(x,z)$ *and* $\overline{\Pi}^S(0,0) > \overline{\Pi}^S(x,z)$ *for* $(x,z) = (0,1), (1,0), (1,1)$.

Proof. Observe that $\overline{\Pi}^N(0,z) = \Pi^N_{NP}(0,z)$ and $\overline{\Pi}^N(1,z) = \Pi^N_P(1,z)$ are the Northern average well-being levels when every N-agent adopts NP and every N-agent adopts P, respectively. Similarly, for the South we have $\overline{\Pi}^S(x,0) = \Pi^S_{NP}(x,0)$ and $\overline{\Pi}^S(x,1) = \Pi^S_P(x,1)$.

Let us first consider the average payoff of the North. Notice that: $\overline{\Pi}^N(0,0) = a_2$, $\overline{\Pi}^N(0,1) = a_2 - c_2$, $\overline{\Pi}^N(1,0) = a_1 - b_1$, $\overline{\Pi}^N(1,1) = a_1 - b_1 - c_1$. Therefore, it is always: $\overline{\Pi}^N(0,0) > \overline{\Pi}^N(0,1)$ and $\overline{\Pi}^N(1,0) > \overline{\Pi}^N(1,1)$. Furthermore, $\overline{\Pi}^N(0,0) > \overline{\Pi}^N(1,0)$ if $a_2 > a_1 - b_1$; such condition is satisfied when $(0,0)$ is a sink. The proof of the part of the proposition concerning well-being in the South is identical.

Remark 1. From the well-being analysis above it follows that, in the case represented in Fig. 1, each hemisphere achieves its highest well-being level in $(0,0)$. Only one of the four possible vertex selected by the dynamics implies, therefore, the maximum well-being level. Notice that the lowest well-being level is achieved in $(1,1)$, while intermediate levels are reached in $(0,1)$ and $(1,0)$.

It is easy to check that if $(0,0)$ does not Pareto-dominate all the other fixed points (in the North and in the South), then the dynamics (1) is trivial, i.e. $\dot{x}$ and $\dot{z}$ are always positive in Q. In such case, the fixed point $(1, 1)$ is globally attracting and Pareto-dominates any other possible state (x, z) in the North and in the South.

5 Comparative Dynamics Analysis: Transferring Negative Externalities to the South

One of the most debated issues in the environmental literature concerns the possibility that introducing stricter environmental policies in the North may lead Northern industries to move polluting productions to the South where ecological regulations are less severe. To what extent such mechanism, known as environmental dumping, takes place in reality is still a matter of investigation in the empirical literature. In the case of worldwide problems like global warming, however, shifting polluting productions to the South may generate negative feedback effects on the North that counterbalance Northern ecological policies. The effects of environmental dumping can be examined by simple comparative dynamics analysis.

In our context, dumping produces an increase in the parameters e_1 and e_2. This does not necessarily modify the relative performance of the strategies P and NP for S-agents (this is the case if the value of the difference $e = e_1 - e_2$ remains constant). However, it is reasonable to think that the increase in e_2 is higher than that in e_1 since S-agents choosing P are more protected against the growth of negative externalities generated by the environmental dumping. This produce a reduction of e, so that the straight line $d - ex - fz = 0$ (where $\dot{z} = 0$) will move downwards. If so, in the South the strategy P becomes better performing than strategy NP as the Northern self-protection activities increasingly transfer negative externalities to the South.

The possible consequences of moving polluting productions to the South can be exemplified by looking at the dynamics showed in Fig. 1. Recall that in this case the fixed point $(0, 0)$ Pareto-dominates all other vertex, while $(1, 1)$ is Pareto-dominated by all of them; furthermore, all vertices are locally attractive. A reduction in e-ceteris paribus- may cause the instability of the fixed points in the edge $Q_{z=0}$ (see Proposition 1) giving rise to the dynamic regime represented in Fig. 2. If so, the fixed point $(0, 0)$ with the highest well-being level is no longer attractive, while that with the lowest well-being level $(1, 1)$ is still attractive. Transferring the negative externalities generated by Northern agents to the South, therefore, might end up decreasing well-being in both hemispheres.

6 Conclusions

Nowadays an increasing number of people make self-protection choices to protect against the deterioration of the environment they live in. This phenomenon is becoming more and more frequent in modern industrialized economies. This observation has recently induced some studies to examine the relationship between

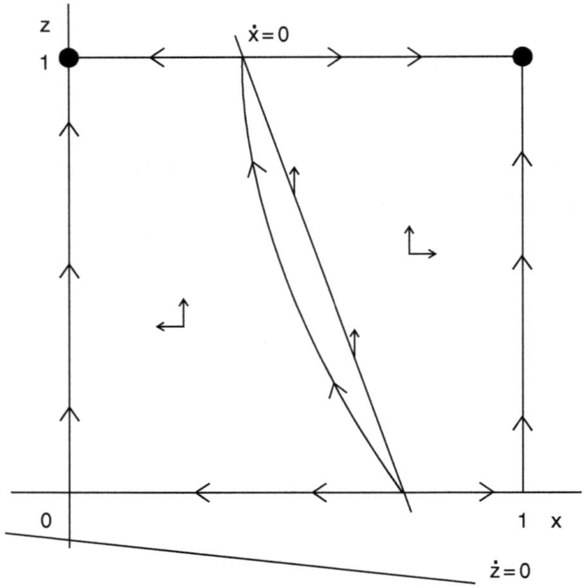

Fig. 2 Dynamic regime with environmental dumping

environmental defensive expenditures and well-being. The basic idea underlying these works is that negative externalities could contribute to a self-enforcing diffusion process of self-protection choices that further increase environmental degradation and reduce individuals' well-being.

The present paper builds on this literature by extending the analysis from a single population to a North–South context. The aim of the paper is to investigate the possible feedback effects that environmental defensive expenditures may generate between the two hemispheres and their impact on welfare in rich and poor countries. We show that the adoption dynamics of self-protection choices are characterized by "imitation" effects, that is all agents in each hemisphere end up choosing the same strategy, thus leading to symmetric Nash equilibria. Furthermore, the diffusion of self-protection choices may give rise to undesirable (i.e. well-being-reducing) increase in self-protection levels. Both hemispheres, in fact, may end up in a situation where populations self-protect "too much": people choose to protect themselves against pollution, but they might be better-off by investing less in self-protection and enjoying a cleaner world.

Finally, we show that transferring the environmental impact of Northern individuals to the South (e.g. transferring polluting activities, production waste, exploitation of natural resources) may end up decreasing welfare in both hemispheres.

We are fully aware that the results emerging from this work may appear provocative, but we believe that they could contribute to shed light on some aspects of the self-protection activities that have generally been neglected in literature.

Acknowledgements The authors would like to thank D. Bromley, R. Lopez, C. Mason, C. Perrings, A. Vercelli and seminar participants at the Workshop MDEF 2008 (Urbino, 25–27/9/2008) for helpful comments and suggestions on a preliminary version of this paper. The usual caveats apply.

References

Antoci, A. (2009). Environmental degradation as engine of undesirable economic growth via self-protection consumption choices. *Ecological Economics, 68*, 1385–1397.

Antoci, A., & Bartolini, S. (1999). *Negative externalities as the engine of growth in an evolutionary context* (Working paper 83.99). Milan: FEEM.

Antoci, A., & Bartolini, S. (2004). Negative externalities and labor input in an evolutionary game. *Environment and Development Economics, 9*, 1–22.

Antoci, A., Galeotti, M., & Rossu, P. (2005). Consumption of private goods as substitutes for environmental goods in an economic growth model. *Nonlinear Anal-Model Control, 10*, 3–34.

Antoci, A., Borghesi, S., & Galeotti, M. (2008). Should we replace the environment? Limits of economic growth in the presence of self-protective choices. *International Journal of Social Economics, 35*, 283–297.

Aronsson, T., Johansson, P. O., & Lofgren, K. G. (1999). On the proper treatment of defensive expenditures. In M. Boman, R. Brannlund, & B. Kriström (Eds.), *Topics in environmental economics*. Dordrecht: Kluwer.

Bartolini, S., & Bonatti, L. (2002). Environmental and social degradation as the engine of economic growth. *Ecological Economics, 43*, 1–16.

Bartolini, S., & Bonatti, L. (2003). Undesirable growth in a model with capital accumulation and environmental assets. *Environment and Development Economics, 8*, 11–30.

Bird, J. (1987). The transferability and depletability of externalities. *Journal of Environmental Economics and Management, 14*, 54–57.

Borgers, T., & Sarin, R. (1997). Learning through reinforcement and replicator dynamic. *Journal of Economic Theory, 77*, 1–14.

Hirsch, F. (1976). *Social limits to growth*. Cambridge: Harvard University Press.

Hueting, R. (1980). *New scarcity and economic growth. More welfare through less production?* Amsterdam: North Holland.

Kadekodi, G. K., & Agarwal, M. M. (2000). Towards integrated environmental accounting: Some possibilities in the Indian context. *Pacific and Asian Journal of Energy, 10*, 211–26.

Leipert, C. (1989). National income and economic growth: The conceptual side of defensive expenditures. *Journal of Economic Issues, 23*, 843–856.

Leipert, C., & Simonis, U. E. (1989). Environmental protection expenditures: The German example. *Rivista Internazionale di Scienze Economiche e Commerciali, 36*, 255–270.

Schlag, K. (1998). Why imitate, and if so, how? A boundedly rational approach to multi-armed bandits. *Journal of Economic Theory, 78*, 130–156.

Sethi, R., & Somanathan, E. (1996). The evolution of social norms in common property resource use. *American Economic Review, 59*, 766–788.

Shogren, J. F., & Crocker, T. D. (1991). Cooperative and non-cooperative protection against transferable and filterable externalities. *Environmental and Resource Economics, 1*, 195–213.

Vincent, J. R. (2000). Green accounting: from theory to practice. *Environment and Development Economics, 5*, 13–24.

Weibull, J. W. (1995). *Evolutionary game theory*. Cambridge: MIT Press.

Structural Change, Economic Growth and Environmental Dynamics with Heterogeneous Agents

Angelo Antoci, Paolo Russu, and Elisa Ticci

1 Introduction

In many developing countries the asset distribution is highly concentrated and the economic agents differ not only by income, but also by their vulnerability to environmental depletion. The poor, especially in rural areas, tend to be more dependent on natural resources and more vulnerable to ecosystem degradation. Three quarters of the poor live in rural areas and more than half of the rural poor depend on breeding and agricultural activities: cultivation of staple food is the main source of calories, income and job for the rural poor (IFAD 2001). Moreover, it is commonly recognized that the rural poor in developing countries significantly rely on the common pool resources of the community they live in (Dasgupta (2001)), while according to World Resources Institute (2005) estimates, around 1 billion of the world poor rely in some way on forests (indigenous people wholly dependent on forests, smallholders who grow farm trees or manage remnant forests for subsistence and income). A meta-analysis of 54 case studies in developing countries found that the poor tend to be more dependent on forest environmental income than better-off households (Vedeld et al. 2004). Natural assets and common or free access resources provide the poor with other additional services: regulating production services such as flood, drought and erosion mitigation, soil renewal, soil fertility or the provision of food, fuelwood and energy and fresh water. Microeconomic studies confirm the relevance of the dependence of the rural population on the community or free access resources (Beck and Nesmith 2001; Cavendish 2000; Falconer 1990; Fisher 2004; Jodha 1986; Narain et al. 2005). On the other hand, the rich have a greater ability to substitute private goods for environmental goods. They are thus able to protect themselves from pollution and to face the depletion of natural capital (United Nations Environment Programme 2004).

Against this background, we analyze a model that considers an economy with two sectors: a traditional resource-based sector that relies on self-employment of

A. Antoci (✉)
Dipartimento di Economia Impresa e Regolamentazione, Università di Sassari, Via Sardegna 58, 07100, Sassari,
e-mail: antoci@uniss.it

G.I. Bischi et al. (eds.), *Nonlinear Dynamics in Economics, Finance and the Social Sciences*, DOI 10.1007/978-3-642-04023-8_2,

poor households and a sector managed by the rich. Physical capital is completely concentrated in the endowments of the rich, while all agents -poor and rich- have access to environmental capital. The polarization of society into two sectors and two classes of agents is clearly an oversimplification, but this assumption makes the model tractable using standard methodology. Moreover, although we consider a highly stylized context, it reflects the ways in which different assets (natural, physical, social, human capital) are typically distributed in several developing countries. Physical capital tends to have a concentrated dispersion across the population because of financial market failures. In absence of perfect information and competition, wealthier individuals and large firms have privileged access to capital market, because they are more endowed with collateral and have a higher ability to exploit scale economies. Conversely, services deriving from environmental resources may be more dispersed and tend to have the characteristics of public goods (in our model all agents have access to environmental capital). In this context, economic agents also differ by feed-back mechanisms and interaction between their production (consumption) choices and environmental dynamics.

In this setting, we show that economic dynamics are path dependent in that the model admits a multiplicity of stable steady states. Furthermore, the model may exhibit a zero-sum game structure. Physical capital endowments allow the Rich to employ wage labor and this possibility is the root of the difference between the rich (labor employers) and the poor (labor force providers) in terms of vulnerability to environmental degradation. The rich are more able to defend themselves from environmental degradation because they can partially substitute natural capital with physical capital or wage labor employment. Thus, the rich may be not disadvantaged by the environmental degradation because they can rely on substitution possibilities as a defensive strategy. To the contrary, they may benefit from the role played by the natural capital scarcity in accelerating labor movement from the traditional to the modern sector. This, in turn, generates incentives to physical capital accumulation. On the other hand, the poor are disadvantaged because they face a reduction in productivity of their labor, namely, in their greatest means of subsistence.

In the history of the development theory, structural change, i.e. the movement of a labor force from the traditional resource-based to the modern sector, is regarded by some economists as a cause and consequence of economic development and growth (see e.g. Lewis 1955; Lucas 2004; Ranis and Fei 1961): growth of the non-resource sectors may permit an unending process of labor productivity growth because they rely on assets (human capital and physical capital) that can expand over time. Saving and investment in physical capital can produce an increase in labor productivity leading to economic expansion. In a dual framework, such vision implies that capital intensive activities are able to sustain a process of economic growth, while the production of the subsistence sector is constrained and cannot overcome a certain threshold because it relies on limited production factors. Therefore a labor shift towards the "modern" sector leads to a structural change associated with an increase of social welfare. Conversely, in our model, structural changes may be "perverse" in the sense of López (2003, 2007), i.e. associated with growing problems of poverty and environmental degradation. Pressures on natural resources can cause a decline in productivity of traditional agricultural activities and the consequent reduction of labor opportunity costs fuels a labor migration from the agricultural sector. The result is a movement of the labor force from the traditional resource-based to the modern sector associated with declining or stagnant wages and with a loss of welfare for labor force.[1]

The remainder of the article is organized as follows. Sections 2 and 3 present the model. Section 4 analyzes the model and investigates some possible dynamics that may emerge and their implications in terms of welfare. Section 5 draws conclusions. A mathematical appendix concludes the paper.

2 The Dual Context

We consider a small open economy[2] with three production factors: labor, a free access renewable natural resource (E) and physical capital (K). In this economy, agents belong to two different populations: the "Rich" (R-agents) and the "Poor" (P-agents). The R-agents accumulate physical capital, hire the labor force and employ all their potential work - represented by a fixed amount of entrepreneurial activity - to produce a storable private good. We call their production "capitalistic sector" or "modern sector". The P-agents are endowed only with labor and they have to choose the distribution of their labor between two activities: working as employees of the Rich in the capitalistic sector or directly exploiting natural resources to produce a non storable good. Let "subsistence sector" or "traditional sector" denote production of the Poor. Given that the Poor cannot invest and accumulate physical capital, we assume that the capital market is completely segmented and is accessible only by the Rich.

The population of the Poor is constituted by a continuum of identical individuals and the size of the population is represented by the positive parameter $\overline{N}$. The P-population's welfare depends on two goods:

1. A non storable good deriving directly from free access renewable natural resources, hereafter referred to as an environmental good.

[1] López points out that indirect factors capable of triggering a perverse structural change are inadequate policies aimed at fostering productivity in the modern sector in addition to a complete neglect of the traditional subsistence sector of the rural poor.

[2] The majority of developing countries are little open economies. In the last two decades, several countries have undertaken trade liberalization reforms and, consequently, the importance of the domestic demand in sustaining economic growth has diminished (at least for trade sectors) because economies are less constrained by limited national demand. To the contrary in open economies, a fundamental factor for economic growth is productive competitiveness that depends on, among other important factors, labour cost. In this sense, Matsuyama's model (Matsuyama 1992) is particularly explicative because it shows how the growth process might be driven by different factors in an open and a closed context: he finds a negative relationship between agricultural productivity and economic growth in open economies, while detecting the inverse links in closed economies.

2. A good (hereafter denoted private good) which can be consumed as a substitute for the services coming from the environmental good.

We assume that the instantaneous utility function of each P-agent is the following

$$U_P(c_P, c_S) = \ln(c_P + ac_S), \tag{1}$$

where:

c_S is the consumption of the produced good as a substitute for the environmental good.
c_P is the consumption deriving from the exploitation of the environmental resource.

According to (1), c_S and c_P are perfect substitutes, with a (constant) rate of substitution equal to $a > 0$. That is, the private good produced by the Rich is able to substitute completely c_P. This is a stylized fact, but it can represent the main components of poor people's welfare: if they work in the subsistence sector in rural areas (fishing, forestry, agriculture or breeding) their living standard strictly depends on their access to and exploitation of E; while if they move to urban zones or they become a wage labor force, they satisfy their needs mainly through the consumption of private goods.

Each P-agent, in each instant of time, employs all his potential labor (that we normalize to unity) in the subsistence sector or in the sector of the Rich. Thus, he cannot rely on alternative income sources at the same time. However, in the absence of inter-sectorial moving costs, significant divergences from the case with employment diversification are not a priori expected. Therefore, for the sake of analytical simplicity, the hypothesis of indivisible labor allocation will be retained.

Let us indicate with N_P and N_R the number of Poor that work, respectively, in the subsistence sector and in the capitalist sector. Consequently, we have $N_P + N_R = \overline{N}$. The aggregate function of production in the traditional sector is given by[3]

$$Y_P = \alpha N_P E.$$

We have assumed that the Poor cannot save and that production is completely exhausted by their consumption. From this equation, it follows that per capita output and consumption of the Poor working in this sector is equal to

$$c_P = \frac{Y_P}{N_P} = \alpha E. \tag{2}$$

The Poor that are hired in the market goods sector receive a real wage equal to w (in terms of the private good produced by the Rich) that is considered as exogenously

[3] This specification was proposed by Schaefer (1957) for fishery and since then it has been widely adopted in literature in modelling natural resources (see e.g. Brander and Taylor 1998a,b; Conrad 1995; López et al. 2007; McAusland 2005; Munro and Scott 1993).

given. By (1), the Poor are indifferent between work in the traditional sector and in the capitalistic sector if and only if

$$c_P = ac_S = aw \tag{3}$$

which can be re-expressed as

$$\frac{1}{a}\alpha E = w. \tag{4}$$

If $\frac{1}{a}\alpha E > w$ (respectively, $\frac{1}{a}\alpha E < w$), then no Poor (respectively all Poor, i.e. $\overline{N}$) would like to work in the capitalistic sector. We assume that E is taken as exogenously given by the Poor, that is, they do not internalize the impact of their production on natural resources; however, we will return to this issue later. In (4), the parameter a determines the difference between the wage in the capitalistic sector and the average output in the traditional sector that allows for the same level of utility.

The population of the Rich is constituted by a continuum of identical individuals and the size of the population is represented by the positive parameter $\overline{M}$. We normalize the size of the R-population by assuming $\overline{M} = 1$. As said, the representative R-agent employs all his fixed potential labor in the modern sector as entrepreneurial activity. Without loss of plausibility, we assume that the marginal product of entrepreneurial labor in the modern sector is higher than the marginal product of labor in the subsistence sector. Therefore, the possibility that the Rich work in the subsistence sector is excluded a priori and the production function of the modern sector can be specified as follows

$$Y_R = \beta K^{\gamma} E^{\delta} (N^D)^{1-\gamma-\delta},$$

where:

- $\gamma > 0$, $\delta \geq 0$ and $\gamma + \delta < 1$ (i.e. the production function satisfies the constant returns to scale assumption).
- K is the physical capital accumulated by the representative R-agent.
- N^D is labor demand by the representative R-agent.
- β is a positive parameter representing (exogenous) technical progress.

3 Economic Dynamics

P and R-agents consider the effect of their choices on the environment as negligible and they do not internalize it; therefore, in their maximization problems they take the evolution of E as given; that is, they behave without taking into account the shadow value of the natural resource and so nobody has an incentive to preserve or restore natural resources. Thus, investment in natural capital does not affect the environmental stock; the dynamics of E is given by the usual logistic function modified for human intervention

$$\dot{E} = E(\overline{E} - E) - \epsilon\alpha N_P E - \eta \overline{Y}_R, \tag{5}$$

where:

- $\overline{E}$ is the carrying capacity of the environmental resource, that is, the maximum stock at which E stabilizes in absence of negative impacts due to P and R-agents' economic activities.
- $\epsilon\alpha N_P E$ is the aggregate environmental impact by the subsistence sector and the parameter $\epsilon > 0$ represents the exploitation of the natural resource by P-agents.
- $\eta > 0$ is a parameter measuring the environmental deterioration caused by the average production $\overline{Y}_R$ of R-agents.

Since there is no investment in natural capital, the R-agent invests in physical capital accumulation everything he saves after consumption expenditures and remuneration of the employed labor force. Therefore the stock of physical capital grows according to the following equation

$$\dot{K} = \beta K^{\gamma} E^{\delta} (N^D)^{1-\gamma-\delta} - wN^D - c_R. \tag{6}$$

Preferences of the Rich are assumed to be representable by a utility function defined over the consumption of the private good. Let the R-agent's instantaneous utility be

$$U_R(c_R) = \ln c_R.$$

Therefore U_R is twice continuously differentiable, strictly increasing and strictly concave, that is $U_R' > 0$ and $U_R'' < 0$. The representative R-agent maximizes his utility by choosing c_R and the labor demand N^D, that is, he solves the following intertemporal optimization problem

$$\underset{c_R, N^D}{Max} \int_0^{\infty} (\ln c_R) e^{-rt} dt$$

under the constraints (5) and (6), where $r > 0$ is the discount rate. The solution to the R-agent's problem is found considering the following current value Hamiltonian function

$$H = \ln c_R + \lambda(\beta K^{\gamma} E^{\delta} (N^D)^{1-\gamma-\delta} - wN^D - c_R) + \theta(E(\overline{E} - E) - \epsilon\alpha N_P E - \eta \overline{Y}_R),$$

where λ and θ are the co-state variables associated to K and E, respectively. It is easy to verify that the dynamics of K, E and λ, do not depend on θ. In fact, we have assumed that agents consider $\epsilon\alpha N_P E$ and $\overline{Y}_R$ as given in the maximization problem above and consequently the resulting dynamics are not optimal; however, the trajectories under such dynamics are Nash equilibria (see Wirl, 1997), in the sense that no (Rich or Poor) agent has an incentive to modify his choices along each trajectory generated by the model as long as the others do not modify theirs. The dynamics generated by the model are found by applying the maximum principle

$$\dot{K} = \frac{\partial H}{\partial \lambda} = \beta K^{\gamma} E^{\delta} (N^D)^{1-\gamma-\delta} - wN^D - c_R,$$

$$\dot{E} = \frac{\partial H}{\partial \theta} = E(\overline{E} - E) - \epsilon \alpha N_P E - \eta \overline{Y}_R,$$

$$\dot{\lambda} = r\lambda - \frac{\partial H}{\partial K} = \lambda \left[r - \beta \gamma K^{\gamma-1} E^{\delta} (N^D)^{1-\gamma-\delta} \right],$$

where c_R, N^D and N_P are determined by the following conditions

$$\frac{\partial H}{\partial c_R} = \frac{1}{c_R} - \lambda = 0 \qquad \text{(i.e. } c_R = \frac{1}{\lambda}\text{)},$$

$$\frac{\partial H}{\partial N^D} = \lambda(\beta(1-\gamma-\delta)K^{\gamma} E^{\delta} (N^D)^{-\gamma-\delta} - w) = 0,$$

that is

$$\beta(1-\gamma-\delta)K^{\gamma} E^{\delta} (N^D)^{-\gamma-\delta} = w. \tag{7}$$

The labor market is perfectly competitive and wage is flexible. The equilibrium value of N_P is given by the labor market equilibrium condition [obtained by equalizing the left sides of (4) and (7)]

$$\frac{\alpha}{a} E = \beta(1-\gamma-\delta)K^{\gamma} E^{\delta} (\overline{N} - N_P)^{-\gamma-\delta}.$$

In particular, we obtain

$$N_P = \overline{N} - \left[\frac{a\beta(1-\gamma-\delta)}{\alpha} \right]^{\frac{1}{\gamma+\delta}} E^{-\frac{1-\delta}{\gamma+\delta}} K^{\frac{\gamma}{\gamma+\delta}} \tag{8}$$

if the right side of (8) is not negative, otherwise $N_P = 0$ (i.e. $\overline{N}$ Poor work in the capitalistic sector). By substituting $N_P = 0$ in (8) and solving it with respect to K we obtain the curve that separates the region where $N_P > 0$ from that where $N_P = 0$ in the plane (E, K)

$$K = L(E) := \left[\frac{\alpha \overline{N}^{\gamma+\delta}}{a\beta(1-\gamma-\delta)} \right]^{\frac{1}{\gamma}} E^{\frac{1-\delta}{\gamma}}, \tag{9}$$

where $\frac{1-\delta}{\gamma} > 1$.

Along and above the curve (9), $N_P = 0$ holds. By substituting N^D with the equilibrium value of $\overline{N} - N_P$ in (7) the equilibrium wage w is found.

Finally, given that (ex-post) $\overline{Y}_R$ is equal to Y_R, the dynamics generated by the model are the following

$$\dot{K} = \beta(\gamma+\delta)K^{\gamma}E^{\delta}(\overline{N} - N_P)^{1-\gamma-\delta} - \frac{1}{\lambda}, \tag{10}$$

$$\dot{E} = E(\overline{E} - E) - \epsilon\alpha N_P E - \eta\beta K^{\gamma}E^{\delta}(\overline{N} - N_P)^{1-\gamma-\delta}, \tag{11}$$

$$\dot{\lambda} = \lambda(r - \beta\gamma K^{\gamma-1}E^{\delta}(\overline{N} - N_P)^{1-\gamma-\delta}), \tag{12}$$

where $N_P = 0$ for (E, K) above (9) while N_P is given by (8) for (E, K) below the curve (9). The following restrictions on variables and parameters hold: K, E, $\lambda > 0$; $a, \alpha, \beta, \gamma, \epsilon, \eta, r, \overline{E}, \overline{N} > 0$; $\delta \geq 0$, $\gamma + \delta < 1$.

4 Analysis of the Model

In this section we analyze the existence and stability of the fixed points (i.e. the stationary states) of the model dynamics, obtained by imposing $\dot{E} = 0$, $\dot{K} = 0$, $\dot{\lambda} = 0$ in the system (10)–(12). Note that, for $\lambda > 0$, equations $\dot{E} = 0$ and $\dot{\lambda} = 0$ depend only on E and K and consequently, solving them, we obtain the fixed point values of E and K. The corresponding value of λ is obtained by solving the equation $\dot{K} = 0$.

4.1 *The Case Without Specialization*

In the case without specialization (i.e. $\overline{N} > N_P > 0$), the condition $\dot{E} = 0$ is satisfied along the graph of the function

$$K = F(E) := E^{\frac{1-\delta}{\gamma}} \left(\frac{\overline{E} - E - \epsilon\alpha\overline{N}}{M(\beta\eta M^{-\gamma-\delta} - \epsilon\alpha)} \right)^{\frac{\gamma+\delta}{\gamma}},$$

where $M := \left(\dfrac{a\beta(1-\gamma-\delta)}{\alpha} \right)^{\frac{1}{\gamma+\delta}}$, and the condition $\dot{\lambda} = 0$ is satisfied along the graph of the function

$$K = G(E) := \left(\frac{\beta\gamma}{r} M^{1-\gamma-\delta} \right)^{\frac{\gamma+\delta}{\gamma}} E^{\frac{2\delta+\gamma-1}{\delta}}.$$

Therefore, the intersections between $F(E)$ and $G(E)$ (occurring below the curve (9)) identify the fixed points under the regime of no specialization. To state the existence and stability results on these fixed points, we define

$$\Omega := \alpha\left(\frac{\eta}{a(1-\gamma-\delta)} - \epsilon\right),$$

$$\Delta := \frac{r}{\beta\gamma\left(\frac{a\beta(1-\gamma-\delta)}{\alpha}\right)^{\frac{1-\gamma}{\gamma}}},$$

$$\overline{N}_1 := \frac{1}{\Delta^{\frac{\gamma}{1-\gamma}}}\left[\frac{\delta a}{\alpha[\eta - \epsilon a(1-\gamma-\delta)]}\right]^{\frac{1-\gamma-\delta}{1-\gamma}},$$

$$\overline{E}_1 := \frac{\left(1+\frac{\delta}{1-\gamma-\delta}\right)}{\left[\left(\overline{N}_1\right)^{\delta}\Delta^{\gamma}\right]^{\frac{1}{1-\gamma-\delta}}} + \alpha\epsilon\overline{N},$$

$$\overline{E}_2 := \frac{\alpha\eta\overline{N}}{1-\gamma-\delta} + \left(\frac{1}{\overline{N}^{\delta}\Delta^{\gamma}}\right)^{\frac{1}{1-\gamma-\delta}}.$$

According to the sign of Ω, two regimes can be distinguished:

1. REGIME DCS (Dirty Capitalistic Sector). *We denote regime DCS (Dirty Capitalistic Sector) as the scenario in which η, the rate of environmental impact caused by the capitalistic sector, is relatively high (ceteris paribus) in comparison to the environmental impact of the traditional sector, measured by ε. That is, $\Omega > 0$ holds, where $\Omega > 0$ if and only if $\frac{\eta}{\epsilon} > a(1-\gamma-\delta)$.*
2. REGIME DTS (Dirty Traditional Sector). *We denote regime DTS (Dirty Traditional Sector) as the scenario in which: $\Omega < 0$.*

Now we can state the following proposition. The proof of such a proposition requires straightforward but tedious calculations; due to space constraints, we will therefore omit it.[4]

[4] The proof is available from the authors on request.

Proposition 1. *In the regime DCS (i.e. $\Omega > 0$), two fixed points with $\overline{N} > N_P > 0$ at most exist. In particular, two fixed points exist if*

$$\overline{N} > \overline{N}_1, \quad \overline{E}_1 < \overline{E} < \overline{E}_2.$$

One fixed point exists if

$$\overline{N} \geq \overline{N}_1, \quad \overline{E} \geq \overline{E}_2.$$

No fixed point exists in the remaining cases.

In the regime DTS (i.e. $\Omega < 0$), one fixed point with $\overline{N} > N_P > 0$ at most exists. In particular, it exists if

$$\overline{E} \geq \overline{E}_2.$$

No fixed point exists in the remaining cases.

In the regime DCS (i.e. $\Omega > 0$), if two fixed points exist, in one of these the curve $G(E)$ intersects $F(E)$ from above in the plane (E, K) (we will indicate such a point with the letter A) while in the other point (which we will indicate with B) the opposite holds; in A the value of E is lower than in B. If only one fixed point is admissible, its configuration is like a point B, namely in it $G(E)$ intersects $F(E)$ from below (see Fig. 6 of the mathematical appendix). In the regime DTS (i.e. $\Omega < 0$), in the unique fixed point the curve $G(E)$ intersects $F(E)$ from above.

Proposition 1 highlights that the fixed points with $\overline{N} > N_P > 0$ exist only when the carrying capacity $\overline{E}$ overcomes certain thresholds ($\overline{E} \geq \overline{E}_1$ if $\Omega > 0$ and $\overline{E} \geq \overline{E}_2$ if $\Omega < 0$). These thresholds are positively correlated to the rate of environmental impact caused by the two sectors (ϵ and η). Thus, if the economic activities are too polluting then stationary points with $\overline{N} > N_P > 0$ do not exist.

Proposition 1 also implies that $\overline{E}$ or $\overline{N}$ can always be found so that two fixed points exist if $\Omega > 0$ and one fixed point exists if $\Omega < 0$, namely the maximum number of admissible stationary points.

Let (E^*, K^*, λ^*) denote the fixed point value of the variables. The stability properties of fixed points depend on the signs of the real parts of the eigenvalues associated to the Jacobian matrix J of the dynamic system (10)–(12) evaluated in (K^*, E^*, λ^*). We define "saddle-point stable" a fixed point that has two eigenvalues with negative real parts, i.e. with a two-dimensional stable manifold. As a matter of fact, under the perfect foresight assumption, if the fixed point has a two-dimensional stable manifold, given the initial values $K(0)$ and $E(0)$ of the state variables K and E, R-agents are able to fix the initial value $\lambda(0)$ of the jumping variable λ so that the growth trajectory starting from $(E(0), K(0), \lambda(0))$ approaches the fixed point. Therefore the fixed point can be reached by growth trajectories. If the fixed point has less than two eigenvalues with negative real parts, then given the initial values $K(0)$ and $E(0)$, a value $\lambda(0)$ does not (generically) exist so that the growth trajectory starting from $(K(0), E(0), \lambda(0))$ approaches the fixed point.

Proposition 2. *The fixed points without specialization ($\overline{N} > N_P > 0$) are characterized by the following stability properties:*

In the regime DCS (i.e. $\Omega > 0$), the fixed point A has always two eigenvalues with positive real parts. The fixed point B is always saddle-point stable if $\gamma + 2\delta - 1 < 0$ while, if $\gamma + 2\delta - 1 > 0$, it can be saddle-point stable or repulsive; however, if $E^* > \frac{1}{2}\left(\overline{E} - \epsilon\alpha\overline{N} - \frac{r\delta}{\gamma}\right)$, it is saddle-point stable.

In the regime DTS (i.e. $\Omega < 0$), the unique fixed point is always saddle-point stable.

Proof. See appendix.

From Proposition 2, it follows that if the gap between the value of the parameter $\overline{E}$ - denoting the carrying capacity - and E^* is not too wide (namely if $E^* > \frac{1}{2}(\overline{E} - \epsilon\alpha\overline{N} - \frac{r\delta}{\gamma})$), the fixed point B is saddle-point stable. As we will see in the following sections, this gap depends on demographic pressure and on the environmental impact of the production of the Poor and the Rich because E^* is decreasing in ϵ, η and $\overline{N}$. As long as the parameters ϵ, η and $\overline{N}$ overcome a certain threshold, the gap is such that the fixed point cannot be reached.

4.2 The Case with Specialization $N_P = 0$

In this context, the condition $\dot{E} = 0$ is satisfied along the graph of the function

$$K = F_0(E) := \frac{E^{\frac{1-\delta}{\gamma}}(\overline{E} - E)^{\frac{1}{\gamma}}}{(\eta\beta\overline{N}^{1-\gamma-\delta})^{\frac{1}{\gamma}}}$$

while the condition $\dot{\lambda} = 0$ is satisfied along the graph of the function

$$K = G_0(E) := \left(\frac{\beta\gamma}{r}\overline{N}^{\frac{1}{\gamma}}\right)^{\frac{1}{1-\gamma}} E^{\frac{\delta}{1-\gamma}}.$$

Therefore the intersections between $F_0(E)$ and $G_0(E)$ identify the fixed points under the regime of perfect specialization in the production of the capitalistic sector.

To state the following proposition, we define

$$\Gamma := \frac{1-\gamma-\delta}{2-2\gamma+\delta},$$

$$\overline{E}_0 := \left(\frac{\overline{N}}{\Gamma}\right)^{\theta}\left(\frac{\left(\frac{\beta\gamma}{r}\right)^{\frac{1}{1-\gamma}}}{\frac{\gamma}{\eta r}(1-\Gamma)}\right)^{\frac{1-\gamma}{2-2\gamma-\delta}},$$

$$\overline{N}_0 := \frac{r\eta}{\gamma}(1-\Gamma)\left(\frac{\beta\gamma}{r}\right)^{\frac{1}{\gamma}}\left(\frac{\alpha\eta\Gamma^{\Gamma}}{a(1-\Gamma)(1-\gamma-\delta)}\right)^{\frac{2\gamma+\delta-1}{1-\gamma}}.$$

With straightforward calculations, we can prove that:[5]

Proposition 3. *Two fixed points with $N_P = 0$ at most exist. In particular, two fixed point exist if*

$$\overline{N} < \overline{N}_0, \quad \overline{E}_0 < \overline{E} < \overline{E}_2.$$

One fixed point exists if

$$\overline{E} \geq \overline{E}_2.$$

No fixed point exists in the remaining cases.

When two fixed points with specialization exist, in one of these points (the fixed point that we will denote with A_0) the graph of $G_0(E)$ intersects that of $F_0(E)$ from above, viceversa in the other fixed point (which we will indicate with B_0) Furthermore, in A_0 the value of E is lower than in B_0. If only one fixed point exists, its configuration is like a point A_0 namely in this point $G_0(E)$ intersects $F_0(E)$ from above (see Fig. 7 of the mathematical appendix).

Proposition 4. *The fixed point A_0 has always two eigenvalues with positive real parts, while B_0 can be saddle-point stable; in particular, it is the case if*

$$E^* > \frac{1}{2}\left(\overline{E} - \frac{r}{\gamma(1-\gamma)}\right).$$

Proof. See appendix.

According to Proposition 4, E^* has to be sufficiently high for saddle-point stability, i.e. $E^* > \frac{1}{2}\left(\overline{E} - \frac{r}{\gamma(1-\gamma)}\right)$. These are sufficient conditions so that the system presents a saddle-point stable stationary state with disappearance of the traditional sector and a complete process of "proletarianization" with all the Poor employed in capitalistic production.

[5] The proof is available from the authors on request.

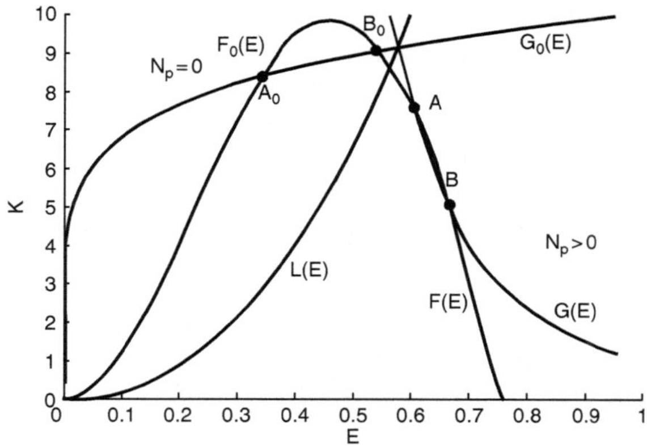

Fig. 1 Four fixed points: A_0 and B_0 with $N_p = 0$, A and B with $N_p > 0$. The parameters' values are: $\alpha = 2$, $\beta = 1$, $\gamma = 0.4$, $\delta = 0.1$, $\epsilon = 0.1$, $\eta = 0.1$, $a = 1$, $r = 0.1$, $\overline{E} = 0.96$, $\overline{N} = 1$

We can also investigate whether the existence of fixed points with $N_P = 0$ is compatible with the existence of fixed points with $N_P > 0$. The following proposition identifies necessary and sufficient conditions for the simultaneous existence of four fixed points A, B, A_0 and B_0.

Proposition 5. *Four fixed points exist -A_0 and B_0 with $N_P = 0$, A and B with $N_P > 0$- if and only if $\overline{N}_0 > \overline{N} > \overline{N}_1$, $max\{\overline{E}_0, \overline{E}_1\} < \overline{E} < \overline{E}_2$ and $\Omega > 0$.*

The proof of this proposition follows from Propositions 1 and 3.

For a numerical example in which four fixed point exist, see Fig. 1. When two saddle-point stable stationary states exist, the choice between B and B_0 depends on the initial conditions. This is a typical example of path dependence: the initial value of E and K determines the fixed point (B or B_0) that the growth trajectory will approach.

4.3 Welfare

The following proposition helps to identify the most significant variables that represent the dynamics of the economy.

Proposition 6. *The stationary state value of consumption c_R^* of the Rich is positively proportional to the stationary state value of physical capital K^*. More precisely, $c_R^* = \dfrac{r(\gamma+\delta)}{\gamma}K^*$ holds. The stationary state values of consumption c_S^* of the Poor working in the capitalistic sector and of consumption c_P^* of the Poor working in the traditional sector are positively proportional to the stationary state value of natural capital E^*. More precisely, $c_S^* = \dfrac{\alpha}{a}E^*$ and $c_P^* = \alpha E^*$ hold.*

This implies that the Rich are able to face effectively environmental degradation through physical capital accumulation. It means that exogenous changes leading to an increase in K^* ensure a growing c_R^*, even if E^* declines. This is not the case for the Poor, whose welfare is positively proportional to E^*.

The above proposition allows to focus on fixed point values of N_P, E and K. From these variables, Poor and Rich agents' welfare can be computed. The following proposition concerns Poor agents' welfare in the context in which two saddle-point stable stationary states coexist, B and B_0.

Proposition 7. *When two saddle-point stable stationary states coexist, B and B_0, then the value of E^* (and consequently P-agents' welfare) is higher in B than in B_0; the value of K^* (and consequently R-agents' welfare) may be higher or lower.*

The proof of such proposition is straightforward. The numerical simulations in Figs. 2–5 show how the fixed point values of K and E change, varying the parameters $\overline{E}$ and γ. In these figures, the continuous (dotted) lines indicate values of E^* and K^* corresponding to saddle-point stable stationary states (respectively, to fixed points with at least two eigenvalues with positive real part). Note that for some values of η and $\overline{E}$, the conditions set in Proposition 5 are satisfied: four fixed points exist and the initial levels of E and K determine whether B or B_0 will be reached. Moreover, as $\overline{E}$ (η) overcomes a minimum (maximum) level, only B_0-type fixed points with full specialization can be approached. Thus, point B_0 can be generated as a final step of an "excessive" depletion of the stock of environmental resources.

Notice that in the numerical examples in Figs. 2–5 when B and B_0 coexist, then P-agents' welfare is higher in B than in B_0 while the opposite holds for R-agents' welfare. Furthermore, observe that varying the parameters $\overline{E}$ and η, Poor agents'

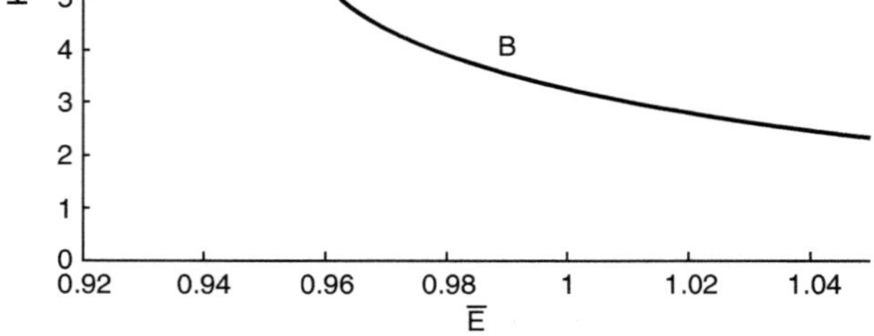

Fig. 2 The value of K, evaluated at the fixed points with $N_p > 0$ and $N_p = 0$ varying $\overline{E}$. *Continuous lines* represent saddle-point stable stationary states

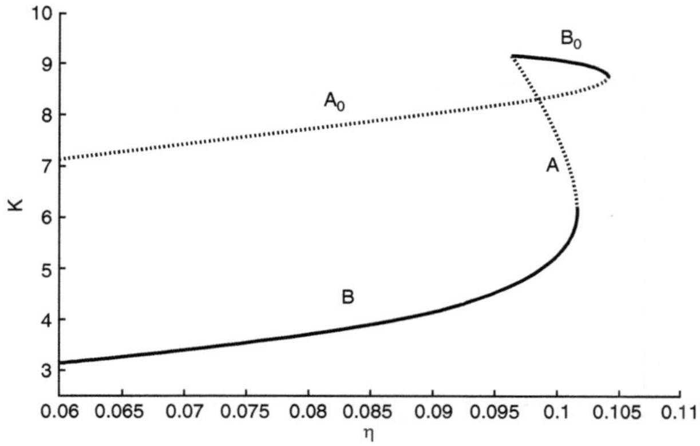

Fig. 3 The value of K, evaluated at the fixed points with $N_p > 0$ and $N_p = 0$ varying η. *Continuous lines* represent saddle-point stable stationary states

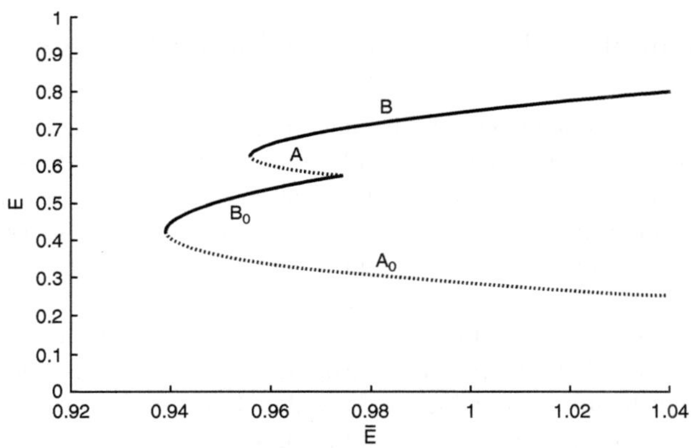

Fig. 4 The value of E, evaluated at the fixed points with $N_p > 0$ and $N_p = 0$ varying $\overline{E}$. *Continuous lines* represent saddle-point stable stationary states

welfare and Rich agents' welfare are inversely correlated, if evaluated at the fixed point without specialization B: a reduction of the endowment of the natural resource (or an increase of the negative impact of the modern sector on the environmental resource) leads to an increase of K^* and to a decrease of E^*. To the contrary, at the fixed point with specialization B_0, a positive correlation is observed. This difference is explained by the fact that along B a perverse structural change occurs in that the reduction of E^* generates a reduction of equilibrium wages associated to an increase of the proportion of Poor employed in the modern sector.

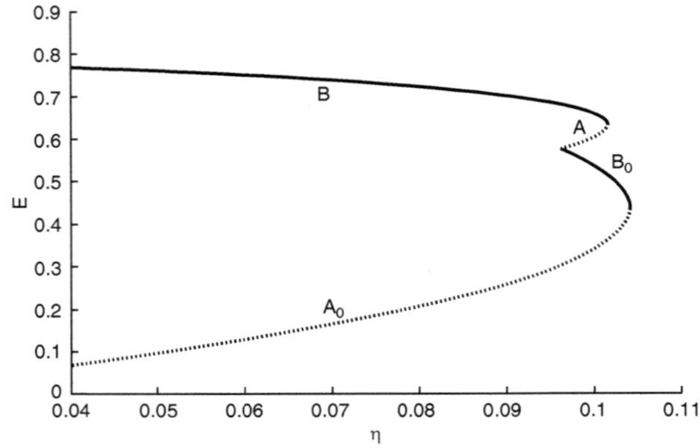

Fig. 5 The value of E, evaluated at the fixed points with $N_p > 0$ and $N_p = 0$ varying η. *Continuous lines* represent saddle-point stable stationary states

5 Discussion of the Results and Concluding Remarks

The bulk of growth models with environmental resources focuses on the relationship between environmental depletion and economic growth or total social welfare, while the links between environmental degradation, economic growth and asset distribution has often been overlooked. Indeed, vulnerability to scarcity or to reduction of natural capital is correlated to asset endowments: it depends on defensive substitution possibilities that, in turn, are affected by the availability of other production factors. Consequently environmental degradation can be expected to have a distributive impact too. This effect can be particularly relevant in developing countries where asset distribution is often highly skewed and the typology of income sources tend to differ across income levels. From this perspective, this article has attempted to apply a less aggregative approach to the study of the links between open access environmental resources, welfare of different population groups, composition and level of output.

The analysis of the model shows that, in contexts with highly concentrated physical capital distribution and free-access renewable natural resources, when physical-capital-intensive activities (i.e. the modern sector in our model) are relatively more polluting or resource demanding than the traditional activities, unexpected results can emerge. A labor shift to these activities can be fuelled not only by advantages in terms of total factor and labor productivity, but also by environmental degradation which, eventually, can lead to a complete specialization in the capital-intensive sector which drives the economy towards B_0, the unique stationary point that is admissible. If the environmental impact produced by these activities is still relatively high but does not overcome a certain threshold, two saddle-point stable stationary states exist: one with specialization in modern sector production (B_0) and one with the presence of both sectors (B). In this case the economic dynamics are path dependent and the selection between these fixed points is affected by the initial level of natural and physical capital. Economies with low natural capital endowments will be more likely to approach the fixed point B_0 and to follow a transition to a complete specialization. It is worth noting that, in such context, the poor obtain a higher welfare level in the stationary state without complete specialization than in the case of a complete process of "proletarianization". Therefore, our model shows that a trade-off between the welfare of the poor and the expansion of modern activities can emerge when environmental externalities and agents' heterogeneity are considered in a joint framework. Conversely, expansion of the modern activities might stimulate counter-intuitive consequences: an immiserizing growth process, namely, an output growth resulting in a further impoverishment of the poor and in a worsening of income distribution. In conclusion, our model suggests that in some contexts[6] the expansion of activities usually regarded as the engine of economic growth and, consequently, necessary (though not sufficient) conditions for poverty reduction, might actually increase poverty and inequality through the erosion of the resources upon which poor people depend.[7]

This trade-off does not emerge in the regime DTS, i.e. when the modern sector produces a relatively lower environmental impact than the traditional sector. In this scenario, for both the poor and the rich the welfare effect of an increase in output production and labor employment of the modern sector is positive.

In conclusion the proposed model shows that environmental degradation may represent a push factor of economic development in an economy polarized into two main classes (the rich and the poor) and characterized by the following stylized facts:

(a) The main income source of the rural poor is self-employment in traditional activities highly dependent on natural resources.
(b) Labour remuneration in rural sector represents the basic opportunity cost for (unskilled) labour in the economy. Thus, given that environmental degradation reduces labour productivity of the rural poor, it may depress wages.
(c) Production of the modern sector managed by the rich is less affected by the depletion of natural resources; they are able to defend themselves by partially substituting natural resources with physical capital accumulation and wage labour employment.

In this context, if the modern sector is sufficiently low-dependent on natural capital (i.e. the natural capital elasticity of the modern sector output is sufficiently low)

---

[6] When income and asset concentration is high and the capitalistic sector is heavily polluting.

[7] Models that predict scenarios with undesirable economic processes are not new in literature. Actually, Antoci and Bartolini (1999, 2004), Antoci et al. (2005, 2008) and Antoci (2009) have proposed models in which negative externalities may constitute an engine of economic growth. In their models, economic growth produces negative externalities that reduce the capacity of natural or social environment to provide free goods. Agents try to defend themselves from welfare losses by increasing their labor supply in order to raise their consumption of private goods that are substitute for free access goods. This, in turn, stimulates economic growth. As a result, defensive strategies generate a growth path along which the production and consumption of private goods are higher than the socially optimal level.

environmental depletion may benefit the modern sector through an increase in low cost labour supply and, in turn, may stimulate physical capital accumulation and expansion of the modern sector. However, if the environmental impact of the modern sector is sufficiently heavy and relatively higher than that of the traditional sector, the structural change is likely to result in an increase in inequality.

Appendix

Proof Proposition 2

Substituting $N_P = \overline{N} - MK^{\frac{\gamma}{\gamma+\delta}} E^{\frac{\delta-1}{\gamma+\delta}}$, the system (10)–(12) becomes

$$\dot{K} = \beta(\gamma+\delta) M^{1-\gamma-\delta} K^{\frac{\gamma}{\gamma+\delta}} E^{\frac{2\delta+\gamma-1}{\gamma+\delta}} - \frac{1}{\lambda},$$

$$\dot{E} = E(\overline{E} - E) + M(\epsilon\alpha - \eta\beta M^{-\gamma-\delta}) K^{\frac{\gamma}{\gamma+\delta}} E^{\frac{\delta-1}{\gamma+\delta}} - \epsilon\alpha\overline{N},$$

$$\dot{\lambda} = \lambda\left(r - \beta\gamma M^{1-\gamma-\delta} K^{-\frac{\delta}{\gamma+\delta}} E^{\frac{2\delta+\gamma-1}{\gamma+\delta}}\right),$$

where $M = \left(\frac{a\beta(1-\gamma-\delta)}{\alpha}\right)^{\frac{1}{\gamma+\delta}}$. Let (K^*, E^*, λ^*) denote the fixed point values of (K, E, λ). Remember that the fixed points without specialization are given by the intersections between the graphs of the functions $K = F(E)$ and $K = G(E)$ occurring below the curve $K = L(E)$ in the plane (E, K). Figure 6 shows all possible configurations of curves $K = F(E)$ and $K = G(E)$; in this figure, the curve $K = L(E)$ is drawn only if $K = F(E)$ and $K = G(E)$ have intersections above it; $E_1 := \frac{1-\delta}{1+\gamma}(\overline{E} - \epsilon\alpha\overline{N})$ indicates the value of E maximizing $F(E)$.

The Jacobian matrix evaluated at the fixed point (K^*, E^*, λ^*) is

$$J^* = \begin{pmatrix} h_K & h_E & h_\lambda \\ f_K & f_E & f_\lambda \\ g_K & g_E & g_\lambda \end{pmatrix}$$

with

$$h_K = r > 0,$$

$$h_E = \frac{r\rho K^*}{\gamma E^*},$$

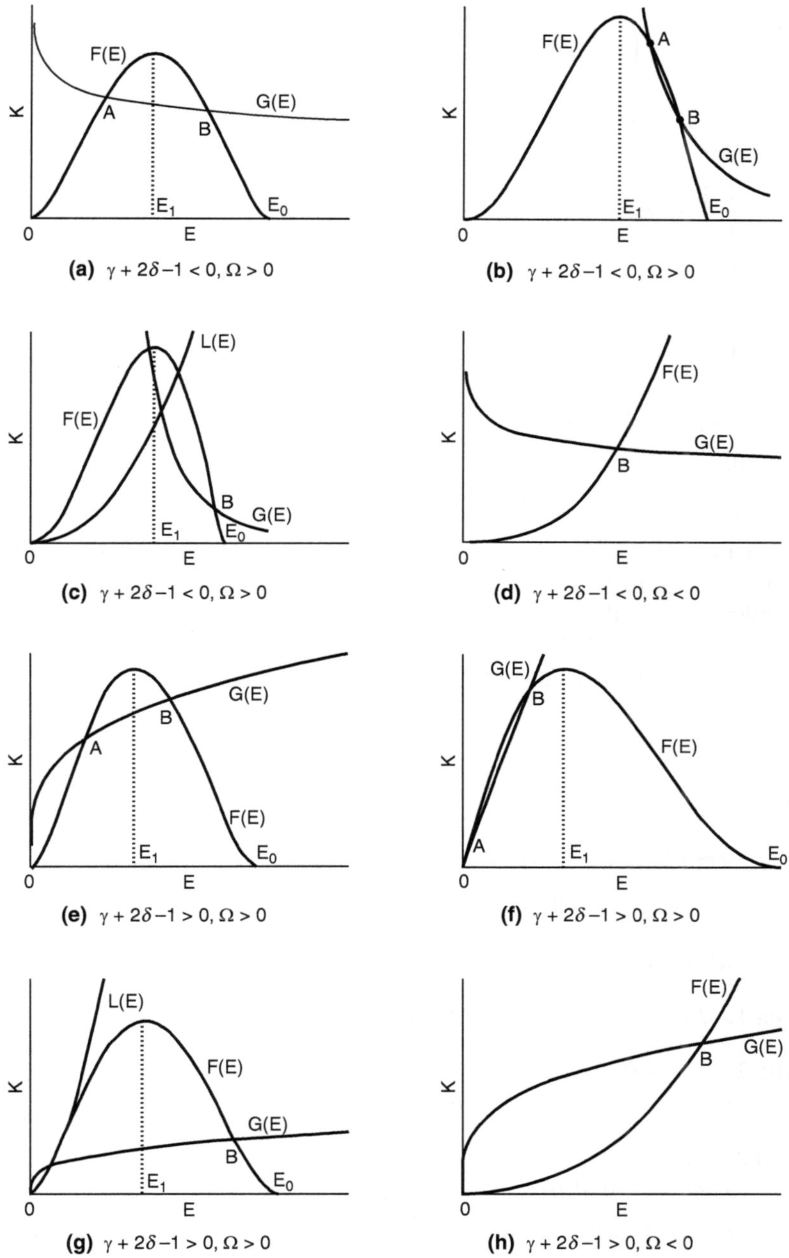

Fig. 6 Fixed points with $N_p > 0$

$$h_\lambda = \frac{1}{(\lambda^*)^2} = \left(\frac{r(\gamma+\delta)K^*}{\gamma}\right)^2 > 0,$$

$$f_K = -\frac{\gamma}{\gamma+\delta}\frac{\Omega E^*(\overline{N} - N_P)}{K^*},$$

$$f_E = \frac{1+\gamma}{\gamma+\delta}(E_1 - E^*),$$

$$f_\lambda = 0,$$

$$g_K = \frac{\gamma\delta}{(\gamma+\delta)^2(K^*)^2} > 0,$$

$$g_E = -\frac{\gamma}{(\gamma+\delta)^2}\frac{\rho}{E^*K^*},$$

$$g_\lambda = 0,$$

where $\rho = \gamma + 2\delta - 1$ and $\Omega = \alpha\left(\frac{\eta}{a(1-\gamma-\delta)} - \epsilon\right)$.

Notice that $sign(h_E) = sign(\rho)$, $sign(g_E) = sign(-\rho)$, $sign(f_E) = sign(E_1 - E^*)$ and $sign(f_K) = sign(-\Omega)$.

In order to study the stability properties of fixed points, we apply the methodology proposed by Wirl (1997). The eigenvalues of the system are the roots of the following characteristic polynomial

$$P(z) = z^3 - tr(J^*)z^2 + wz - |J^*|,$$

where

$$tr(J^*) = h_K + f_E + g_\lambda, \qquad |J^*| = h_\lambda(f_K g_E - f_E g_K),$$
$$w = -h_\lambda g_K + h_K f_E - h_E f_K.$$

The following results can be easily proved.

Lemma 1. *If $E^* < E_1$, then $tr(J^*) > 0$.*

Lemma 2. *If $\Omega > 0$, then $|J^*| < 0$ in A and $|J^*| > 0$ in B.*
If $\Omega < 0$, then $|J^*| > 0$ in the unique admissible fixed point.

Lemma 3. *If $\rho < 0$, then $w < 0$.*
If $\rho > 0$ and $\Omega < 0$, then $w < 0$.
If $\rho > 0$ and $\Omega > 0$, then $E^* > \frac{1}{2}\left(\overline{E} - \epsilon\alpha\overline{N} - \frac{r\delta}{\gamma}\right)$ is a sufficient condition for $w < 0$.

It is now possible to discuss the stability properties of A and B, in the regime $\Omega > 0$, and of the unique admissible fixed point in the regime $\Omega < 0$. As explained in the main text, a fixed point (K^*, E^*, λ^*) is said "saddle-point stable" if J^* admits two eigenvalues with negative real parts.

Stability Analysis of *A*

By Lemma 2, $|J^*| < 0$ holds in A; therefore, A is either a saddle with two positive eigenvalues or a sink. Conditions for local attractivity are (see Wirl, 1997): $tr(J*) < 0$, $|J^*| < 0$ and $w > 0$. Figure 6 shows that A may assume two possible configurations. In the cases (a) and (b), $\rho < 0$ holds; thus, from Lemma 3, it follows that $w < 0$, therefore A is not attractive. In the cases (e) and (f), $E^* < E_1$ holds in A; this implies, by Lemma 1, that $tr(J*) > 0$. Thus A cannot be attractive. In short, the fixed point A is always a saddle with two positive eigenvalues.

Stability Analysis of *B* and of the Unique Fixed Point in the Regime $\Omega < 0$

In B and in the unique fixed point in the regime $\Omega < 0$, $|J^*| > 0$ holds; therefore, such a fixed point is either a source or a saddle point with a two-dimensional stable manifold (Wirl 1997). Wirl finds that a positive determinant and a negative coefficient w are sufficient conditions for saddle-point stability. Given Lemmas 2 and 3, this happens when $\rho < 0$ (Fig. 6, cases a–d) or when $\rho > 0$ and $\Omega < 0$ (Fig. 6, case h). If $\rho > 0$ and $\Omega > 0$, the sign of w is not univocally determined. Consequently, in this case, B may be repulsive or saddle-point stable. However, by Lemma 3, $E^* > \frac{1}{2}\left(\overline{E} - \epsilon\alpha\overline{N} - \frac{r\delta}{\gamma}\right)$ is a sufficient condition for saddle-point stability (Fig. 6, cases e–g); this completes the proof of Proposition 2.

Proof of Proposition 4

In the regime $N_P = 0$, the dynamic system (10)–(12) becomes

$$\dot{K} = \beta(\gamma+\delta)K^{\gamma}E^{\delta}\overline{N}^{1-\gamma-\delta} - \frac{1}{\lambda},$$
$$\dot{E} = E(\overline{E} - E) - \beta\eta K^{\gamma}E^{\delta}\overline{N}^{1-\gamma-\delta},$$
$$\dot{\lambda} = \lambda(r - \beta\gamma K^{\gamma-1}E^{\delta}\overline{N}^{1-\gamma-\delta}).$$

In order to study the stability properties of fixed points, we calculate the Jacobian matrix J_0^* evaluated at a fixed point (K^*, E^*, λ^*) with $N_P = 0$

$$J_0^* = \begin{pmatrix} h_{0K} & h_{0E} & h_{0\lambda} \\ f_{0K} & f_{0E} & f_{0\lambda} \\ g_{0K} & g_{0E} & g_{0\lambda} \end{pmatrix}$$

with

$$\begin{aligned}
h_{0K} &= r(\gamma+\delta) > 0, \\
h_{0E} &= \frac{r\delta(\gamma+\delta)K^*}{\gamma E^*} > 0, \\
h_{0\lambda} &= \frac{r^2(\gamma+\delta)^2(K^*)^2}{\gamma^2} > 0, \\
f_{0K} &= -r\eta < 0, \\
f_{0E} &= \overline{E}(1-\delta) - (2-\delta)E^*, \\
f_{0\lambda} &= 0, \\
g_{0K} &= \frac{\gamma(1-\gamma)}{(\gamma+\delta)(K^*)^2} > 0, \\
g_{0E} &= -\frac{\gamma\delta}{(\gamma+\delta)K^*E^*} < 0, \\
g_{0\lambda} &= 0.
\end{aligned}$$

The eigenvalues of the system are the roots of the following characteristic polynomial

$$P(z) = z^3 - tr(J_0^*)z^2 + wz - |J_0^*|,$$

where

$$tr(J_0^*) = h_{0K} + f_{0E}, \qquad |J_0^*| = h_{0K}(f_{0K}g_{0E} - f_{0E}g_{0K}),$$
$$w = -h_{0\lambda}g_{0K} + h_{0K}f_{0E}f_{0K}.$$

Let us first consider $tr(J_0^*)$. Figure 7 shows all possible configurations of the fixed points with $N_P = 0$. The fixed points correspond to the intersections between the graphs of the functions $K = F_0(E)$ and $K = G_0(E)$, occurring above the curve $K = L(E)$ in the plane (E, K).[8] Notice that $f_{0E} > 0$ if $E^* < E_M := \frac{\overline{E}(1-\delta)}{2-\delta}$, where E_M is the value of E maximizing $F_0(E)$. Being $E^* < E_M$ in A_0, $f_{0E} > 0$ and $tr(J_0^*) > 0$ hold in A_0 (see cases a-c in Fig. 7).

In Fig. 7a, $E^* < E_M$ holds in B_0; therefore $f_{0E} > 0$ and $tr(J_0^*) > 0$. In Fig. 7b, $E^* > E_M$ holds in B_0; therefore $f_{0E} < 0$ and the sign of $tr(J_0^*)$ is not univocally determined.

[8] In Fig. 7, the curve $K = L(E)$ is not drawn when no intersection between $K = F_0(E)$ and $K = G_0(E)$ occurs below it.

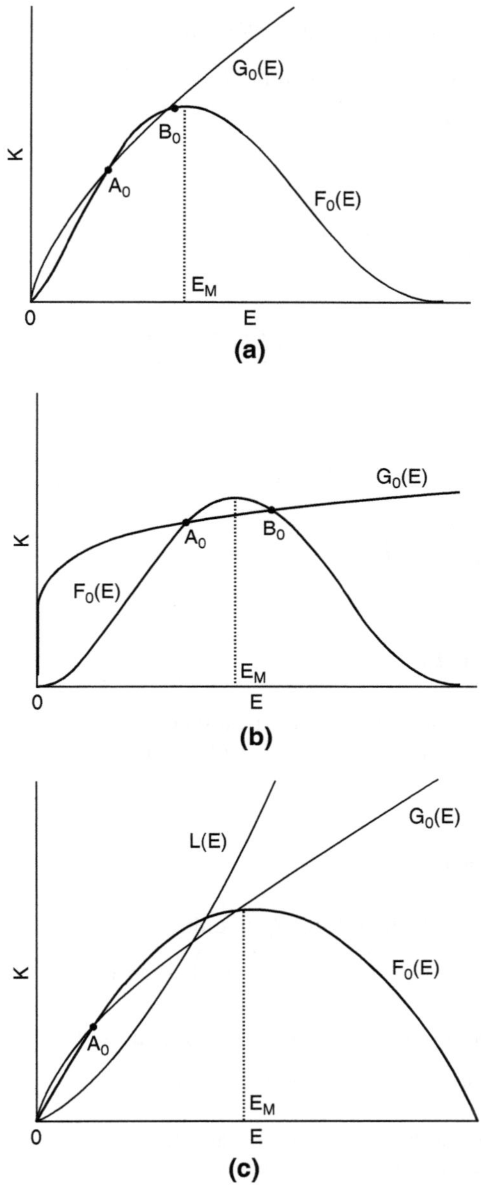

Fig. 7 Fixed points with $N_p = 0$

Let us now analyze the sign of $|J_0^*|$. We can observe that $F_0' > G_0'$ in A_0, while $F_0' < G_0'$ in B_0, where $F_0' = -\frac{f_{0E}}{g_{0K}}$ and $G_0' = -\frac{g_{0E}}{g_{0K}}$. It follows that $|J_0^*| < 0$ in A_0 while $|J_0^*| > 0$ in B_0.

Finally, let us consider

$$w = -\frac{r^2(\gamma+\delta)}{\gamma(1-\gamma)} + r(\gamma+\delta)(\overline{E}(1-\delta) - (2-\delta)E^*) + \frac{\delta\eta r^2(\gamma+\delta)K^*}{\gamma E^*}.$$

Replacing[9]

$$K^* = \frac{\gamma E^*(\overline{E} - E^*)}{r\eta} \tag{13}$$

we obtain

$$w = r(\gamma+\delta)\left\{-\frac{r}{\gamma(1-\gamma)} + \overline{E} - 2E^*\right\} < 0$$

if $E^* > \frac{1}{2}\left(\overline{E} - \frac{r}{\gamma(1-\gamma)}\right)$.

Stability Analysis of A_0

$|J_0^*| < 0$ holds in A_0; therefore A_0 may be a saddle point with two eigenvalues with positive real parts or a sink. Given that $tr(J_0^*) > 0$, local attractivity is excluded.

Stability Analysis of B_0

In B_0 we have $|J_0^*| > 0$; therefore B_0 is either a source or a saddle-point stable stationary state. If $E^* > \frac{1}{2}\left(\overline{E} - \frac{r}{\gamma(1-\gamma)}\right)$, then $w < 0$ and consequently the fixed point cannot be repulsive (see Wirl, 1997). That is, $E^* > \frac{1}{2}\left(\overline{E} - \frac{r}{\gamma(1-\gamma)}\right)$ is a sufficient condition for saddle-point stability.

Acknowledgements The authors would like to thank Ramón López for the insightful conversations and discussions on the topics dealt with in this work. We are also grateful to Simone Bertoli, Giovanni Andrea Cornia, Javier Escobal and Alessandro Vercelli as well as the audience of conferences in Ascona, Urbino and Wageningen for their helpful suggestions and comments on a preliminary version. The usual caveats apply.

---

[9] Formula (13) is obtained from equations $\dot{E} = 0$ and $\dot{\lambda} = 0$.

References

Antoci, A. (2009). Environmental degradation as engine of undesirable economic growth via self-protection consumption choices. *Ecological Economics, 68*, 1385–1397.

Antoci, A., & Bartolini, S. (1999). *Negative externalities as the engine of growth in an evolutionary context* (Fondazione Eni Enrico Mattei Working paper 83.99). Milan

Antoci, A., & Bartolini, S. (2004). Negative externalities and labor input in an evolutionary game. *Environment and Development Economics, 9*, 591–562.

Antoci, A., Galeotti, M., & Russu, P. (2005). Consumption of private goods as substitutes for environmental goods in an economic growth model. *Nonlinear Anal-Model Control, 10*, 3–34.

Antoci, A., Borgesi, S., & Galeotti, M. (2008). Should we replace the environment? Limits of economic growth in the presence of self-protective choices. *International Journal of Social Economics, 35*, 283–297.

Beck, T., & Nesmith, C. (2001). Building on poor people's capacities: the case of common property resources in India and West Africa. *World Development, 29*, 119–133.

Brander, J. A., & Taylor, M. S. (1998a). The simple economics of Easter Island: a Ricardo–Malthus model of renewable resource use. *American Economic Review, 88*, 119–138.

Brander, J. A., & Taylor, M. S. (1998b). Open access renewable resources: Trade and trade policy in a two-country model. *Journal of International Economy, 44*, 181–209.

Cavendish, W. (2000). Empirical regularities in the poverty-environment relationship of rural households: Evidence from Zimbabwe. *World Development, 28*, 1979–2003.

Conrad, J. M. (1995). Bioeconomic model of the fishery. In D. Bromley (Ed.), *Handbook of environmental economics*. New York: Springer.

Dasgupta, P. (2001). *Human well-being and natural environment*. New York: Oxford University Press.

Falconer, J. (1990). *The major significance of minors forest products*. Rome: FAO.

Fisher, M. (2004). Household welfare and forest dependence in Southern Malawi. *Environment and Development Economics, 9*, 135–154.

IFAD. (2001). *The challenge of ending rural poverty, Rural Poverty Report 2001*. New York: Oxford University Press.

Jodha, N. S. (1986). Common property resources and the rural poor. *Economic and Political Weekly, 21*, 1169–1181.

Lewis, W. A. (1955). *The theory of economic growth*. London: Allen & Unwin.

López, R. E. (2003). The policy roots of socioeconomic stagnation and environmental implosion: Latin America 1950–2000. *World Development, 31*, 259–280.

López, R. E. (2007). Structural change, poverty and natural resource degradation. In G. Atkinson, S. Dietz, & E. Neumayer (Eds.), *Handbook of sustainable development*. Cheltenham: Edwards Elgar.

López, R. E., Anriquez, G., & Gulati, S. (2007). Structural change and sustainable development. *Journal of Environmental Economic Management, 53*, 307–322.

Lucas, R. (2004). Life earnings and rural–urban migration. *Journal of Political Economy, 112*, 29–59.

Matsuyama, K. (1992). Agricultural productivity, comparative advantage, and economic growth. *Journal of Economic Theory, 58*, 317–334.

McAusland, C. (2005). Learning by doing in the presence of an open access renewable resource: Is growth sustainable? *Natural Resources Modeling, 18*, 41–68.

Munro, G. R., & Scott, A. D. (1993). The economics of fisheries management. In A. V. Kneese & J. L. Sweeney (Eds.), *Handbook of natural resource and energy economics* (Vol. III). Amsterdam: North-Holland.

Narain, U., Shreekant, G., & vant Veld, K. (2005). *Poverty and the environment-exploring the relationship between household incomes, private assets and natural assets* (Working paper 134). Centre for Development Economics, Delhi School of Economics, Delhi.

Ranis, G., & Fei, J. C. H. (1961). A theory of economic development. *American Economic Review, 51*, 533–565.

Schaefer, M. B. (1957). Some considerations of population dynamics and economics in relation to the management of marine fisheries. *Journal of the Fisheries Research Board of Canada, 5*, 669–681.

United Nations Environment Programme. (2004). *Human well-being, poverty and ecosystem services. Exploring the links*. Nairobi: Author.

Vedeld, P., Angelsen, A., Sjaastad, E., & Kobugabe-Berg, G. (2004). *Counting on the environment: Forest incomes and the rural poor* (Paper no.98). Washington, DC: World Bank Environment Department.

Wirl, F. (1997). Stability and limit cycles in one-dimensional dynamic optimisations of competitive agents with a market externality. *Journal of Evolutionary Economics, 7*, 73–89.

World Resources Institute in collaboration with United Nations Development Programme, United Nations Environment Programme, and World Bank. (2005). *World resources 2005. The wealth of the poor: Managing ecosystems to fight poverty*. Washington, DC: Author.

Bifurcations and Chaotic Attractors in an Overlapping Generations Model with Negative Environmental Externalities

Angelo Antoci, Ahmad Naimzada, and Mauro Sodini

1 Introduction

We analyze an overlapping generations model with the following features. There exists a continuum of identical individuals whose welfare depends on leisure, on the stock E of a free access environmental good and on the consumption C of a private good. The private good is produced by a continuum of identical perfectly competitive firms via a constant returns technology (represented by a Cobb–Douglas production function); the representative firm uses physical capital K and labour L of the representative individual as productive inputs. Each economic agent considers as negligible the negative impact of his choices on the environmental good; this implies that the choices of each agent generate negative externalities on the others.

Following Zhang (1999), Antoci et al. (2007) and Itaya (2008) (among the others), we assume that individuals' utility function is non separable in E and C; that is, the marginal utility of C depends on the value of E; in particular, we consider both the cases in which marginal utility increases (i.e. C and E are *substitutes*) and decreases (i.e. C and E are *complements*) when the value of E decreases.[1]

The assumption of non separability of the utility function allows us to analyze possible feedback effects on consumption and economic growth generated by environmental degradation. In particular, we show that the dynamics can admit at most one steady state when C and E are *complements* while at most three steady states can exist when C and E are *substitutes*. In the context in which C and E are

[1] It is obvious that the marginal utility deriving from the consumption of private goods may be reduced by environmental deterioration; for example, to drink a cup of coffee in front of an uncontaminated seaside is better than in front of a polluted one. Living in a house on an uncontaminated river is obviously better than in a house placed on a polluted and smelling one. However, environmental degradation may also increase the marginal utility deriving from the consumption of some private goods; the relevance of substitutive consumption is stressed by the literature on environmental defensive expenditures (see, e.g. Antoci 2009; Antoci and Bartolini 1999, 2004; Antoci et al. 2008; Hueting 1980; Leipert 1989; Leipert and Simonis 1988; Shogren and Crocker 1991).

A. Naimzada (✉)
Università di Milano-Bicocca, Dipartimento di Metodi Quantitativi Piazza Ateneo Nuovo, 1 20126 Milano, Italy,
e-mail: ahmad.naimzada@unimib.it

G.I. Bischi et al. (eds.), *Nonlinear Dynamics in Economics, Finance and the Social Sciences*, DOI 10.1007/978-3-642-04023-8_3,

substitutes, we find that *local indeterminacy* – i.e. the existence of an infinite number of (Nash) equilibrium orbits approaching the same steady state – can occur and that environmental externalities play a key role in generating it. More precisely, we show that starting from a context in which the environment doesn't enter in individuals' utility function and dynamics are not indeterminate, then indeterminacy may occur (ceteris paribus) introducing E in the utility function.[2] Furthermore, by numerical analysis, we obtain examples of long-run indeterminacy; in particular, we show that a chaotic attractor may arise; in such case, starting from the same initial value K_0, there exist a continuum of initial values L_0 that lead the economy to reach the attractor. So the long run evolution of the economy depends on the choice of L_0. We also give examples in which a chaotic attractor and an attracting steady state coexist giving rise to global indeterminacy;[3] in such multistability regime, starting from the same initial value K_0, there exist a continuum of initial values L_0 that lead the economy to approach the steady state and a continuum that lead the economy towards the chaotic attractor. Finally, in our examples, we find that when global indeterminacy occurs then, along the orbits approaching the chaotic attractor, the values of L and K are higher than along the orbits reaching the steady state, while individuals' welfare is lower. This Pareto-dominance result implies that along the orbits approaching the chaotic attractor the economy experiments a process of undesirable economic growth: individuals' welfare would be higher by choosing lower labour and capital accumulation levels.[4]

The paper has the following structure. Section 2 introduces the set up of the model and the associated dynamic system. Section 3 deals with the existence and local stability of the normalized stationary state. Sections 4 and 5 shows, via numerical simulations, some complex dynamic regimes generated by the model and the evolution of individuals' welfare along the orbits followed by the economy. Section 6 concludes.

2 The Model

We consider a standard overlapping generations economy. Time is discrete: $t = 1, 2, 3, \ldots, \infty$; there exist a continuum of individuals who live for two periods of time and two generations of individuals (*young* and *old*) coexist in each period

[2] In economic literature, local indeterminacy is usually generated via the effect of positive externalities arising from production activity (see, e.g. Benhabib and Farmer, 1999; Bennet and Farmer, 2000; Cazzavillan et al., 1998; Cazzavillan, 2001; Grandmont et al., 1998; Reichlin, 1986). However, some works focus on the role played by negative externalities as engine of local indeterminacy; see, for example, Chen and Lee (2007), Itaya (2008), Meng and Yip (2008).

[3] The term "global indeterminacy" (see, among the others, Grandmont et al., 1998; Krugman, 1991; Matsuyama, 1991; Pintus et al., 2000 and, more recently, Benhabib et al., 2008 and Coury and Wen, 2009) refers to the situation where, starting from the same initial value of the state variable, there exist different equilibrium paths that approach different attractors.

[4] This result confirms analogous results obtained under the assumption of substitutability between C and E (see, e.g. Antoci, 2009; Antoci and Bartolini, 2004; Antoci et al., 2005, 2007, 2008; Bartolini and Bonatti, 2002, 2003).

of time t. Individuals work when they are young and consume the private good when they are old.[5] The private good is produced by a continuum of perfectly competitive firms.

In each period t, the representative young individual has to allocate his time endowment L^* (L^* is a fixed parameter) between leisure and labour supply L_t ($L^* \geq L_t \geq 0$) to the representative firm, remunerated at the wage rate W_t. The remuneration $L_t W_t$ is entirely invested in productive capital K_{t+1} (i.e. $K_{t+1} = L_t W_t$) that the individual will rent to the representative firm in time t at the interest factor R_{t+1}. The sum obtained, $W_t L_t R_{t+1}$, allows him to buy and consume the quantity $C_{t+1} = W_t L_t R_{t+1}$ of the good produced by the firm ($W_t L_t$ and $W_t L_t R_{t+1}$ are expressed in unities of the consumption good).

2.1 The Utility Function

We assume the following utility function:

$$U(L^* - L_t, C_{t+1}, E_{t+1}) = \ln(L^* - L_t) + \frac{Q}{1+\theta} \frac{(PC_{t+1}E_{t+1}^{\varepsilon})^{1-\sigma} - 1}{1-\sigma},$$

where E_{t+1} indicates the stock of the free access environmental good at time $t+1$; $\frac{1}{1+\theta}$ is the discount factor, P and Q are positive scale parameters that will be used to apply the "normalized steady state" technique, ε and σ are positive parameters, $\sigma \neq 1$. The parameter σ denotes the inverse of the intertemporal elasticity of substitution in consumption. Notice that if $\sigma \in (0, 1)$, then C_{t+1} and E_{t+1} are substitutes, while if $\sigma > 1$ they are complements, in that:

$$\frac{\partial^2 U(l_t, C_{t+1}, E_{t+1})}{\partial C_{t+1} \partial E_{t+1}} \lesseqgtr 0$$

for $\sigma \lesseqgtr 1$. This function is concave in $L^* - L_t$ and C_{t+1}; it is not assumed to be jointly concave in $L^* - L_t$, C_{t+1} and E_{t+1} in that in the decentralized competitive market economy on which we focus, the variable E_{t+1} is not a choice variable for each economic agent.

2.2 The Production Function

We assume constant returns to scale; in particular, the representative firm has the following Cobb–Douglas production function:

[5] This assumption is adopted in several overlapping generations models (see, among the others, Duranton, 2001 and Zhang, 1999). It simplifies our analysis by abstracting from the consumption-saving choices of agents.

$$Y = AF(K_t, L_t) = AL_t^{1-\alpha} K_t^{\alpha} = A\, L_t k_t^{\alpha},$$

where $k_t := K_t/L_t$ and A is a positive parameter representing (exogenous) technological progress.

2.3 Economic Agents' Choices

The economy is assumed perfectly competitive and so, in each period t, the representative firm maximizes the profit function:

$$AF(K_t, L_t) - W_t L_t - R_t K_t \tag{1}$$

taking the wage rate W_t and the interest factor R_t as exogenously given. As usual, this assumption gives rise to the following first order conditions:

$$W_t = A(1-\alpha)k_t^{\alpha}, \tag{2}$$

$$R_t = A\alpha k_t^{\alpha-1}. \tag{3}$$

The representative individual maximizes the objective function:

$$\max U(L_t, C_{t+1}, E_{t+1})$$

under the constraints:

$$C_{t+1} = R_{t+1} W_t L_t, \tag{4}$$

$$L_t \in \left[0, L^*\right]. \tag{5}$$

In our perfectly competitive economy, W_t and R_{t+1} are considered as exogenously given. Furthermore, we assume that the representative individual, at time t, is able to perfectly foresee the value of E_{t+1}. However, E_{t+1} is considered as exogenously determined in that the representative individual considers as negligible the impact of his choices on the environmental quality.

Under these assumptions, the first order condition for an interior solution (it always holds $0 < L_t < L^*$) of the representative individual's choice problem is

$$-\frac{1+\theta}{L^* - L_t} + Q\frac{[PR_{t+1}W_t E_{t+1}^{\varepsilon}]^{1-\sigma}}{L_t^{\sigma}} = 0. \tag{6}$$

By substituting (1) and (2) in (6) we obtain

$$-\frac{1+\theta}{L^* - L_t} + Q\frac{[P(\alpha(1-\alpha)A^2 k_t^{\alpha} k_{t+1}^{\alpha-1}) E_{t+1}^{\varepsilon}]^{1-\sigma}}{L_t^{\sigma}} = 0. \tag{7}$$

2.4 Dynamics

We assume that the environmental stock E_{t+1} in time $t+1$ depends negatively on the average production level in time t, that is

$$E_{t+1} = \overline{E} - \eta A F(\overline{K}_t, \overline{L}_t) = \overline{E} - \eta A \overline{L}_t \overline{k}_t^{\alpha}, \tag{8}$$

where $\overline{E}$ is a positive parameter representing the endowment of the environmental good, i.e. the value that E_{t+1} would assume in absence of the negative impact of production. $\overline{L}_t$, $\overline{K}_t$ and $\overline{k}_t$ indicate the economy wide average values of L_t, K_t and k_t in time t, respectively; $AF(\overline{K}_t, \overline{L}_t)$ represents average production. The positive parameter η measures the impact of production on the environmental stock. We assume that each economic agent considers $\overline{L}_t$ and $\overline{K}_t$ as exogenously determined. However, being all economic agents identical, ex-post $\overline{L}_t = L_t$, $\overline{K}_t = K_t$ and $\overline{k}_t = k_t$. So, in this model, the choices of the representative individual are not optimal and generate negative externalities. However, the orbits followed by the economy are Nash equilibria, in that no single individual has interest to modify his choices if also the others don't revise theirs.

By plugging (8) in (7) and taking into account that, by (2), it holds

$$K_{t+1} = L_{t+1}k_{t+1} = L_t W_t = L_t A(1-\alpha)k_t^{\alpha}$$

the dynamic system representing the dynamics of the economy is

$$-\frac{1+\theta}{L^* - L_t} + Q\frac{[P(\alpha(1-\alpha)A^2 k_t^{\alpha} k_{t+1}^{\alpha-1})(\overline{E} - \eta A(L_t k_t^{\alpha}))^{\varepsilon}]^{1-\sigma}}{L_t^{\sigma}} = 0, \tag{9}$$

$$k_{t+1}L_{t+1} = A(1-\alpha)k_t^{\alpha} L_t. \tag{10}$$

3 Steady States of Dynamics

3.1 The Normalized Steady State

The system (9)–(10) defines k_{t+1} and L_{t+1} as functions of k_t and L_t. In this section, we study the stability of fixed points of such discrete dynamic system. Since our model contains a large number of parameters, to make clear the study we use the geometrical–graphical method developed by Grandmont et al. (1998) that allows us to characterize the stability properties of the steady states of this dynamic system. We impose some conditions on parameters under which a fixed point $(k_{s.s}, L_{s.s}, E_{s.s})$ with $k_{s.s} = L_{s.s} = E_{s.s} = 1$ exists. This allows us to analyze the effects on stability due to changes in parameters' values being sure that the fixed point doesn't disappear. Without loss of generality, we pose $L^* = 2$.

By requiring that $k_{s.s} = L_{s.s} = E_{s.s} = 1$ [by (9)–(10)] we obtain the following conditions on parameters' values:

$$A = \frac{1}{1-\alpha}, \quad \overline{E} = \frac{1-\alpha+\eta}{1-\alpha}, \quad P = P^* := \frac{1-\alpha}{\alpha} \quad \text{and } Q = Q^* := 1+\theta. \tag{11}$$

Using conditions (11), the dynamic system (9)–(10) can be explicitly written as

$$k_{t+1} = \left[\frac{k_t^{\alpha}(1+\omega-\omega L k^{\alpha})^{\varepsilon} L^{\frac{\sigma}{\sigma-1}}}{(2-L)^{\frac{1}{\sigma-1}}}\right]^{\frac{1}{1-\alpha}}, \tag{12}$$

$$L_{t+1} = L_t k_t^{\alpha}\left[\frac{(2-L)^{\frac{1}{\sigma-1}}}{k_t^{\alpha}(1+\omega-\omega L k^{\alpha})^{\varepsilon} L^{\frac{\sigma}{\sigma-1}}}\right]^{\frac{1}{1-\alpha}}, \tag{13}$$

where $\omega := \frac{\eta}{1-\alpha} \in (0, +\infty)$.[6] Such system always admits the normalized steady state defined above; to look for other steady states, notice that the steady states of system (12)–(13) are characterized by the following conditions:

$$k = 1,$$

$$g(L) := \left[\frac{(1+\omega-\omega L)^{\varepsilon} L^{\frac{\sigma}{\sigma-1}}}{(2-L)^{\frac{1}{\sigma-1}}}\right]^{\frac{1}{1-\alpha}} = 1. \tag{14}$$

Notice that $g(L)$ is defined in the interval $[0, 2)$ and that $g(1) = 1$. By straightforward calculations it is easy to check that at most one fixed point (the normalized one) exists if $\sigma < 1$ (i.e. if C and E are complements) while at most three fixed points exist if $\sigma > 1$ (i.e. if C and E are substitute). The next section provides some examples of dynamics admitting one, two or three fixed points.

3.2 The Stability Properties of the Normalized Steady State and Indeterminacy

In our model, productive capital K_t represents a state variable, so its initial value K_0 is given. Differently from K_t, the variable L_t is a "jumping" variable in that it represents the representative individual's labor input, chosen taking into account of the average labour input in the economy, the expected environmental quality and the accumulated productive capital. Consequently, individuals have to choose the initial value L_0 (and consequently the initial value of $k_t = K_t/L_t$). If the normalized

[6] Notice that $L_{t+1} = L_t k_t^{\alpha}/k_{t+1}$.

steady state is a saddle and K_0 is near enough to 1, then there exists an unique initial value of L_t, L_0, such that the orbit passing through (k_0, L_0) approaches the fixed point. When the fixed point is a sink, given the initial value K_0, then there exists a continuum of initial values L_0 such that the orbit passing through (k_0, L_0) approaches the fixed point; consequently, the orbit the economy will follow is "indeterminate" in that it depends on the choice of the initial value L_0. The following results show how indeterminacy depends on the parameters σ and ε of the model.

The Jacobian matrix of (12)–(13), evaluated at the normalized steady state, is

$$J^N = \frac{1}{1-\alpha}\begin{pmatrix} -\frac{\alpha(\alpha-1+\eta\varepsilon)}{(\alpha-1)} & \frac{\varepsilon\eta(\sigma-1)+(1+\sigma)(\alpha-1)}{(\sigma-1)(\alpha-1)} \\ \frac{\alpha(\varepsilon\eta+\alpha(\alpha-1))}{(\alpha-1)} & \frac{(\sigma-1)(\alpha^2-\alpha+\varepsilon\eta)+2(\alpha-1)}{(\sigma-1)(\alpha-1)} \end{pmatrix}$$

with:

$$Det(J^N) = \frac{2\alpha}{(1-\sigma)(1-\alpha)}, \tag{15}$$

$$Tr(J^N) = \frac{2}{(1-\sigma)(1-\alpha)} + \frac{\eta}{1-\alpha}\varepsilon. \tag{16}$$

Let us assume $\sigma \in (0,1)$. In such context, it is easy to check that varying σ in the interval $(0,1)$, the point $\left(\frac{2}{(1-\sigma)(1-\alpha)}, \frac{2\alpha}{(1-\sigma)(1-\alpha)}\right)$ [see (15) and (16)] describes in the plane $(Tr(J), Det(J))$ (see Fig. 1) a half line T_1 with slope α $(0 < \alpha < 1)$ starting on the right of the line AC, in the first orthant of the plane. This implies that, if $\varepsilon = 0$ (i.e. E doesn't enter the utility function), the point $\left(Tr(J^N), Det(J^N)\right)$ always belongs to the region "saddle" in Fig. 2, for every $\sigma \in (0,1)$. Now, we fix the value of σ, i.e. we fix a point p_1 in T_1; for $\varepsilon = 0$, the points p_1 and $\left(Tr(J^N), Det(J^N)\right)$ are coincident; by increasing ε, the point $\left(Tr(J^N), Det(J^N)\right)$ moves on the right of p_1, along an horizontal line. This

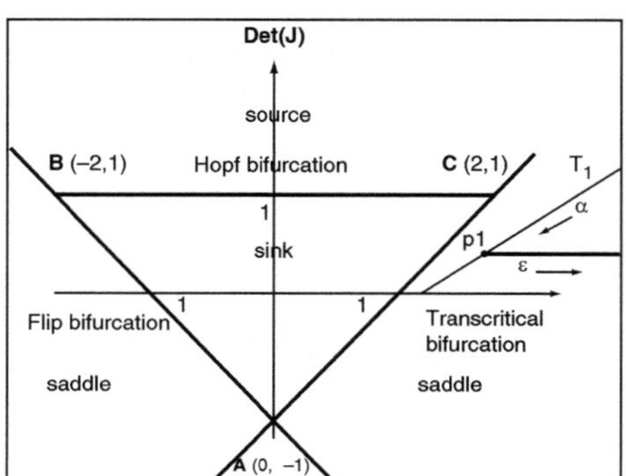

Fig. 1 Stability analysis of the normalized fixed point in case

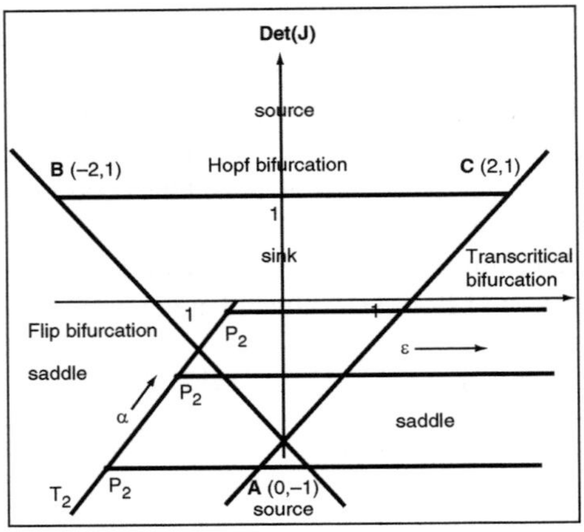

Fig. 2 Stability analysis of the normalized fixed point in case

implies that the point $\left(Tr(J^N), Det(J^N)\right)$ always belongs to the region "saddle" in Fig. 2, for every $\sigma \in (0,1)$ and $\varepsilon \geq 0$.

Under the substitutability assumption $\sigma > 1$, the point $\left(\frac{2}{(1-\sigma)(1-\alpha)}, \frac{2\alpha}{(1-\sigma)(1-\alpha)}\right)$ describes in the plane $(Tr(J), Det(J))$ (see Fig. 2) a half line T_2 with slope α lying in the third orthant of the plane $(Tr(J), Det(J))$ and approaching the origin of it for $\sigma \to \infty$. This implies that, if $\varepsilon = 0$, the point $\left(Tr(J^N), Det(J^N)\right)$ belongs to the region "sink" in Fig. 2 for σ high enough. Notice that, by following the same steps as in case $\sigma \in (0,1)$, it is easy to see that starting from a point p_2 of T_2 in the region "saddle", the point $\left(Tr(J^N), Det(J^N)\right)$ can enter in the region "sink" by increasing ε, if σ is high enough. This implies that an increase of the dependence on E of individuals' welfare can be a source of indeterminacy. However, it is worth to stress that, in our model, indeterminacy can also occur in case $\varepsilon = 0$ (but for higher values of σ).

Notice that Hopf bifurcations are ruled out. A flip bifurcation occurs (in case $\sigma > 1$) when the line AB in Fig. 2 is crossed, along which it holds $Tr(J) + Det(J) + 1 = 0$; so the bifurcation value of ε is

$$\varepsilon_{flip} = \left[-(1-\alpha) - \frac{2(\alpha+1)}{(1-\sigma)}\right]\frac{1}{\eta}.$$

A transcritical bifurcation arises (in case $\sigma > 1$) when the line AC in Fig. 2 is crossed, along which it holds $Det(J) - Tr(J) - 1 = 0$; so the bifurcation value of ε is

$$\varepsilon_{tr} = \left[1 - \frac{2}{1-\sigma}\right]\frac{1-\alpha}{\eta}.$$

3.3 Global Analysis and Numerical Simulations

Because of analytic complexity of the equation describing the evolution of the economy, we cannot find closed form expressions of non-normalized fixed points. Consequently we fix $\alpha = 0.17$, $\eta = 0.11$ and $\sigma = 8$, and use ε as bifurcation parameter.[7] Figure 3a refers to the case $\varepsilon = 7.2$; in such case, there exists only the normalized fixed point, corresponding to the intersection between the graph of $g(L)$ [see (14)] and the vertical line $k = 1$; this point is the unique attractor of the system and in Fig. 3b three orbits approaching it are represented.

By increasing the value of ε up to $\varepsilon = 8.2$, two other fixed points arise (see Fig. 4a), $(k, L) = (1, L_1)$ and $(1, L_2)$, with $L_2 > L_1 > 1$, via a *fold* bifurcation (the bifurcation value is $\varepsilon = 7.995$); $(1, 1)$ and $(1, L_2)$ are attracting while $(1, L_1)$ is a saddle whose stable manifold separates the basins of attraction of $(1, 1)$ and $(1, L_2)$ (see Fig. 4b).

Trajectories near to $(1, L_2)$ follow to the classical period doubling route to chaos while the normalized point preserves its stability (see Fig. 4c, $\varepsilon = 8.50$). Note that, since the initial value of labour input L is a jumping variable, in the dynamic regimes represented in Fig. 4b–c,[8] the long run growth path of the economy is indeterminate in that a slightly different initial choice of L may determine a rather different long run behaviour of the economy. This type of indeterminacy differs from that relative to the orbits approaching the same attracting fixed point in that, in such case, indeterminacy occurs with respect to transient dynamics only.

A further increase of ε causes the disappearance of the strange attractor ($\varepsilon = 8.53$) and the normalized point becomes the unique attractor of the system.

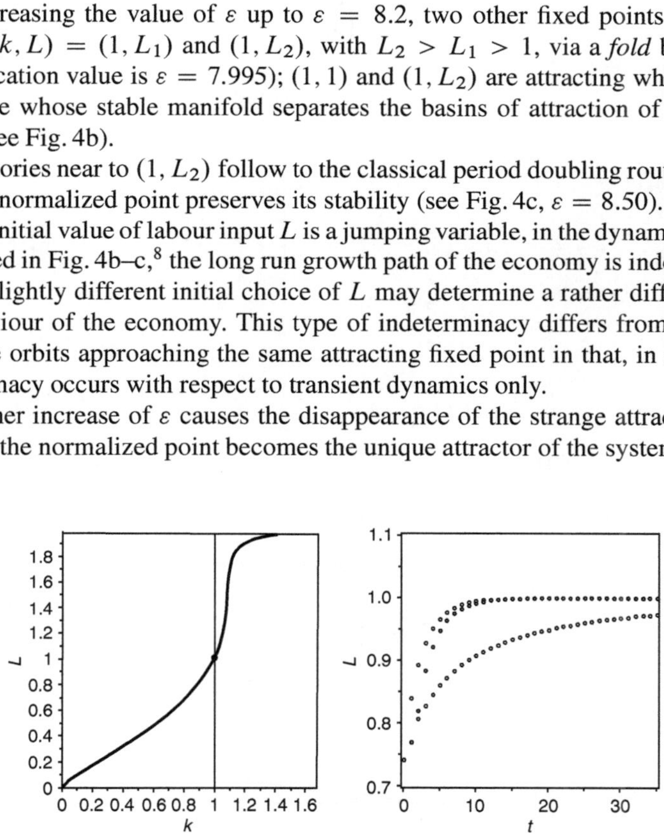

(a) (b)

Fig. 3 Example in which the normalized fixed point is the unique attractor

[7] All the numerical simulations we have made suggest that similar results are obtained by using α, η or σ as bifurcations parameters.

[8] For initial conditions in the white area, the system diverges to infinity (same convention for Fig. 4b–c). Points with white interior part are saddle.

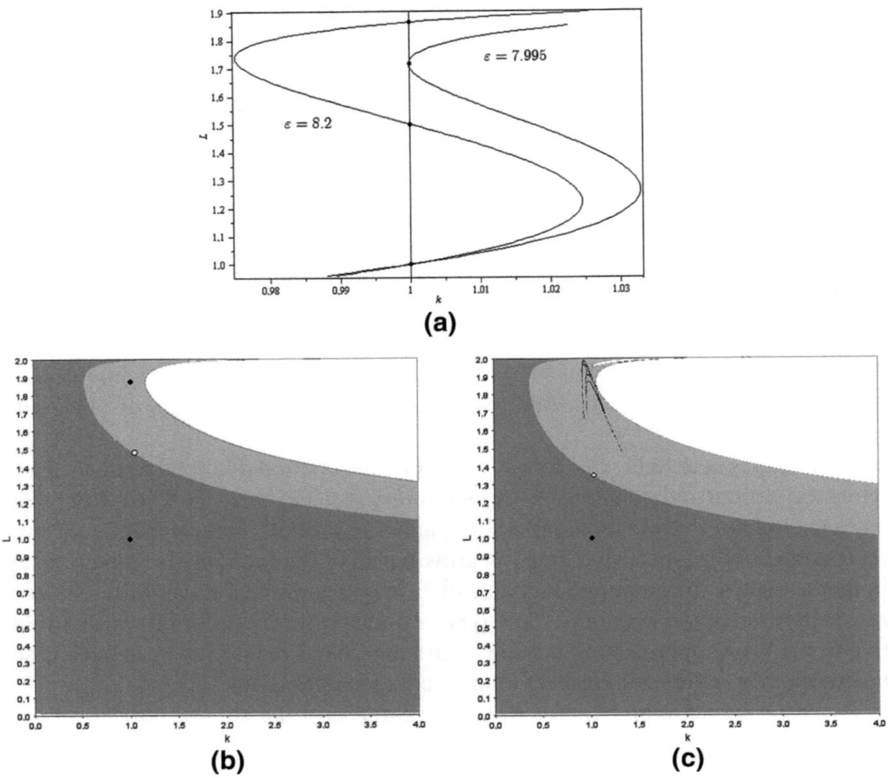

Fig. 4 Example in which three fixed points exist

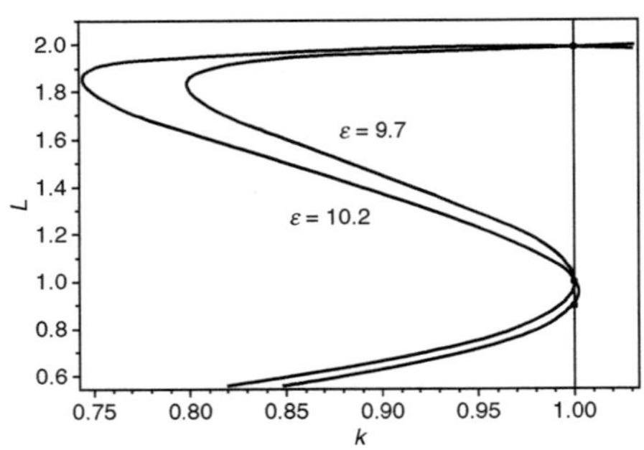

Fig. 5 Example in which three fixed points exist and the normalized fixed point is a saddle

Finally, observe that, for $\varepsilon = 9.7$, $(1, 1)$ and $(1, L_1)$ are coincident while, increasing ε, a switching between $(1, 1)$ and $(1, L_1)$ occurs (see Fig. 5): $(1, L_1)$ and $(1, 1)$, with $L_1 < 1$, become respectively an attracting point and a saddle. The overlapping bifurcation diagrams in Fig. 6 (with initial conditions (4, 1.52) and (0.2, 0.2)) show the evolution of the attractors.

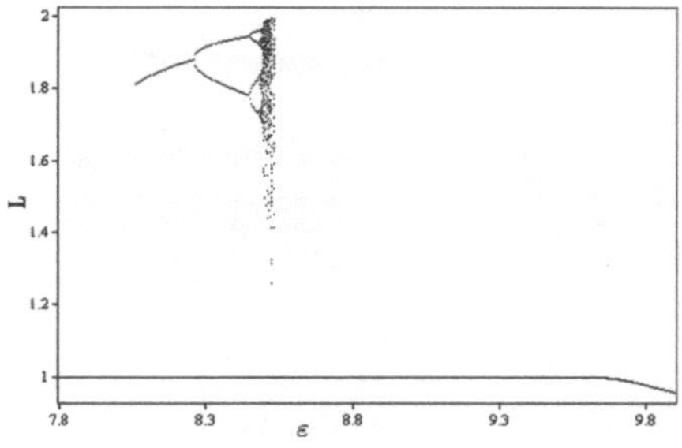

Fig. 6 Overlapping bifurcation diagrams varying the parameter

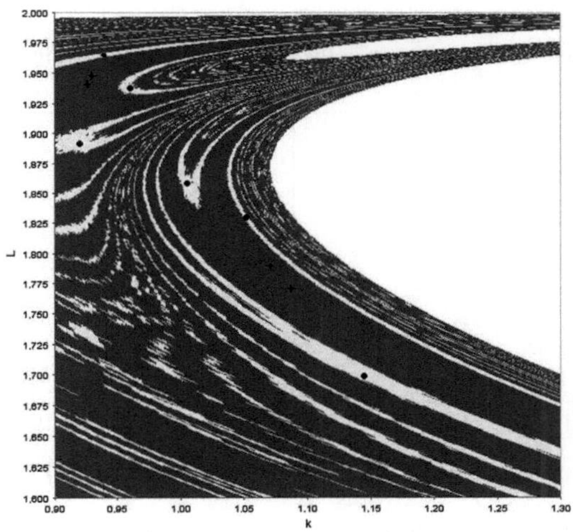

Fig. 7 The fractal nature of the basins of attraction of period-6 and period-4 cycles

The system shows a further interesting dynamic phenomenon non evidenced by the previous figures. For $\varepsilon = 8.452$ a period-6 cycle is born and a coexistence of three attractors arises up to $\varepsilon = 8.4587$. Figure 7 evidences the fractal nature of the basins of attraction of period-6 and period-4 cycles. The magnification of the superior part of bifurcations diagram in Fig. 6[9] describes the evolution of the period-6 cycle that via period doubling bifurcations generates a chaotic attractor ($\varepsilon = 8.465$) with several periodic windows while period-4 cycle preserves the stability. Finally the chaotic attractor dies and all the trajectories in that area converge to the period-4 cycle.

[9] The initial condition for period-4 cycle is (3.5, 1.1). To analyse the evolution of the other attractor, we used the continuation algorithm of EFChaos.

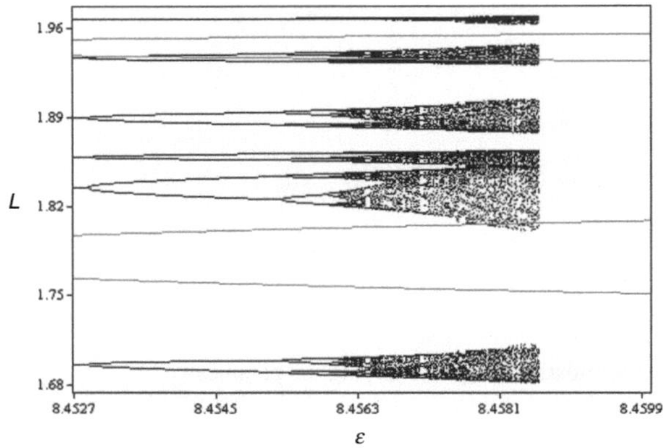

Fig. 8 An enlargement of the upper part of the bifurcation diagram in Fig. 6

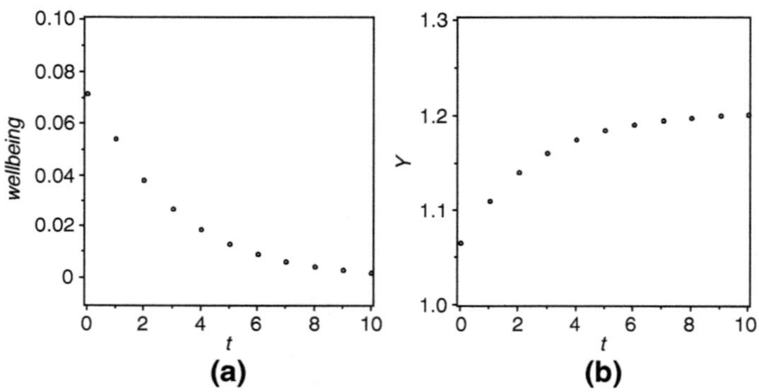

Fig. 9 Wellbeing and output evaluated along an orbit approaching the normalized fixed point

4 Wellbeing

In this section we analyze welfare (i.e. the value of the utility function) of each generation along equilibrium orbits by some numerical exercises. Figures 9a–b show the behavior of wellbeing and output Y along an orbit approaching the attracting normalized steady state ($\alpha = 0.17, \eta = 0.11, \sigma = 8, \varepsilon = 7.5$). Notice that output and wellbeing are inversely correlated and so the economy experiments undesirable economic growth: the increase in production and consumption of output is not able to compensate the negative effects of environmental degradation.

In the case showed in Fig. 10 ($\alpha = 0.17, \eta = 0.11, \sigma = 8, \varepsilon = 8.2$) there exist two attracting steady states, the normalized one and a steady state with higher labour and output levels.

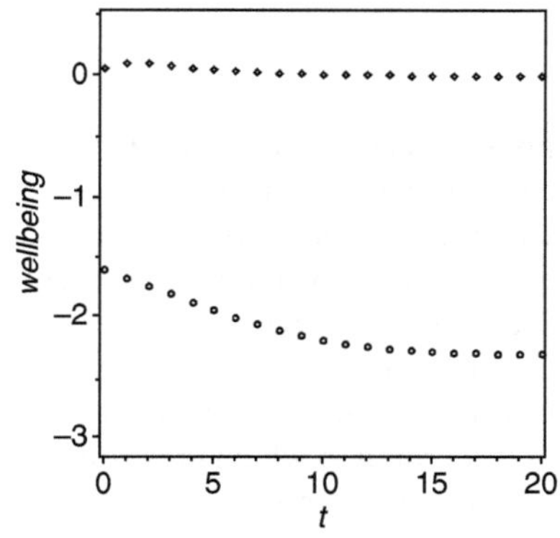

Fig. 10 Time evolution of wellbeing

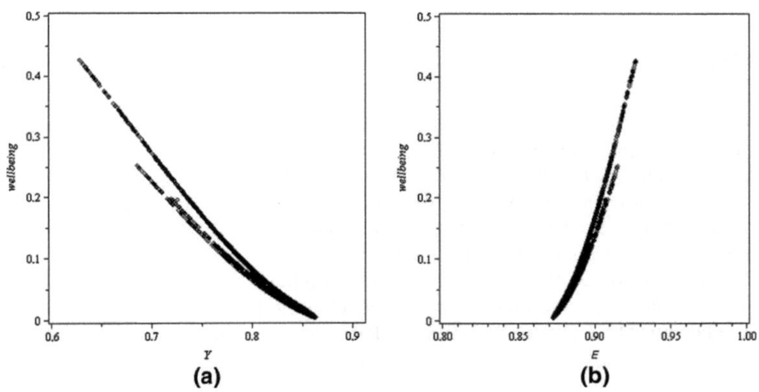

Fig. 11 Wellbeing, output Y and stock E of the environmental resource evaluated along orbits approaching a strange attractor

In such context, wellbeing is evaluated along an orbit approaching the normalized steady state (that with the higher wellbeing level) and along one approaching the other attracting steady state.

In Fig. 11a–b wellbeing, output Y and the stock of the environmental resource E are evaluated along orbits approaching a strange attractor ($\alpha = 0.17$, $\eta = 0.11$, $\sigma = 8$, $\varepsilon = 8.5$). Notice that E is positively correlated with welfare while the opposite holds for Y. In such context, the normalized fixed point is characterized by levels of output and labour lower than those corresponding to the orbits plotted in Fig. 11a–b and it Pareto-dominates them.

5 Conclusions

By analyzing an overlapping generations model in which the utility function is non separable with respect to private goods and free access environmental goods, we have shown that undesirable growth paths may emerge in the context in which private and environmental goods are substitutes; that is, growth paths may exist that are Pareto-dominated by other orbits characterized by lower private production and consumption levels. In such context, we have also showed that two types of indeterminacy may occur; a *short run indeterminacy*, associated (as usual) to the existence of an attracting fixed point, according to which the orbit that will be followed by the economy during *transient dynamics* is not predictable (because it depends on the choice of the initial value of labour input), and a *long run indeterminacy* according to which the initial choice of the labour input deeply affect long run dynamics.

References

Antoci, A. (2009). Environmental degradation as engine of undesirable economic growth via self-protection consumption choices. *Ecological Economics, 68*, 1385–1397.

Antoci, A., & Bartolini, S. (1999). *Negative externalities as the engine of growth in an evolutionary context* (Fondazione Eni Enrico Mattei Working paper 83.99). Milan

Antoci, A., & Bartolini, S. (2004). Negative externalities and labor input in an evolutionary game. *Environment and Development Economics, 9*, 591–562.

Antoci, A., Galeotti, M., & Russu, P. (2005). Consumption of private goods as substitutes for environmental goods in an economic growth model. *Nonlinear Anal-Model Control, 10*, 3–34.

Antoci, A., Galeotti, M., & Russu, P. (2007). Undesirable economic growth via economic agents' self-protection against environmental degradation. *Journal of The Franklin Institute, 344*, 377–390.

Antoci, A., Borgesi, S., & Galeotti, M. (2008). Should we replace the environment? Limits of economic growth in the presence of self-protective choices. *International Journal of Social Economics, 35*, 283–297.

Bartolini, S., & Bonatti, L. (2002). Environmental and social degradation as the engine of economic growth. *Ecological Economics, 43*, 1–16.

Bartolini, S., & Bonatti, L. (2003). Undesirable growth in a model with capital accumulation and environmental assets. *Environment and Development Economics, 8*, 11–30.

Benhabib, J., & Farmer, R. E. (1999). Indeterminacy and sunspots in macroeconomics. In J. B. Taylor & M. Woodford (Eds.), *Handbook of macroeconomics*. Amsterdam: North-Holland.

Benhabib, J., Nishimura, K., & Shigoka, T. (2008). Bifurcation and sunspots in the continuous time equilibrium model with capacity utilization. *Journal of Economic Theory, 4*, 337–355.

Bennet, R. L., & Farmer, R. E. (2000). Indeterminacy with non-separable utility. *Journal of Economic Theory, 93*, 118–143.

Cazzavillan, G., Lloyd-Braga, T., & Pintus, P. (1998). Multiple steady states and endogenous fluctuations with increasing returns to scale in production. *Journal of Economic Theory, 80*, 60–107.

Cazzavillan, G. (2001). Indeterminacy and endogenous fluctuations with arbitrarily small externalities. *Journal of Economic Theory, 101*, 133–157.

Chen, B. L., & Lee, S. F. (2007). Congestible public goods and local indeterminacy: A two-sector endogenous growth model. *Journal of Economic Dynamics and Control, 31*, 2486–2518.

Coury, T., & Wen, Y. (2009). Global indeterminacy in locally determinate real business cycle models. *International Journal of Economic Theory, 101*, 133–157.

Duranton, G. (2001). Endogenous labour supply, growth and overlapping generations. *Journal of Economic Behavior and Organization, 44*, 295–314.

Grandmont, J. M., Pintus, P., & de Vilder, R. (1998). Capital-labor substitution and competitive nonlinear endogenous business cycles. *Journal of Economic Theory, 80*, 14–59.

Hueting, R. (1980). *New scarcity and economic growth. More welfare through less production?* Amsterdam: North Holland.

Itaya, J.-I. (2008). Can environmental taxation stimulate growth? The role of indeterminacy in endogenous growth models with environmental externalities. *Journal of Economic Dynamics and Control, 32*, 1156–1180.

Krugman, P. (1991). History versus expectations. *Quarterly Journal of Economics, 106*, 651–667.

Leipert, C. (1989). National Income and economic growth: The conceptual side of defensive expenditures. *Journal of Economic Issues, 23*, 843–856.

Leipert, C., & Simonis, U. E. (1988). Environmental damage – environmental expenditures: Statistical evidence on the federal republic of Germany. *International Journal of Social Economics, 15*, 37–52.

Matsuyama, K. (1991). Increasing returns, industrialization, and indeterminacy of equilibrium. *Quarterly Journal of Economics, 106*, 587–597.

Meng, Q., & Yip, C. K. (2008). On indeterminacy in one-sector models of the business cycle with factor-generated externalities. *Journal of Macroeconomics, 30*, 97–110.

Pintus, P., Sands, D., & deVilder, R. (2000). On the transition from local regular to global irregular fluctuations. *Journal of Economic Dynamics and Control, 24*, 247–272.

Reichlin, P. (1986). Equilibrium cycles in an overlapping generations economy with production. *Journal of Economic Theory, 40*, 89–102.

Shogren, J. F., & Crocker, T. D. (1991). Cooperative and non-cooperative protection against transferable and filterable externalities. *Environmental and Resource Economics, 1*, 195–213.

Zhang, J. (1999). Environmental sustainability, nonlinear dynamics and chaos. *Economic Theory, 14*, 489–500.

Stock Dynamics in Stage Structured Multi-agent Fisheries

En-Guo Gu and Fabio Lamantia

1 Introduction

The sustainable use of public renewable resources is a crucial issue for the long run survival of mankind. With a rapidly increasing population and a quick economy development, overexploitation of worldwide renewable resources seriously affects their ability to renew themselves and therefore their sustainable use (Food and Agriculture Organization, 2004; Garcia & Grainger, 2005). To complicate the problem further, the modelling of commercial exploitation of renewable resources represents an extremely challenging task, as it always involves nonlinear interaction among many different components (biological, economic, social) as well as uncertainty. In particular the issue of fishery management with chaotic and catastrophic dynamics has been thoroughly discussed in Rosser (2002a). Many researchers have investigated the dynamic of an exploited biomass regarded as a single species (Bischi & Lamantia, 2007; Bischi, Kopel, & Szidarovszky, 2005; Clark, 1990; Fan & Wang, 1998; Gu, 2007). Recently also the evolution of an exploited stage-structured single species has been addressed (see Jing and Ke (2004); Song and Chen (2002); Gao, Chen, and Sun (2005)). This analysis is of particular significance especially for those many species whose individuals have different economic value at different ages. For example, little eels are often called "soft gold," for their high economic value, so many agents are interested in harvesting only little eels. But the case is exactly the opposite for those species whose immature individuals have negligible economic value, so that exploiters want to harvest only the mature population and let the immature population grows, so that it can acquire a greater value. Often also public regulators try to direct the harvesting activity toward a target stage, for instance by limiting the use of trawl with too small meshes in order to protect the infant population.

When multi-agents (societies, countries, or exploiters) compete for public resource exploitation, their strategic interaction can be modeled within the setting

F. Lamantia (✉)
Department of Business Sciences, University of Calabria, Via P. Bucci 3C, Rende (CS), Italy,
e-mail: lamantia@unical.it

G.I. Bischi et al. (eds.), *Nonlinear Dynamics in Economics, Finance and the Social Sciences*, DOI 10.1007/978-3-642-04023-8_4,

of (Cournot) oligopoly games, which have attracted economists' interest in recent years as (apparently) simple models capable of originating complex dynamics (see Rosser (2002b)). In fishery economics game-theoretic models are more and more employed (see, amongst many others, Bischi and Lamantia (2007); Clark (1990); Hanneson (1995); Mckelvey (1997); Mesterton-Gibbons (1993); Szidarovszky and Okuguchi (1998, 2002)).

In recent papers on the dynamic of an exploited stage-structured species, decisions on harvesting do not presuppose strategic interaction among agents, as they are based on concepts such as maximum sustainable yield (Jing & Ke, 2004) or constant effort exploitation (Gao, Chen, & Sun 2005; Song & Chen 2002). Instead in this paper we derive a total harvesting function within a game-theoretic framework, as outlined below. First we assume that the population of exploiters is ex-ante subdivided into two fractions: one fraction $0 \leq \gamma \leq 1$ of the population acts as a cooperative venture and consequently tries to maximize the overall profit of the coalition; on the other hand, each exploiter in the complementary fraction of the population, $(1 - \gamma)$, behaves as a selfish profit maximizer (referred to as a "defector" in the following). These assumptions generalize those given in Bischi, Kopel, and Szidarovszky (2005), as cooperators and defectors are assumed to coexist and the two limiting cases of all cooperators and all defectors considered in that paper are here obtained as limiting case, given by $\gamma = 1$ and $\gamma = 0$, respectively. A fishery model with interaction between cooperators and defectors has been proposed in Bischi, Lamantia, and Sbragia (2004), where, differently from the present paper, the fraction of agents in the two groups dynamically changes according to an evolutive process.

Assuming that a representative agent of each type (cooperator or defector) harvests at each time period exactly the quantity prescribed by its Nash equilibrium strategy (see Sethi and Somanathan (1996)), we derive a total harvesting function for each age class by each representative agent. Note that in this model, agents do not aim at preserving the resource for future generations, as they are not enforced by an authority to do so. In fact from the point of view of intergenerational altruism, we can say that all agents are non cooperators. In any case this kind of model is interesting, since it can shed some light on the dynamics of unregulated harvesting and provide hints on the type of control that a regulator can pose to enable resource conservation.

Although we formulate the game theoretical model with m exploitable age classes, we consider its dynamic version for the case of a fishery where a single species is subdivided into two exploitable stages (young and mature). In this way, we keep the model tractable still capturing its main features. Moreover the biological model without harvesting (proposed in Gao and Chen (2005)) has the simplest asymptotic behavior, namely convergence to a globally stable fixed point, which can correspond to extinction of the species or to the natural carrying capacity, according to the parameter values. Consequently the complex dynamic we find is entirely induced by human harvesting.

All in all we consider three main cases for the discrete time dynamical systems at hand, i.e., the exclusive harvesting of young individuals, mature ones and the complete model with harvesting from both stages. Since, in general, landed resource from different stages can be sold in different markets, those goods are sold in two (independent) markets in the complete model. For each case we carry out separate analysis, first studying the existence of positive equilibria and then exploring, mainly numerically, their local stability and asymptotic behavior. In order to answer questions related to the crisis of extinction, we use a global dynamic approach based on numerical and geometric methods to analyze the topological structure of the set of initial conditions which generate acceptable or feasible time path. We believe that an important parameter in this model is represented by the fraction of cooperators γ, which can be increased by an authority whenever conservative consciousness of exploiters is developed. Therefore we mainly focus on the impact of this parameter on the long-run dynamic of the biomass.

The paper is organized as follows. In Sect. 2, a game-theoretical model with multi-agents is outlined, leading to the formulation of a total harvest function. In Sect. 3 the dynamical model of resource exploitation is derived, under different assumptions on the harvested stage. In Sect. 4 existence of positive equilibria and their local stability are studied. Section 5 focuses on some global dynamic issues, mainly studying the impact of the cooperative rate on the structure of the feasible set. Section 6 concludes.

2 The Game-Theoretical Model

Let us assume that n players can harvest in a fishery subdivided in m age classes. The landing from each age class is sold in a different market, as it has, in general, its own economic value. The inverse demand function of harvested stock for the jth market, $j = 1, \ldots, m$ is given by

$$p_j = a_j - b_j H_j, \quad H_j = \sum_{i=1}^{n} h_{i,j}, \tag{1}$$

with maximal price and marginal demand given respectively by $a_j, b_j > 0$. Total harvesting in the j-th age class is denoted by $H_j(t)$ and so $h_{i,j}(t) = x_j(t) q_{i,j} e_{i,j}$ is the amount of resource in age class j harvested by agent i at time period t. We assume that individual harvesting for age class j is proportional to the resource stock level in the class, a specific catchability coefficient (related to the adopted technology) and the exerted harvesting effort, denoted respectively by $x_j(t)$, $q_{i,j}$, $e_{i,j}$. On the cost side we assume that each player's harvesting cost depends on the harvesting effort. In the easiest case, namely the linear one, the cost function of player i is given by

$$C_i = C_i(e_1, \ldots, e_m) = \left(\sum_{j=1}^{m} c_{i,j} e_{i,j} \right) - c_i^f, \tag{2}$$

where c_i^f is a fixed cost and $c_i > 0$ represents a variable cost or a technological parameter.

As we assume that the technology levels are fixed, the only decision variable is given by the harvesting effort to exert for each age class. Moreover we assume that $n\gamma$ agents act as one player, so forming a cooperative venture. Consequently they try to maximize the overall profit of coalition (here $0 \leq \gamma \leq 1$ is the fraction of agents with cooperative attitude toward exploitation); the remaining $(1-\gamma)n$ agents behave as selfish profit maximizers.

Let $h_{i,j}^c$ and $h_{v,j}^d$ represent the quantities harvested (and sold) on market j by cooperators $i, i = 1, 2, \cdot, n\gamma$, and defectors $v, v = 1, 2, \cdots, (1-\gamma)n$ respectively. Total harvest (and supply) on market j is $H_j = H_j^c + H_j^d = \sum_{i=1}^{n\gamma} h_{i,j}^c + \sum_{i=1}^{n(1-\gamma)} h_{i,j}^d$. Therefore, the expected profit of ith cooperator and defectors are

$$\pi_i^T = \sum_{j=1}^{m} h_{i,j}^T \left(a_j - b_j H_j\right) - C_i = \sum_{j=1}^{m} h_{i,j}^T \left[a_j - b_j \left(\sum_{i=1}^{n\gamma} h_{i,j}^T + \sum_{i=1}^{n(1-\gamma)} h_{i,j}^T \right) \right] - C_i, \tag{3}$$

where $T \in \{c, d\}$ denotes the type of the agent (cooperator or defector) and C_i is given in (2). Consequently each agent decides how much effort $e_{i,j}$ to exert for each subclass j, by solving m optimization problems. Formally there are m first order conditions for each agent, but if we construct the model in this way, it is straightforward to observe that this maximization problem is equivalent to m distinct maximization problems (one for each age class), because it is $\frac{\partial^2 \pi_i}{\partial e_{i,p} e_{i,q}} = 0$ for classes $p \neq q$.

The defectors solve the optimization problem $\max_{e_{i,j}^d} \pi_i^d$, which leads, assuming interior optimum, to the following m FOCs

$$\frac{\partial \pi_i^d}{\partial e_{i,j}^d} = (a_j - b_j H_j) q_{i,j}^d x_j - b_j q_{i,j}^d x_j h_{i,j}^d - c_{i,j}^d = 0, \quad j = 1, \ldots, m. \tag{4}$$

Instead, each cooperator determines $e_{i,j}^c$ by solving the optimization problem $\max_{e_{i,j}^c} \pi$, where $\pi = \sum_{i=1}^{n\gamma} \pi_i^c$ denotes the total profit of the cooperative venture. Assuming interior optimum also in this optimization problem, the first order conditions are

$$\frac{\partial \pi}{\partial e_{i,j}^c} = (a_j - b_j H_j) q_{i,j}^c x_j - b_j q_{i,j}^c x_j H_j^c - c_{i,j}^c = 0, \quad j = 1, \ldots, m. \tag{5}$$

By employing conditions (4) or (5), each agent decides her harvesting activity according to the corresponding Nash equilibrium level, as proposed in Sethi and Somanathan (1996).

Dividing equation (4) by $q_{i,j}^d x_j$ and then adding it for all i, we have

$$n(1-\gamma)(a_j - b_j H_j) - b_j H_j^d - \sum_{i=1}^{n(1-\gamma)} \frac{c_{i,j}^d}{q_{i,j}^d} \frac{1}{x_j} = 0. \tag{6}$$

Adding equation (5) for all i and then dividing it by $(\sum_{i=1}^{n\gamma} q_{i,j}^c)x_j$, it results

$$(a_j - b_j H_j) - b_j H_j^c - \frac{\sum_{i=1}^{n\gamma} c_{i,j}^c}{\sum_{i=1}^{n\gamma} q_{i,j}^c} \frac{1}{x_j} = 0. \tag{7}$$

Adding equation (6) and (7), we obtain the total harvesting function for age class j as

$$H_j(x_j) = \frac{1}{(n+2-n\gamma)b_j}\left[(n+1-n\gamma)a_j - \left(\sum_{i=1}^{n(1-\gamma)} \frac{c_{i,j}^d}{q_{i,j}^d} + \frac{\sum_{i=1}^{n\gamma} c_{i,j}^c}{\sum_{i=1}^{n\gamma} q_{i,j}^c}\right)\frac{1}{x_j}\right], \tag{8}$$

which is meaningful provided that $x_j > H_j(x_j) \geq 0$. As conditions ensuring meaningful harvesting are not obvious in the general case, we discuss them in the next section, under the same assumptions on agents' homogeneity carried out for the dynamic model.

3 Dynamic of the Stage-Structured Single Species with Harvesting

To reduce the dimension of the dynamical system to a bidimensional one, we consider only two age classes. Without harvesting, the single-species population model with two stages can be expressed by the following biological growth law (see Gao and Chen (2005); Tang and Chen (2002)):

$$\begin{cases} \dot{x}(t) = re^{-N(t)}y(t) - d_1 x(t) - \delta x(t), \\ \dot{y}(t) = \delta x(t) - d_2 y(t), \end{cases} \tag{9}$$

where x and y are the population densities of immature and mature, respectively. $N(t) = x(t) + y(t)$, $re^{-N(t)}$ is the birth rate of mature population, $d_1 > 0, d_2 > 0$ are the death rate constant of immature and mature respectively, and $r > d_1 + d_2$. The maturity rate $\delta > 0$ determines the mean length of the juvenile period.

According to (9), we can investigate the corresponding discrete population model:

$$\begin{cases} x_{n+1} = re^{-N_n} y_n + (1 - d_1 - \delta)x_n \\ y_{n+1} = \delta x_n + (1 - d_2)y_n \end{cases} \quad n \in N, \tag{10}$$

where $N_n = x_n + y_n$, N denotes the set of non-negative integers. Let $\alpha = 1 - d_1 - \delta$, $\mu = 1 - d_2$. For ecological reasons, we assume that $0 < \alpha < 1$ so that we have $0 < \delta < 1,\ 0 < \mu < 1,\ 0 < \alpha + \delta < 1$. System (10) yields

$$\begin{cases} x_{n+1} = re^{-N_n} y_n + \alpha x_n \\ y_{n+1} = \delta x_n + \mu y_n \end{cases} \quad n \in N. \tag{11}$$

Clearly, $E_0 = (0, 0)$ is the trivial equilibrium of system (11), corresponding to the extinction of the species. There exists also a unique positive equilibrium

$$E^* = \left(\frac{1-\mu}{\delta + 1 - \mu} ln \frac{r\delta}{(1-\delta)(1-\mu)}, \frac{\delta}{\delta + 1 - \mu} ln \frac{r\delta}{(1-\delta)(1-\mu)} \right), \tag{12}$$

provided that $R_0 = \frac{r\delta}{(1-\delta)(1-\mu)} > 1$, which can be referred to as a natural *carrying capacity* equilibrium.

For the local and global stability of equilibria E_0 and E^*, we recall the following theorems (see Gao and Chen (2005)):

Theorem 1. *E_0 is locally asymptotically stable if $R_0 < 1$, and unstable if $R_0 > 1$; E^* is locally asymptotically stable if $R_0 > 1$.*

Theorem 2. *E_0 is globally asymptotically stable if $R_0 < 1$; E^* is globally asymptotically stable if $R_0 > 1$.*

As remarked in Gao and Chen (2005), $R_0 = \frac{r\delta}{(1-\delta)(1-\mu)}$ is the intrinsic net reproductive number, or net reproductive rate. From theorem (2), if $R_0 > 1$, then the positive equilibrium E^* exists and is globally asymptotically stable, so that, on average, individuals replace themselves before they die, leaving constant the unharvested population.

Now we introduce harvesting, as in the game theoretic framework described in the previous section. The dynamics of harvested stock can be represented by the two-dimensional system:

$$\begin{cases} x_{n+1} = re^{-N_n} y_n + \alpha x_n - \Theta_1 H_1(x_n) \\ y_{n+1} = \delta x_n + \mu y_n - \Theta_2 H_2(y_n) \end{cases} \quad n \in N, \tag{13}$$

where Θ_1, Θ_2 are two binary variables taking on the values $\{0, 1\}$. The reason to introduce variables Θ_i is that of modelling the protection of a given age class by the government so that if $\Theta_i = 0$ then no fishing from class i is allowed. Moreover we can study (at least numerically) what happens where $\Theta_1 = 1$ and $\Theta_2 = 1$.

Denoting by $A_j = [n(1-\gamma)+1]\, a_j$, $B_j = [n(1-\gamma)+2]\, b_j$, $\beta_j = \sum_{i=1}^{n(1-\gamma)} \frac{c_{i,j}^d}{q_{i,j}^d} + \frac{\sum_{i=1}^{n\gamma} c_{i,j}^c}{\sum_{i=1}^{n\gamma} q_{i,j}^c}$, $j = 1, \ldots, m$, then we have

$$H_j(x_j) = \frac{A_j}{B_j} - \frac{\beta_j}{B_j x_j}, \tag{14}$$

which is nonnegative if $x_j \geq \frac{\beta_j}{A_j}$. However, it is important to remark that the proposed total harvesting function is meaningful provided that $x_j > H_j(x_j) \geq 0$. When $B_j > \frac{A_j^2}{4\beta_j}$ then condition $x_j > H_j(x_j)$ holds true, and so harvesting is meaningful whenever $x_j \geq \frac{\beta_j}{A_j}$. On the other hand when $B_j \leq \frac{A_j^2}{4\beta_j}$, then it is easy to show that a sufficient condition for $x_j > H_j(x_j)$ is that the biomass in age class j is above a survival threshold, namely $x_j > \frac{1}{2}\left(\frac{A_j}{B_j} + \sqrt{\frac{A_j^2 - 4B_j\beta_j}{B_j^2}}\right)$, which also implies nonnegative harvesting. For these reasons we will assume, also in numerical simulations, that $H_j(x) = \min\left[x, \max\left[0, \frac{A_j}{B_j} - \frac{\beta_j}{B_j x}\right]\right]$.

To simplify the notation, we set henceforth $(x_n, y_n) = (x, y)$, and we denote by $'$ the unit-time advancement operator.

Note that for $\Theta_1 = 1$ and $\Theta_2 = 0$, i.e., when only immature stock is harvested, the two-dimensional dynamical systems in (13) can be represented as iterated point mapping[1]

$$T_1 : \begin{cases} x' = rye^{-(x+y)} + \alpha x - H_1(x), \\ y' = \delta x + \mu y, \end{cases} \tag{15}$$

whereas for $\Theta_1 = 0$ and $\Theta_2 = 1$ the system in (13) becomes

$$T_2 : \begin{cases} x' = rye^{-(x+y)} + \alpha x, \\ y' = \delta x + \mu y - H_2(y). \end{cases} \tag{16}$$

Finally for $\Theta_1 = 1$ and $\Theta_2 = 1$ we have the dynamics of a species where no fishing limitation at all is imposed by the government, i.e.,

$$T_3 : \begin{cases} x' = rye^{-(x+y)} + \alpha x - H_1(x), \\ y' = \delta x + \mu y - H_2(y). \end{cases} \tag{17}$$

For simplicity from now on, we assume that all agents of the same type $T \in \{c, d\}$ (cooperators or defectors) are homogeneous, i.e., $c_i^T = C^T$, $q_i^T = q^T$, $h_{i,j}^T = h_j^T$, $e_{i,j}^T = e_j^T$, so that we can also write $\beta = n(1-\gamma)\frac{C^d}{q^d} + \frac{C^c}{q^c}$.

Under these assumptions, by (3) and (14), profits of a representative agent of type $T \in \{c, d\}$ from harvesting x_j from stage j can be written as

$$\pi_j^T = \frac{h_j^T}{n(1-\gamma)+2}\left[a_j + \frac{\beta_j}{x_j}\right] - \left(C^T e_j^T + c^f\right), \tag{18}$$

which are positive for all x_j, provided that the coefficients in the cost function are sufficiently low.

[1] When only one age class is harvested, we suppress the subscripts in A, B, β, as no confusion arises.

An interesting characterization can be given in particular, when all agents (cooperators and defectors) are homogeneous, i.e., if $q^c = q^d = q$ and $C^c = C^d = C$. In this case condition $x_j > \frac{\beta_j}{A_j}$ (ensuring nonnegative harvesting) reduces to $x_j > \frac{C}{a_j q}$. From homogeneity within groups of agents and from (8), we have that $\frac{\partial H_j}{\partial \gamma} = \frac{n\left[2C^d - q^d\left(\frac{C^c}{q^c} + a_j x_j\right)\right]}{bq^d x_j (2+n(1-\gamma))^2} < 0$ if and only if $x_j > \frac{2q^c C^d - q^d C^c}{a_j q^c q^d}$. Again, when cooperators and non-cooperators are homogeneous this last condition reduces to $x_j > \frac{C}{a_j q}$, which coincides with the condition on non-negativity of harvesting. To Conclude, agent homogeneity is sufficient to ensure that as the cooperative level γ is increased a lower total harvesting for age class j is achieved. This point is particularly important, since more cooperation among agents is equivalent to lower harvesting in each stage class.

4 Fixed Points and Local Stability

We begin our analysis on equilibria in the benchmark cases (15) and (16) where only one age class is exploited. Then we briefly consider the general case of harvesting in both stages (17). As we showed earlier, when biomass level in a given age class is too low, no harvesting takes place so that the model reduces to the one considered in Gao and Chen (2005), whose equilibrium analysis has been previously outlined. Hence, for the analysis of the three main cases (15), (16), and (17), the harvesting function is taken as in (14), i.e., without constraints, unless otherwise stated.

4.1 Exclusive Harvesting of Infant Population

Let us consider the case of exclusive harvesting of infant population. The positive steady state $E(\bar{x}, \bar{y})$ of the system (15) satisfies the following equation:

$$1 - \alpha + \frac{A}{Bx} - \frac{\beta}{Bx^2} = \frac{r\delta}{1-\mu} \exp\left(-\frac{1+\delta-\mu}{1-\mu} x\right). \tag{19}$$

Condition (19) says that the fixed point coordinate $\bar{x}$ is determined from the intersection points of an exponential function, $\psi(x) = \frac{r\delta}{1-\mu} \exp\left(-\frac{1+\delta-\mu}{1-\mu} x\right)$ with the function $\phi(x) = 1 - \alpha + \frac{A}{Bx} - \frac{\beta}{Bx^2}$. Clearly, the function $\psi(x)$ is strictly decreasing and convex, with $\psi(0) = \frac{r\delta}{1-\mu} > 0$ and asymptotic to the horizontal axis ($\lim_{x\to+\infty} \psi(x) = 0$). On the other hand, the function $\phi(x)$ is unimodal, being $\phi'(x) = \frac{2\beta - Ax}{Bx^3}$, $\phi'(x) > 0$ if $0 < x < \frac{2\beta}{A}$ and $\phi'(x) < 0$ if $x > \frac{2\beta}{A}$, and with $\lim_{x\to 0^+} \phi(x) = -\infty$ and $\lim_{x\to+\infty} \phi(x) = 1 - \alpha > 0$. Notice that $\phi(\widehat{x}) = 0$ at $\widehat{x} = \frac{-A + \sqrt{A^2 + 4B\beta(1-\alpha)}}{2B(1-\alpha)} > 0$. From the geometric properties of $\psi(x)$ and $\phi(x)$, it

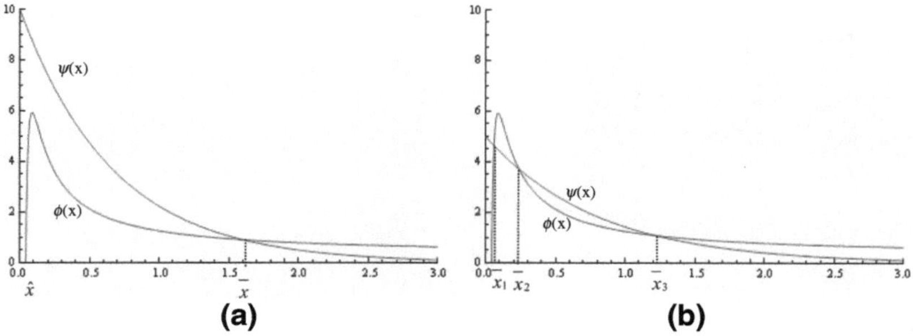

Fig. 1 Sketches of the existence of the positive equilibrium for the system (15). Parameters are given as $r = 20$, $\alpha = 0.7$, $\mu = 0.6$, $a = 3$, $b = 1.5$, $n = 100$, $C^c = C^d = 0.1$, $q^c = q^d = 0.75$, $\gamma = 1$. (**a**) $\delta = 0.2$: a unique equilibrium $(\overline{x}, \overline{y})$ exists; (**b**) $\delta = 0.1$: three equilibria with coordinates $\overline{x}_1, \overline{x}_2, \overline{x}_3$ exist

follows that system (15) has at least one positive steady state $E = (\bar{x}, \bar{y})$ with coordinates $\bar{x}$ and $\bar{y}$ satisfying $\bar{x} > \widehat{x}$ and $\bar{y} > \frac{\delta \widehat{x}}{1-\mu}$ respectively (see Fig. 1a), provided that if $B > \frac{A^2}{4\beta}$ then $\bar{x} \geq \frac{\beta}{A}$, or if $B \leq \frac{A^2}{4\beta}$ then $\bar{x} > \frac{1}{2}\left(\frac{A}{B} + \sqrt{\frac{A^2-4B\beta}{B^2}}\right)$ so that harvesting at equilibrium is meaningful, as explained at the end of Sect. 3. Moreover multiple equilibria can be obtained (see Fig. 1b). In this case for the positive equilibrium $E = (\bar{x}, \bar{y})$ it must also be that $\bar{x} < \frac{2\beta}{A}$ and $\bar{y} < \frac{2\beta\delta}{A(1-\mu)}$. Sufficient conditions for uniqueness of the positive equilibrium can be stated. For instance, it is easy to verify that condition $\frac{r\delta}{(1-\mu)(1-\alpha)} < 1$ ensures that only one equilibrium exists. Condition $\psi(0) \leq \phi\left(\frac{2\beta}{A}\right)$, i.e., $\frac{r\delta-(1-\alpha)(1-\mu)}{1-\mu} < \frac{A^2}{4B\beta}$, ensures the uniqueness of a positive equilibrium $E = (\bar{x}, \bar{y})$ such that $\widehat{x} < \bar{x} < \frac{2\beta}{A}$ and $\frac{\delta\widehat{x}}{1-\mu} < \bar{y} < \frac{2\delta\beta}{A(1-\mu)}$.

In general, the stability conditions of the positive fixed points can not be analytically given, as the analytical expression of positive fixed points can not be gained and conditions on trace and determinants of the Jacobian matrix are quite involving. To give some insights on the stability of equilibria, we rely on numerical analysis, in particular to the so-called bifurcation diagrams, showing the possible long-term values (equilibria/fixed points, periodic or chaotic orbits) of the system as a function of one or two bifurcation parameters. As propensity to cooperate can be increased by improving the conservative consciousness of exploiters, we mainly focus on the impact of parameter γ on biomass for each age class.

To begin with, let us consider a situation with $n = 100$ agents, whose catchability coefficients and marginal costs are equal, disregarding the fact that they cooperators and defectors. We set $C^c = C^d = 0.5$, $q^c = q^d = 0.75$ and we assume that the landed resource (only infant population in this case) is sold in a market with maximum price $a_1 = a = 3$ and marginal demand $b_1 = b = 1.5$. Biological parameters, following Gao and Chen (2005), are given by $\alpha = 0.7$, $\delta = 0.2$, $\mu = 0.6$, and $r = 60$. Moreover we considered the initial conditions (i.c.) $x_0 = y_0 = 5$. As the net reproductive rate in this case is $R_0 = 37.5$, we know that, without harvesting, the system converges asymptotically to the carrying capacity equilibrium E^*

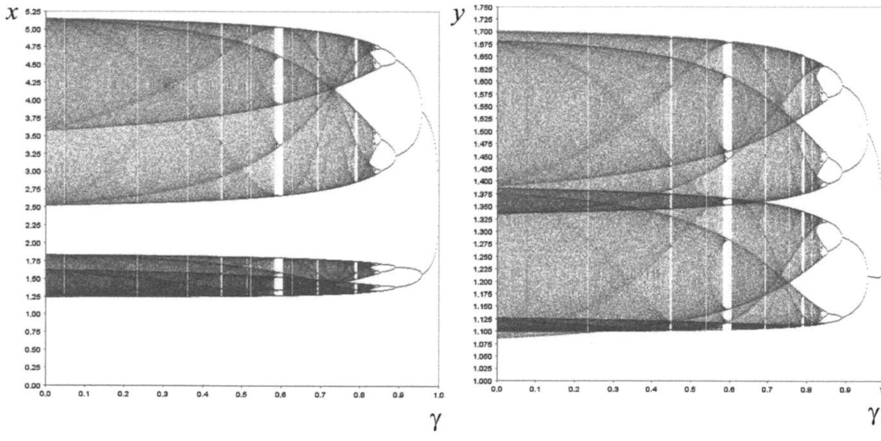

Fig. 2 Bifurcation diagrams of the system (15), with i.c. $x_0 = y_0 = 5$; $n = 100$; $C^c = C^d = 0.5$; $q^c = q^d = 0.75$; $a = 3$; $b = 1.5$; $\alpha = 0.7$; $\delta = 0.2$; $\mu = 0.6$; $r = 60$; $\gamma \in [0, 1]$

given in (12). With harvesting it is possible to show, at least numerically, that for any $\gamma \in [0, 1]$ a unique steady state exists.

As we showed earlier, in this case total harvesting is a decreasing function of the rate of cooperators γ, so we are interested in understanding how a change in the "propensity to cooperate" parameter γ influences the long run behavior of the system. In Fig. 2a, b two bifurcation diagrams for the state variables x and y are presented with bifurcation parameter $\gamma \in [0, 1]$. It is interesting to observe that when all agents are cooperators ($\gamma = 1$) the system still converges in the long run to an equilibrium point. However, as long as a very small amount of agents begins to defect, i.e., $\gamma < 1$, then the steady state looses stability through a flip bifurcation. For low values of γ, a two pieces chaotic attractor exists, which vanishes through a sequence of period halving bifurcations as γ is increased to the value $\gamma = 1$. Many other numerical experiments confirm the complex dynamic arising when only the infant population is exploited and $\gamma < 1$.

However, the fact that all agents are cooperators does not necessarily implies that the system converges to a fixed point. In fact another source of instability for this model is a decreasing marginal demand b. Let us consider again the parameters as in Fig. 2, with $\gamma \in [0, 1]$ and let us take the marginal demand $b \in [1.1, 3]$. The corresponding double parameters bifurcation diagram, depicted in Fig. 3 clearly shows the typical period doubling route to chaos, obtained by decreasing γ and/or b (see white area). Indeed for this parameter constellation, it is always possible, for any level of γ, that a stable fixed point flip bifurcates to a chaotic attractor as long as marginal demand b is reduced, or equivalently as long as the demand curve gets more inelastic, being the elasticity $E_d = -1 + \frac{a}{bq}$. This result is similar to the one obtained in Onozaki, Sieg, and Yokoo (2000). We incidentally observe that for essential goods, like food, price elasticity is usually low, so that complex behavior is reasonable for the model at hand.

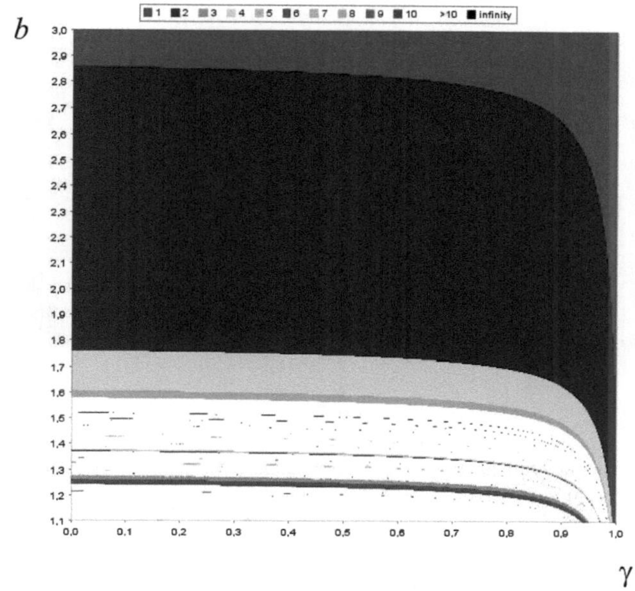

Fig. 3 Two-parameters bifurcation diagram for $b \in [1.1, 3]$ and $\gamma \in [0, 1]$ and other parameters as in Fig. 2

We end this section by showing a last bifurcation diagram, with all parameters as in Fig. 2, but $n = 50$ and $r = 2000$. Also in this case the unique equilibrium (12) is stable when no harvesting takes place. When all agents cooperate ($\gamma = 1$), a generic trajectory is attracted by a stable 8, that flip bifurcates to higher order cycles and a one piece chaotic attractor as soon as we decrease γ. It is interesting to observe the existence of a window with a three pieces chaotic attractor for $0.8566 \approx \gamma \approx 0.8801$ with a sudden jump in biomass of both age classes and a window with a period five cycle for $0.8127 \approx \gamma \approx 0.8177$ (see Fig. 4). From our numerical experiments, we can infer that, on average, resource variability for both age classes is decreasing in γ. This last case will be compared with the one obtained by harvesting only adult population.

4.2 Exclusive Harvesting of Adult Population and the Complete Case

The positive steady state $E(\bar{x}, \bar{y})$ of the system (16) satisfies the following equation:

$$\frac{(1-\alpha) f(y)}{y} = r \exp(-y - f(y)), \tag{20}$$

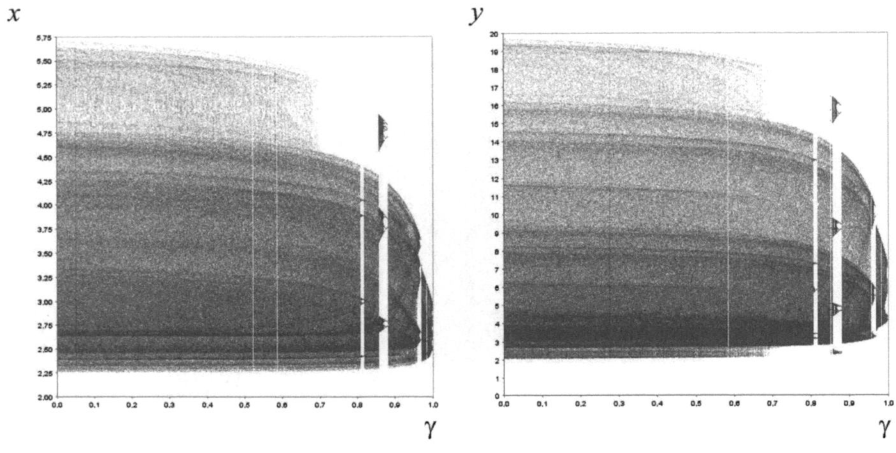

Fig. 4 Bifurcation diagrams for $\gamma \in [0, 1]$ of the system (15), with $n = 50$ and $r = 2000$ and all other parameters as in Fig. 2

where $f(y) = \frac{1}{\delta}\left[(1-\mu)y + \frac{A}{B} - \frac{\beta}{By}\right]$. Condition (20) says that the fixed point coordinate $\bar{y}$ is determined from the intersection points of the function $\eta(y) = r\exp(-y - f(y))$ with the function $\xi(y) = \frac{1-\alpha}{\delta}\left[(1-\mu) + \frac{A}{By} - \frac{\beta}{By^2}\right]$. Clearly, $\eta(y)$ is strictly decreasing, convex, and asymptotic to coordinate axes, being $\lim_{y\to 0^+}\eta(y) = +\infty$ and $\lim_{y\to+\infty}\eta(y) = 0$. On the other hand $\xi(y)$ is again an unimodal function, with $\xi'(y) = \frac{(1-\alpha)}{\delta}\frac{2\beta - Ay}{By^3}$, $\xi'(y) > 0$ if $0 < y < \frac{2\beta}{A}$ and $\xi'(y) < 0$ if $y > \frac{2\beta}{A}$, with $\lim_{y\to 0^+}\xi(y) = -\infty$ and $\lim_{y\to+\infty}\xi(y) = \frac{(1-\alpha)(1-\mu)}{\delta}$. Similarly as the previous case, we observe that $\xi(\widehat{y}) = 0$ at $\widehat{y} = \frac{-A+\sqrt{A^2+4B\beta(1-\mu)}}{2B(1-\mu)} > 0$. Therefore, we can state the existence for system (16) of one positive steady state $E(\bar{x}, \bar{y})$ with coordinate $\bar{y}$ satisfying $\widehat{y} < \bar{y}$ and $\bar{x} = f(\bar{y})$. Also in this case we can numerically verify the multiple equilibria can be obtained.

Again it is not possible to give here analytical stability conditions for the general case. Therefore we explore numerically the behavior of model (16) as main parameters are changed.

Let us reconsider parameters as in Fig. 4 but now for the model (16) where only adult population is exploited. This case is represented in Fig. 5, where, similarly to the case in the previous section, on average variability is decreasing in γ and periodic windows with low period cycles exist. For $\gamma \in (\widetilde{\gamma}, 1]$, with $\widetilde{\gamma} \approx 0.84382$, the only attractor of the model is a stable two-cycle. This stable two-cycle undergoes a (supercritical) Neimark–Sacker (NS) bifurcation at $\gamma = \widetilde{\gamma}$, through which a two-piece quasi periodic attractor becomes stable. Chaotic attractors are then created by sequences of flip bifurcations as γ is further reduced. In our numerical experiments, we did not observe NS bifurcations for the model of exclusive harvesting of infant population (15), so we conjecture that this is a typical feature of harvesting of the adult stage only.

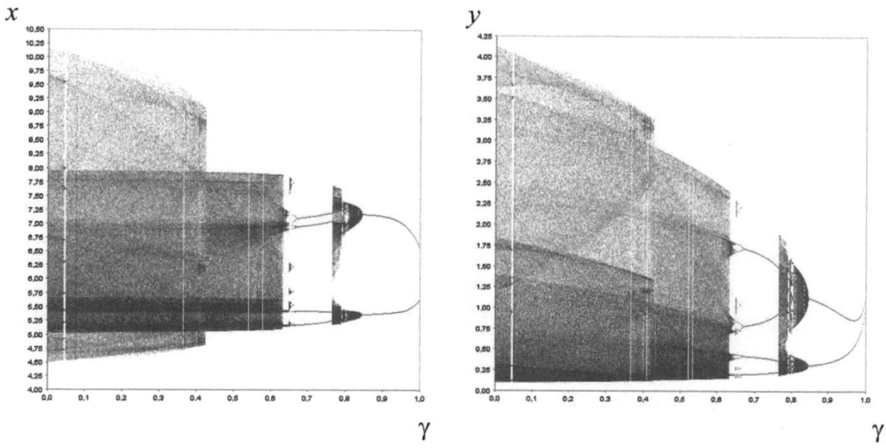

Fig. 5 Bifurcation diagrams for $\gamma \in [0, 1]$ of the system (16), with all parameters as in Fig. 4

Now we consider a case with heterogeneity between cooperators and defectors. As cooperative behavior could be regarded as a virtuous one, it can be either the case that defectors are punished by the government (for instance by imposing taxes on their catches) or that cooperators activity is less controlled by the authority. As a consequence, harvesting by cooperators can be cheaper and/or more effective, i.e., respectively $C^c < C^d$ and/or $q^c > q^d$. We investigate such a case with $n = 10$ agents who harvest only adult fish population, with $C^c = 0.1, C^d = 0.5, q^c = 0.3,$ and $q^d = 0.2$. Market demand (for the adult population) is specified by $a_2 = a = 3$, $b_2 = b = 1.5$. Biological parameters given by $r = 5,\ \alpha = 0.2,\ \delta = 0.6$, and $\mu = 0.9$. A unique positive fixed point $E(\bar{x}, \bar{y})$ exists for all γ and it is stable for all $\gamma \in [0, \gamma_c]$, where $\gamma_c = 0.9485$. As the value of the cooperative rate γ crosses the bifurcation value γ_c, the positive fixed point $E(\bar{x}, \bar{y})$ loses its stability through a flip bifurcation (see Fig. 6a, b). Further cascades of flip bifurcations lead to a chaotic attractor as the cooperative rate γ is further increased. For values of γ near its maximum level 1, diverging trajectories are obtained, as described in the next section. This result is the opposite of the one obtained when cooperators and defectors are homogeneous, in the sense that in this case the increment of the cooperative rate γ has a destabilizing role for the fixed point.

Now we briefly consider the case when both age-classes can be harvested. The corresponding dynamical system is given in (17). To analyze equilibria, we begin by considering the system (17) with $H_j(.)$ defined as in (14), so that a positive steady state $E(\bar{x}, \bar{y})$ of the system (17) satisfies the following equation:

$$\frac{A_1}{yB_1} - \frac{\beta_1}{B_1 y f(y)} + (1-\alpha)\frac{f(y)}{y} = r\ \exp(-y - f(y)) \tag{21}$$

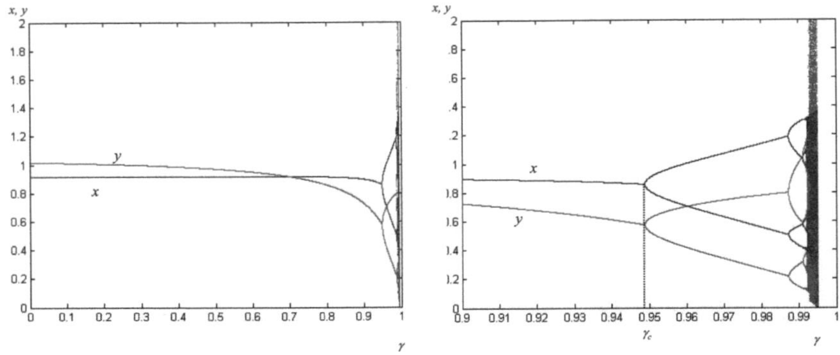

Fig. 6 From a simple attractor to chaos with agents' heterogeneity. Parameters are given as $n = 10$; $C^c = 0.1$; $C^d = 0.5$; $q^c = 0.3$; $q^d = 0.2$; $a = 3$; $b = 1.5$; $\alpha = 0.2$; $\delta = 0.6$; $\mu = 0.9$; $r = 5$; i.c. $x_0 = y_0 = 5$. (**a**) $\gamma \in [0, 1]$; (**b**) zoom in with $\gamma \in [0.9, 1]$

where $f(y) = \frac{1}{\delta}\left(y(1-\mu) + \frac{A_2}{B_2} - \frac{\beta_2}{B_2 y}\right)$ has already been studied in the previous paragraph. So the fixed point is an intersection between the strictly decreasing exponential function $\eta(y)$, already considered in the previous paragraph , with the function $\phi(y) = \frac{A_1}{yB_1} - \frac{\beta_1}{B_1 y f(y)} + (1-\alpha)\frac{f(y)}{y}$. In this case, we have again that $\lim_{y\to 0^+}\phi(y) = -\infty$ and $\lim_{y\to+\infty}\phi(y) = \frac{(1-\alpha)(1-\mu)}{\delta}$, but now there also exists a unique real number $\widehat{y} = \frac{-A_2+\sqrt{(A_2)^2+4B_2\beta_2(1-\mu)}}{2B_2(1-\mu)} > 0$ such that the line $y = \widehat{y}$ is a vertical asymptote for $\phi(y)$, with $\lim_{y\to\widehat{y}^-}\phi(y) = +\infty$ and $\lim_{y\to\widehat{y}^+}\phi(y) = -\infty$. Also considering the constraints on harvesting, we have that the dynamical system (17) has at least two equilibria (x_1^*, y_1^*) and (x_2^*, y_2^*), with $y_1^* < \widehat{y} < y_2^*$ and $x_1^* < f(\widehat{y}) < x_2^*$ provided that at the lower equilibrium levels x_1^* and y_1^*, biomass at each stage is above the threshold specified in (3). Depending on parameters, other cases can be treated similarly.

As for the previous cases, also here it is necessary to rely on numerical simulations. However we do not present new figures, as the dynamical scenarios are similar to the ones previously described. In particular we remark that for low levels of reproductive capacity r, the generic trajectory of system (17) diverges to minus infinity as long as harvesting takes place. In this case harvesting will so deplete the resource that it becomes extinct in finite time.

5 Some Insights on Global Dynamics

In this section, we use a global dynamic approach based on numerical methods to analyze the topological structure of the set of initial conditions which generate acceptable or feasible time path. Following Gu (2007), we shall refer to the set of points leading to a whole trajectory contained in $R_+^2 = \{(x, y)|x \geq 0, y \geq 0\}$ as the feasible set.

In the numerical examples described in this section, we reconsider the case of agents' heterogeneity depicted in Fig. 6, i.e., biological parameters are given as $r = 5$, $\alpha = 0.2$, $\delta = 0.6$, and $\mu = 0.9$, players' parameters as $n = 10$, $C^c = 0.1$, $C^d = 0.5$, $q^c = 0.3$, and $q^d = 0.2$ and market's parameters as $a = 3$ and $b = 1.5$. As we remarked in the previous section, in this case defectors bear higher costs for their activity, for instance because the government wants to punish them for their attitude.

We focus on the impact of changes in the cooperative ratio on the extent of the feasible set. The influence of changes in the cooperative rate is of significant interest to policy makers, since the cooperative rate can be increased by developing the conservative consciousness of exploiters. Furthermore, we shall also investigate the influence of changes in the cooperative rate on the structure of attractors (stable equilibrium or cycles).

We first investigate the benchmark case (16), i.e., only mature biomass is harvested. Intuitively speaking, the high cooperative rate should prevent overexploitation and lead to more conservation of biomass. However, apparently counterintuitive results are possible when only adult population is harvested. This can be explained by the assumed agents' asymmetry, which advantages the cooperative attitude. Let us consider the cases in Fig. 7, where the phase space is depicted for different levels

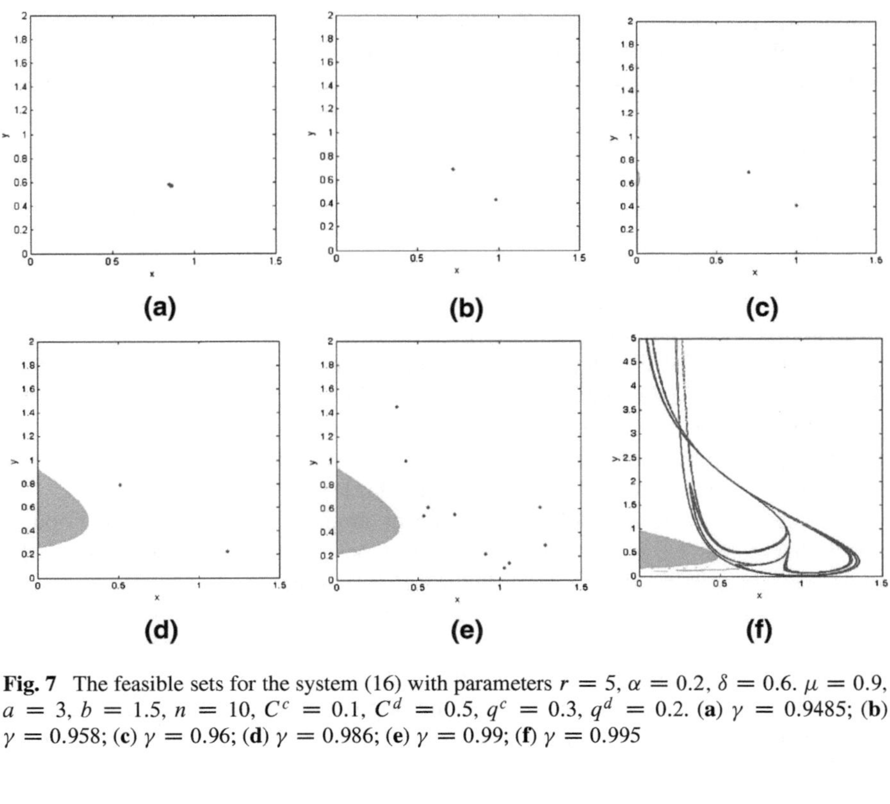

Fig. 7 The feasible sets for the system (16) with parameters $r = 5$, $\alpha = 0.2$, $\delta = 0.6$. $\mu = 0.9$, $a = 3$, $b = 1.5$, $n = 10$, $C^c = 0.1$, $C^d = 0.5$, $q^c = 0.3$, $q^d = 0.2$. (**a**) $\gamma = 0.9485$; (**b**) $\gamma = 0.958$; (**c**) $\gamma = 0.96$; (**d**) $\gamma = 0.986$; (**e**) $\gamma = 0.99$; (**f**) $\gamma = 0.995$

of γ. As the cooperative rate γ is increased, the feasible set shrinks. For example, the feasible set as shown in Fig. 7a, b is first quadrant R_+^2 for $\gamma < 0.958$. That is, for a value of γ up to a certain level, the steady state (see Fig. 7a) and, after the flip bifurcation (at $\gamma_c = 0.9485$ as described in the previous section), the two-cycle (see Fig. 7b) and higher order cycles (see the 10-cycle in Fig. 7e) are global attractors in R_+^2. As γ reaches the value $\gamma_d = 0.958$, one cyan tongue, which characterizes extinction of the species in finite time, appears in the first quadrant and its size increases as γ is increased from 0.96 to 0.995 (see Fig. 7c–f). In this case, the feasible set for $\gamma > 0.958$ is obtained by subtracting from the first quadrant R_+^2 one tongue, which represents the basin of attraction of diverging trajectories. Moreover chaotic attractors arise as the cooperation rate is close to its maximum level [see Fig. 7f where $\gamma_e = 0.995$].

From this analysis we draw the conclusion that, under particular circumstances, an increment in the fraction of agents who cooperate can lead to a reduction in stability, both in the sense of a destabilization of simple attractors (equilibrium or cycles) with birth of chaotic attractors and shrinking in the basin of attraction of feasible trajectories. In fact, the shrinking of the feasible set may cause higher probability of extinction. Moreover, despite a seemingly large feasible set [see again Fig. 7f], the boundary of the chaotic attractor is quite close to the boundary of its basin of attraction (white region). In this case, small perturbations in the biomass (e.g., an external event of modest proportions happens) might lead to the extinction of the species in finite time. So, although this dynamic is feasible over time, a small displacement of its trajectory could lead to extinction of the species. As shown in Fig. 7, in this example as the parameter γ is increased, the attractor changes from a steady state to a cycle to chaotic oscillation and also its basin of attraction acquires a more complex structure, with the appearing of tongues of the basin of diverging trajectories. However we remark that, in general, the complexity of the attractor and of its basin are uncorrelated phenomena in discrete dynamical systems.

As it is visible in Fig. 7f, the chaotic attractor is near the tongue of the unfeasible set, so that when γ is further increased, a contact between them occurs and a global bifurcation called final bifurcation (or boundary crisis) happens. After this contact, the disappearance of the chaotic attractor takes place so that a generic trajectory is unfeasible, i.e., extinction of the species in finite time occurs. We recall that the global bifurcations causing qualitative changes on the structure of the feasible set can be explained by using the concept of critical curves for a non-invertible (many to one) map (Mira, Gardini, Barugola, & Cathala, 1996).

For the same parameters but under the assumption that only infant population is harvested, we found from numerical experiments that for all $\gamma \in [0, 1]$, any trajectory starting in R_+^2 exits this set after a finite number of iterations. So the feasible set is always an empty set. Under this parameter constellation, harvesting the infant population leads to the extinction of the species, whereas the harvesting of adult population results sustainable for most levels of γ. In fact, for values of γ before the boundary crisis, the population might fluctuate, but never becomes extinct as long as the initial fish stock is in the feasible set (see again Fig. 7).

6 Conclusions

Many researchers have studied the exploitability of common resources based on game theoretical models. Most of them assume that all agents maximize their individual profit competing for common resource or maximize their whole profit, i.e., they act as one agent. However, in the real world, agents may have different level of consciousness for protecting the resource, and thus they may take different lines of action. Furthermore, the underlying dynamical model is often based on exploitation of a single species with no age structure. However, many species may have different economic value in different ages, so that it is important to explicitly consider this point.

In this paper, we have formulated a dynamical model assuming harvesting of different stages and multi-agent exploiters (cooperators and defectors). For this model we have analyzed the existence and stability of positive equilibria, which characterize the sustainable use of the renewable resource, under different exploitable stages. The underlying biological model without harvesting, proposed in Gao and Chen (2005), has the simplest asymptotic behavior, i.e., convergence to a carrying capacity. By contrast, we found a rich dynamic behavior as soon as harvesting takes place. In particular, when all agents have the same economic parameters, then the more they compete, the higher harvesting takes place. As a consequence the dynamic complexity increases as soon as more agents defect, with a destabilizing influence on sustainability of resource exploitation. Similar results are obtained by reducing the demand elasticity of biomass.

We have also analyzed cases with agent heterogeneity. In particular we have considered a situation in which defectors are punished for their behavior (e.g., they are more heavily taxed or controlled by the authority). As a consequence the system can be destabilized if many cooperators are present. For this example we carried out a global study of the model showing a possible route to the extinction of the species through a so called final bifurcation. Under these conditions, harvesting the infant population should be forbidden, as it always leads to the extinction of population in finite time.

Acknowledgements The authors acknowledge helpful comments of an anonymous referee. The usual disclaimer applies. This Work is Supported by National Natural Science Foundation of China (10871209). The first author is grateful for the support from the key NSF of SCUFN (YZZ06027).

References

Bischi, G. I. & Lamantia, F. (2007) Harvesting dynamics in protected and unprotected areas. *Journal of Economic Behavior and Organization, 62*, 348–370.

Bischi, G. I., Lamantia, F., & Sbragia, L. (2004) Competition and cooperation in natural resources exploitation: An evolutionary game approach. In C. Carraro & V. Fragnelli (Eds.), *Game practice and the environment* (pp. 187–211). Edward Elgar Publishing, Cheltenham.

Bischi, G. I., Kopel, M., & Szidarovszky, F. (2005) Expectation-stock dynamics in multi-agent fisheries. *Annals of Operations Research, 137*, 299–329.

Clark, C. W. (1990) *Mathematical bioeconomics: the optimal management of renewable resource*. New York, USA: Wiley.

Fan, M. & Wang, K. (1998) Optimal harvesting policy for single population with periodic coefficients. *Mathematical Biosciences, l52*, 165–177.

Food and Agriculture Organization, *The state of world fisheries and aquaculture*. Sofia: FAO.

Garcia, S. & Grainger, J. R. (2005) Gloom and doom? The future of marine capture fisheries. *Philosophical Transactions of the Royal Society B, 360*, 21–24.

Gao, S. & Chen, L. (2005) Dynamic complexities in a single-species discrete population model with stage structure and birth pulses. *Chaos, Solitons & Fractals, 22*, 519–527.

Gao, S., Chen, L., & Sun, L. (2005) Optimal pulse fishing policy in stage-structured models with birth pulses. *Chaos, Solitons & Fractals, 25*, 1209–1219.

Gu, E.G. (2007) Global analysis of an ecological populational model with an external interference. *Chaos, Solitons & Fractals, 32*, 224–233.

Jing, W. and Ke, W., The optimal harvesting problems of a stage-structured population. *Applied Mathematics and Computation, 148*, 235–247.

Hanneson, R. (1995) Sequential fishing: cooperative and non cooperative equilibria. *Natural Resource Modeling, 9*, 51–59.

Mckelvey, R. (1997) Game theoretic insight into the international management of fisheries. *Natural Resource Modeling, 10*, 129–171.

Mesterton-Gibbons, M. (1993) Game-theoretic resource modelling. *Natural Resource Modeling, 7*, 93–147.

Mira, C., Gardini, L., Barugola A., & Cathala, J.C. (1996) *Chaotic dynamics in two-dimensional noninvertible maps*. Singapore: World Scientific

Onozaki, T., Sieg G., & Yokoo, M. (2000) Complex dynamics in a cobweb model with adaptive production adjustment, *Journal of Economic Behavior & Organization, 41*, 101–115.

Rosser, Jr. J. B. (2002a) Implications for fisheries policy of complex ecologic-economic dynamics. *Nonlinear Dynamics, Psychology, and Life Sciences, 6*(2), 103–120.

Rosser, Jr. J.B., The development of complex oligopoly dynamics theory. In T. Puu & I. Sushko (Eds.), *Oligopoly dynamics: models and tools* (pp. 15–29). Berlin: Springer.

Sethi, R. & Somanathan, E. (1996) The evolution of social norms in common property resource use. *The American Economic Review, 86*, 766–788.

Song, X. & Chen, L. (2002) Modelling and analysis of a single-species system with stage structure and harvesting, *Mathematical and Computer Modelling, 36*, 67–82.

Szidarovszky, F. & Okuguchi, K. (1998) An oligopoly model of commercial fishing. *Seoul Journal of Economics, 11*, 321–330.

Szidarovszky, F. & Okuguchi, K. (2002) A dynamic model of international fishing. *Seoul Journal of Economics, 13*, 4711–476.

Tang, S. Y. & Chen, L. S. (2002) Density-dependent birth rate, birth pulses and their population dynamic consequences. *Journal of Mathematical Biology, 64*, 169–184.

International Environmental Agreement: A Dynamical Model of Emissions Reduction

Marta Elena Biancardi

1 Introduction

Over the last two decades, the interest in international environmental problems such as climate change, ozone depletion, marine pollution has grown immensely and it has driven an increased sense of interdependence between countries.

Cooperation among different countries appears necessary and this results in International Environmental Agreements (IEA) such as Helsinky and Olso Protocol signed in 1985 and 1994; Montreal Protocol on the reduction of CFCs that deplete the ozone layer, signed in 1987; Kyoto Protocol, on the reduction of greenhouse gases causing global warming, signed in 1997. In these IEAs, the number of signatories varies considerably and this justifies the increasing interest of many authors to explain why IEAs are ratified only by a fraction of the potential signatories and to suggest strategies to increase their number.

Economists have emphasized two important aspects: agreements must be profitable (there must be gains to all signatory countries), agreements must be self-enforcing (in the absence of any international authority, there must be incentives for countries to join and to remain in an agreement). So, the participation of countries in an international agreement, to improve the quality of the environment, is a complex question for different reasons. First, countries are sovereign and their participation to IEA is voluntary, there is no supra national authority that forces countries to participate to an agreement, as well as there is no international environmental judicial system powerful enough to guarantee compliance to an IEA. Second, each country may have an incentive to free-ride, in fact while the costs for reducing emissions are carried out exclusively by the country that is taking action, the benefits of a reduction in emissions are shared by all countries, so that each country has the incentive to wait for the others to reduce their emissions.

Literature has focused on IEA's stability concepts in order to obtain some conclusions on the size that can be expected and to explain why some IEAs are large

M.E. Biancardi
Facoltà di Economia - Università di Foggia Via IV Novembre, 1 71100 Foggia,
e-mail: m.biancardi@unifg.it

G.I. Bischi et al. (eds.), *Nonlinear Dynamics in Economics, Finance and the Social Sciences*, DOI 10.1007/978-3-642-04023-8_5,

and others are not. Stable IEA means that no individual signatory country has any incentive to leave the IEA and no non-signatory country has an incentive to join the IEA.

Both Cooperative and Non-Cooperative game theory have been used to study coalition formation.

In the Cooperative Game framework, Chander and Tulkens (1995) define γ-core concept starting from the classical core concept. Basically the coalitional stability idea behind the γ-core assumes that if a single player deviates from the grand coalition, which is the coalition ratified by all potential signatories, this will lead to a complete disintegration of the coalition so that we end up in the non-cooperative Nash equilibrium in which all players act as singletons. The authors, using the above concept and implementing transfers to solve the heterogeneity of the countries, reach the stability of the grand coalition which represents the full cooperative solution. This approach supposes the existence of a large number of countries that are predisposed to sign the agreement, from which the naming "grand coalition" approach. It leads to an optimistic view on the size of the stable coalition.

In the Non-Cooperative Game framework the concept of Internal and External stability has been applied to obtain the size of a coalition. The idea is to check for which size of a coalition an individual country is indifferent either remaining in the coalition or leaving it. Carraro and Siniscalco (1993), Hoel (1992), de Zeeuw (2008) show that if signatories act in a Cournot fashion with respect to non-signatories then the size of a stable coalition is very small. If countries act in a Stackelberg fashion, where signatories are the leaders and non-signatories are the followers, a stable IEA can have any number of signatories between two and the grand coalition (see Barrett, 1994; Diamantoudi and Sartzetakis, 2006; Rubio and Ulph, 2006). This approach, known as the "small coalition" approach, leads to a pessimistic result about the coalitions which can emerge.

The mechanism of the γ-core is too strong, since it assumes that the initial coalition falls apart completely, but the mechanism of internal and external stability is too weak, since it assumes that only a deviation takes place.

Recent developments in game theory advocate the concept of *farsighted* stable coalitions against previous notions of stability which are myopic and don't reflect the complexity and foresight of countries' decisions about agreements.

When an agent contemplates leaving a coalition, it compares the welfare it enjoys as a member of the coalition with the welfare it will enjoy once it leaves. The agent implicitly assumes that once it deviates, no one else will want to deviate. But this is not always the case. In fact, it is possible that another country may wish to leave the coalition and so on. Thus, the agent must compare the starting situation with the outcome at the end of the process, after a number of deviations. The final outcome can be characterized as such only if no more countries wish to leave and no more countries wish to join.

The concept of farsightedness inspired a series of works in an abstract environment such as Chwe (1994), in which the Largest Consistent Set captures this notion. An outcome is stable and it is in the Largest Consistent Set if and only if deviations from it do not occur because the deviation itself or potential further deviations are not unanimously preferred to the original outcome, by the coalition considering the deviation.

The concept of farsightedness inspired also a series of papers in which this notion of stability is applied in the context of IEA, such as Diamantoudi and Sartzetakis (2002), Eyckmans (2001), de Zeeuw (2008).

This literature shows that farsightedness allows both large and small stable coalitions and so this concept reconciles the cooperative and non-cooperative approaches. All papers quoted above study the stability of an IEA in a static context while dynamic aspects are ignored, but environmental problems and in particular abatements processes, are usually dynamic as well as the evolution of the stock pollutant. In most models, it is assumed that countries reduce emissions in one step, but it is not realistic and also not rational. The analysis of the stability of an IEA needs to incorporate the dynamic of the stock pollutant and of behavioural reactions of agents. For these reasons we propose a non-cooperative game theoretic analysis of the IEAs with a stock pollutant in a dynamic setting. Other authors as, for example, Rubio and Casino (2005) and de Zeeuw (2008) have applied differential games and optimal control methodologies to analyse environmental problems. In particular, Rubio and Casino (2005) analyse the internal and the external stability of environmental agreements in a dynamic framework, when environmental damages are associated with a stock externality. Coalition formation has been designed as a two stages game in which, in the first stage each country decides to join or not the coalition and in the second stage signatories and non signatories play an emissions differential game. Authors calculate open loop equilibrium and show, by a numerical simulation, that a bilateral coalition is the unique self enforcing IEA. In de Zeeuw (2008) a model of abatement is proposed as a difference game, because the state transition is given as a difference equation. The feedback Nash equilibrium is found and, in order to study the stability of an IEA, the concept of dynamic farsighted stability is introduced showing as large and small stable coalitions can occur.

In this paper we propose an optimal control model with the objective to reduce pollution at the lowest costs. Players determine their abatement levels in a dynamic setting defined in continuous time. In the absence of cooperation, the abatement process is slower and has higher costs than in case some countries cooperate. In the differential game proposed, Open Loop Nash equilibria and Feedback Nash equilibria are calculated in order to determine the optimal paths of the abatement levels and of the stock pollutant. The results obtained are the same and depend on the parameter p which is defined in the cost function and which can be seen as a measure of the environmental awareness of countries. Stability conditions, such as internal and external stability or farsighted stability, are applied showing that different answers about the size of a stable IEA can be obtained.

The paper is organized as follows. In Sect. 2 we describe the model; in Sect. 3 the open loop Nash equilibria of the differential game are calculated and in Sect. 4 the analysis of the stability is proposed. In Sect. 5 Feedback Nash equilibria are obtained showing that they agree with the ones obtained using open loop strategies. Some concluding remarks are given in Sect. 6.

2 The Model

Let us assume that n identical countries decide to abate emissions in order to reduce the environmental pollution. Initially the accumulated emissions are at a level s_0 and each country i chooses an abatement level, $a_i(t)$, $i = 1, 2, \ldots, n$, where $a_i(t) \geq 0$, like an investment in green technology or a structural change. The dynamic of accumulated emissions is given by the following differential equation

$$\dot{s}(t) = L - \sum_{i=1}^{n} a_i(t) - k\, s(t) \quad s(0) = s_0. \tag{1}$$

L represents a constant source of pollutant, resulting from business-as-usual, without environmental concern, it is the baseline emissions, i.e. the level of emissions when the countries do not abate and it is assumed to be constant over time. In order to reduce accumulated emissions, abatement has to compensate the constant source of pollutant and so this implies that L will affect the abatement levels. k is a positive rate of pollution decay by natural processes.

Since $s(t) \geq 0$ the following constraint must be satisfied

$$0 \leq a_i(t) \leq \frac{L}{n} \quad \forall\, i = 1, 2, \ldots, n \tag{2}$$

so, we suppose, by the symmetry, that a single country is allowed to abate only a fraction of the emissions produced by itself.

We assume that players minimize a cost function $c_i(a_i(t))$ which is the sum of two terms: the abatement costs and the costs due to remaining pollution. It is very common in literature to consider this kind of cost functions in which the two terms can be linear or quadratic function. In this model we consider the first term quadratic and the second one linear. So, the cost function for each country is

$$c_i(a_i(t)) = \frac{1}{2} a_i(t)^2 + \frac{1}{2} p\, s(t). \tag{3}$$

A major role is played by the parameter $p > 0$; it can be seen as a measure of the environmental awareness of countries, i.e. it denotes the relative weight attached to the damage costs as compared to the abatement costs. By symmetry, p is the same for every country.

We follow the non-cooperative game theory approach in order to describe the formation of coalitions because countries, in international negotiations care only about their self interest and moreover because the predictions coming from the cooperative game theory, about abatement levels and global costs, do not fulfill. As most of the literature quoted into introduction, we discuss a class of non-cooperative games of coalition formation, where all players announce simultaneously and independently their decision to form coalitions and their abatement levels. This assumption excludes the possibility that one of the countries has any advantage in the game (see Carraro and Siniscalco, 1998). Among the simultaneous games, we considers open membership games in which any player is free to join or leave a coalition. Accordingly, players cannot specify in advance the coalition they wish to form, rather players announce a message and coalitions are formed by all players who make the same announcement. So, we have a two stages game, in which in the first stage each country decides whether to join or not the agreement while in the second stage each country chooses his abatement level.

The game is solved in a backward order. Let us assume that, as the outcome of the first stage game, there are m signatory countries ($i = 1, \ldots, m$) and $n - m$ nonsignatory countries ($j = m + 1, \ldots, n$). So, we consider a simple structure in which there is only one coalition while the other countries play as individual outsiders.

In the second stage, non signatory countries choose their abatement levels acting non-cooperatively in order to minimize the present value of their costs taking as given the strategy of the other countries; signatory countries choose their abatement levels acting non-cooperatively against nonsignatories in order to minimize the present value of the aggregate costs of the m signatories. Signatories also take as given the strategies of nonsignatories. The optimal abatement levels and accumulated emission paths are given by the Nash equilibria of a differential game. Consequently it is possible to obtain the equilibrium present value of the costs of a signatory country $C_i(m)$ and of a nonsignatory country $C_j(m)$. In the first stage, countries have the choice between acceding to an IEA or remaining a non signatory. The Nash equilibria of a game, as the one described above, correspond to internal and external stability conditions, introduced by D'Aspremont et al. (1983) in the context of cartel formation and applied by Carraro and Siniscalco (1993) to IEAs. Internal and External Stability have been proposed for static emission games, but can be extended to dynamic games.

A coalition of size m is internally stable if no member has an incentive to leave the coalition because the costs for an outsider in respect to a coalition of size $m - 1$ are larger than the costs for a member of an m-*sized* coalition. External stability means that no country has an incentive to join a coalition of size m because the cost for a member of a coalition of size $m + 1$ is higher than for an outsider to a coalition of size m. We have the following definition:

Definition 1. A coalition of size m is said to be stable if it is internally stable, i.e. $C_i(m) \leq C_j(m - 1)$ and externally stable, i.e. $C_i(m + 1) \geq C_j(m)$ where $i = 1, \ldots, m$ and $j = m + 1, \ldots, n$.

The concept captures the idea of voluntary participation since it can be shown that if a coalition is internally stable it is also profitable, that is, all participants receive more than in the status quo. A deeper investigation shows that the internal and external stability definition is too weak, since this assigns a myopic behaviour to players assuming that no further deviations take place. Moreover key results of this stability concept are that the number of signatories falls short of the grand coalition and the equilibrium coalition is usually rather small. For most model, in this traditional approach, the result about the stability is rather grim but small coalitions which are stable can exist. In particular in our model , using the internal and the external stability, we will see that if $m = 1$ we obtain the standard Nash equilibrium between individual countries which is not externally stable and so no stable. So, it is quite realistic to start from the situation in which all countries are singleton and to arrive to the formation of small stable coalitions. In fact, the instability of the coalition of size one, induces countries to join an agreement and we will reach the conclusion that coalitions of size 2 and 3 are self-enforcing. In these cases, in fact, cooperation decreases the costs and so, if there are only two or three signatories in the agreement they lose leaving the coalition. On the other hand, we also we will find that nonsignatories cannot gain joining the agreement, so that it is stable.

To overcome all these drawbacks, the concept of *farsightedness* has been introduced in literature, see, e.g. Chwe (1994). The concept of farsightedness is sufficiently rich to allow for a set of large and small stable coalitions.

A country belonging to a coalition of size m decides to abandon the coalition if its current cost $C_i(m)$ is higher than the cost he should pay leaving the coalition. Nevertheless, by the farsighted approach, he will not simply compare its actual cost with the outsider cost $C_j(m-1)$, but he will take into account the possibility that if he leaves the coalition then other coalition members may find it convenient to abandon the coalition, too. So, a disgregation process of the coalition can arise and then a country which decides to abandon a coalition of size m must compare its cost as a member of the coalition with the cost that it should pay as an outsider of the remaining coalition at the end of this disgregation process. If no country has an incentive to leave a coalition of size m, behaving in a farsighted way, then the coalition is said to be internally farsighted stable. A similar definition is given for the external farsighted stability.

The following definition, based upon the concept of the Largest Consistent Set by Chwe (1994) and proposed by Diamantoudi and Sartzetakis (2002) in order to study the farsighted coalitional stability, captures the above ideas and it shows that it is possible to construct a farsighted coalition, step by step, in a recursive way.

We consider a set σ which is the collection of all farsighted stable coalitions, where a coalition is called farsighted stable if it is both internally farsighted and externally farsighted stable. The construction of σ can appear complicated because of its recursive nature. But this recursivity defines some kind of consistency requirement for deviation. The recursivity is resolved assuming that coalitions of size 1 belongs to the set σ of farsighted stable coalitions.

Definition 2. A coalition of size m is internally farsighted stable if a finite sequence of coalitions of size $m-1, \dots, m-s$, where $s \in 1, 2, \dots, m$, such that the coalition of size $m-s$ is in σ and $C_i(m-j) > C_j(m-s)$ $\forall j = 0, 1, \dots, s-1$, doesn't exist.

A coalition of size m is externally farsighted stable if a finite sequence of coalitions of size $m+1, \dots, m+s$, where $s \in 1, 2, \dots, n-m$, such that the coalition of size $m+s$ is in σ and $C_i(m+s) < C_j(m+j)$ $\forall j = 0, 1, \dots, s-1$, doesn't exist.

3 The Open Loop Nash Equilibria of the Differential Game

In order to obtain a self enforcing IEA, we calculate the open loop Nash equilibria of the abatements differential game. Let us assume that $\delta > 0$ is the discount rate, assumed common to all countries. Given the abatement levels of outsiders, signatories commit to a level of abatement that minimize the sum of the costs present value of the countries in the agreement

$$\min_{a_i} \sum_{h=1}^{m} \int_0^{+\infty} e^{-\delta t} \left(\frac{1}{2} a_h(t)^2 + \frac{1}{2} p\, s(t) \right) dt \tag{4}$$

which is equivalent to

$$\max_{a_i} \sum_{h=1}^{m} \int_0^{+\infty} - e^{-\delta t} \left(\frac{1}{2} a_h(t)^2 + \frac{1}{2} p\, s(t) \right) dt. \tag{5}$$

Given the abatement levels of cooperators, nonsignatories commit to a level of abatement that minimize the discounted present value of the costs which is equivalent to

$$\max_{a_j} \int_0^{+\infty} - e^{-\delta t} \left(\frac{1}{2} a_j(t)^2 + \frac{1}{2} p\, s(t) \right) dt. \tag{6}$$

In both cases, countries face the same dynamics

$$\dot{s}(t) = L - \sum_{i=1}^{m} a_i(t) - \sum_{j=m+1}^{n} a_j(t) - k\, s(t) \quad s(0) = s_0 \tag{7}$$

with the constraint on the control variables given by (2).

Let us define the current value of the Hamiltonian in the standard way

$$H_i = - \sum_{h=1}^{m} \left(\frac{1}{2} a_h^2 + \frac{1}{2} p\, s \right) + \lambda_i \left(L - \sum_{h=1}^{m} a_h - \sum_{j=m+1}^{n} a_j - ks \right), \; i = 1, \ldots, m,$$

$$H_j = - \left(\frac{1}{2} a_j^2 + \frac{1}{2} p\, s \right) + \lambda_j \left(L - \sum_{i=1}^{m} a_i - \sum_{j=m+1}^{n} a_j - ks \right), \; j = m+1, \ldots, n,$$

where λ_i and λ_j are the adjoint variables. We obtain the following set of necessary conditions for an interior open-loop equilibrium

$$a_i = -\lambda_i, \quad i = 1, \ldots, m, \tag{8}$$

$$a_j = -\lambda_j, \quad j = m+1, \ldots, n, \tag{9}$$

provided these expressions stay within the interval $[0, L/n]$.

Note that the second derivative of H with respect to a_i and a_j is $\partial^2 H/\partial a^2 = -1 < 0$, so that (8) and (9) satisfy the second order condition for the maximum of a function.

The adjoint equations are

$$\dot{\lambda}_i = (\delta + k)\lambda_i + \frac{1}{2}mp, \quad i = 1, \ldots, m, \tag{10}$$

$$\dot{\lambda}_j = (\delta + k)\lambda_j + \frac{1}{2}p, \quad j = m+1, \ldots, n, \tag{11}$$

and the transversality conditions

$$\lim_{t\to+\infty} e^{-\delta t}\lambda_i = 0, \quad i = 1, \ldots, m, \tag{12}$$

$$\lim_{t\to+\infty} e^{-\delta t}\lambda_j = 0, \quad j = m+1, \ldots, n. \tag{13}$$

Because of the symmetry assumption, the n equations given by (10) and (11) reduce to two. Solving them and using the transversality conditions (12) and (13) which are sufficient because the Hamiltonian functions are strictly concave in s, we obtain

$$\lambda_i = -\frac{mp}{2(\delta + k)} \quad \text{and} \quad \lambda_j = -\frac{p}{2(\delta + k)}.$$

The constraint on the control variables given by (2), (8) and (9) lead to the following optimal abatement levels

$$a_i = \begin{cases} \dfrac{mp}{2(\delta + k)} & \text{if } 0 \le \dfrac{mp}{2(\delta + k)} \le \dfrac{L}{n} \qquad i = 1, \ldots, m, \\[2ex] \dfrac{L}{n} & \text{if } \dfrac{mp}{2(\delta + k)} > \dfrac{L}{n}, \end{cases}$$

for a signatory country;

$$a_j = \begin{cases} \dfrac{p}{2(\delta + k)} & \text{if } 0 \le \dfrac{p}{2(\delta + k)} \le \dfrac{L}{n} \qquad j = m+1, \ldots, n, \\[2ex] \dfrac{L}{n} & \text{if } \dfrac{p}{2(\delta + k)} > \dfrac{L}{n}, \end{cases}$$

for a non signatory country.

Note that if $0 \le \frac{p}{2(\delta+k)} \le \frac{L}{n}$ the abatement levels are always higher under cooperation otherwise they are the same for cooperators and defectors. In particular, if $0 \le \frac{mp}{2(\delta+k)} \le \frac{L}{n}$ the abatement level a_i, due to each member of the coalition is positively correlated with the size m of the coalition. It means that if the number of countries that sign the agreement increases then the abatement level increases; a non signatory country, instead, considers only its own abatement level which does not depend on the number of the signatory countries. Both the abatement levels a_i and a_j are positively correlated with p and so, if the relative weight of the damage costs compared with the abatement costs increases, also the abatement levels of signatories and of nonsignatories increase too, confirming that the parameter p is also an indicator of the environmental awareness of the countries. Moreover a_i and a_j are inversely correlated with the natural decay rate k and with the discount rate δ. This means that if k increases and so, if there is a fast natural decay of the accumulated emissions, then abatement level decreases, in the same manner if δ increases and so, if countries don't give value to the future because they are less farsighted, the abatement levels decreases, too.

In order to simply the computations and to compare the abatement levels and the discounted present value of the costs, let

$$r = \frac{2L(\delta + k)}{np}$$

a parameter which represents the combined effects of the values assumed by L, δ, k, of the number of the countries and of the relative weight p.

We have distinguished three different cases, depending on the value of the parameter r, with regard to the number of cooperators m.

Case I

If $r \geq m$ then

$$a_i = \frac{mp}{2(\delta + k)} \quad i = 1, \dots, m \quad \text{and} \quad a_j = \frac{p}{2(\delta + k)} \quad j = m+1, \dots, n. \tag{14}$$

Optimal abatement levels are constant and signatories abate m times more than nonsignatories, so the abatement levels of nonsignatories are smaller than of signatories. The accumulated emission path is

$$s(t) = s_0 e^{-kt} + \frac{1}{k}\left[L - \frac{m^2 p}{2(\delta + k)} - \frac{(n-m)p}{2(\delta + k)}\right](1 - e^{-kt}) \tag{15}$$

which is a positive, increasing and concave function if $s_0 < \frac{1}{k}\left[L - \frac{m^2 p}{2(\delta+k)} - \frac{(n-m)p}{2(\delta+k)}\right]$ otherwise it is a decreasing and convex one. Moreover, in both cases, for $t \to +\infty$, $s(t)$ approaches asymptotically the value $\frac{1}{k}\left[L - \frac{m^2 p}{2(\delta+k)} - \frac{(n-m)p}{2(\delta+k)}\right]$, which represents the steady state value of accumulated emissions for the first case. It is easy to prove that this value decreases as the number of countries that sign the agreement increases.

Case II

If $1 \leq r < m$ then

$$a_i = \frac{L}{n} \; i = 1, \ldots, m \quad \text{and} \quad a_j = \frac{p}{2(\delta + k)} \; j = m+1, \ldots, n; \tag{16}$$

Optimal abatement levels are constant but while cooperators stop producing emissions, outsiders abate the same quantity of the first case. The accumulated emissions path is

$$s(t) = s_0 e^{-kt} + \frac{(n-m)}{k} \left[\frac{L}{n} - \frac{p}{2(\delta + k)} \right] (1 - e^{-kt}) \tag{17}$$

which is a positive, increasing and concave function if $s_0 < \frac{(n-m)}{k} \left[\frac{L}{n} - \frac{p}{2(\delta+k)} \right]$ otherwise it is a decreasing and convex one. Moreover, in both cases, for $t \to +\infty$, $s(t)$ approaches asymptotically the value $\frac{(n-m)}{k} \left[\frac{L}{n} - \frac{p}{2(\delta+k)} \right]$, which represents the steady state value of accumulated emissions for the second case. Also here, this value decreases if the cooperators increase.

Case III

If $r < 1$ then

$$a_i = \frac{L}{n} \; i = 1, \ldots, m \quad \text{and} \quad a_j = \frac{L}{n} \; j = m+1, \ldots, n. \tag{18}$$

Optimal abatement levels are constant and both cooperators and defectors stop producing emissions. In this case the role of a signatory and of a non signatory is the same. The accumulated emissions path is

$$s(t) = s_0 e^{-kt} \tag{19}$$

which is a positive, decreasing and convex function and for $t \to +\infty$, it approaches zero.

4 Stability

We want to apply the conditions of myopic and farsighted stability proposed in the above sections, in order to obtain the size of stable coalitions.

First of all we need to calculate $C_i(m)$ and $C_j(m)$; substituting the obtained optimal control paths of the pollution stock and of the abatement levels in (4) and in (6), we have the following results.

Case I

If $r \geq m$ then

$$C_i(m) = \frac{m^2 p^2}{8\delta(\delta + k)^2} + \frac{p}{2\delta(\delta + k)} \left[L - \frac{m^2 p}{2(\delta + k)} - \frac{(n-m)p}{2(\delta + k)} \right] + \frac{p s_0}{2(\delta + k)}, \tag{20}$$

$$C_j(m) = \frac{p^2}{8\delta(\delta+k)^2} + \frac{p}{2\delta(\delta+k)}\left[L - \frac{m^2 p}{2(\delta+k)} - \frac{(n-m)p}{2(\delta+k)}\right] + \frac{ps_0}{2(\delta+k)}. \tag{21}$$

Case II

If $1 \le r < m$ then

$$C_i(m) = \frac{L^2}{2\delta n^2} + \frac{p(n-m)}{2\delta(\delta+k)}\left[\frac{L}{n} - \frac{p}{2(\delta+k)}\right] + \frac{ps_0}{2(\delta+k)}, \tag{22}$$

$$C_j(m) = \frac{p^2}{8\delta(\delta+k)^2} + \frac{p(n-m)}{2\delta(\delta+k)}\left[\frac{L}{n} - \frac{p}{2(\delta+k)}\right] + \frac{ps_0}{2(\delta+k)}. \tag{23}$$

It is very easy to prove that, both in the first and in the second case, the present value of the costs of a nonsignatory country is lower than the discounted present value of the costs of a signatory for all m in the interval $[2, n]$, i.e.

$$C_i(m) > C_j(m).$$

If $m > 1$ the following inequality is satisfied

$$C_j(m) < C_j(1).$$

It means that the present value of the costs of the outsiders are also lower then the present value of the costs in the Nash equilibrium between individual countries ($m = 1$). The result will be that the coalition of size one will be instable.

Moreover it is possible to prove that both the discounted present value of the costs of signatories and the discounted present value of the costs of nonsignatories decrease if the number of signatories increase. The result will be that signatories always do better withdrawing from the agreement whenever the number of signatories is higher than 3 and so that a low level of cooperation can be expected.

Case III

If $r < 1$ then

$$C_i(m) = C_j(m) = \frac{L^2}{2\delta n^2} + \frac{ps_0}{2(\delta+k)}. \tag{24}$$

In this case the costs are the same indifferent if there is participation to a coalition. Therefore in the following we definitively assume that $r \ge 1$ and in our analysis we consider only the first and the second case.

4.1 Myopic Stability

In order to have the internal stability of a coalition of size $m > 1$ (it is not possible to prove the internal stability of a coalition of size $m = 1$) we need

$$C_i(m) \leq C_j(m-1). \tag{25}$$

We start supposing that m satisfies the constraint given by the first case, i.e. $r \geq m$; obviously $r \geq m-1$ and so C_i and C_j are given respectively by (20) and (21). Then (25) becomes

$$\frac{m^2p^2}{8\delta(\delta+k)^2} + \frac{p}{2\delta(\delta+k)}\left[L - \frac{m^2 p}{2(\delta+k)} - \frac{(n-m)p}{2(\delta+k)}\right] + \frac{ps_0}{2(\delta+k)}$$

$$\leq \frac{p^2}{8\delta(\delta+k)^2} + \frac{p}{2\delta(\delta+k)}\left[L - \frac{(m-1)^2 p}{2(\delta+k)} - \frac{(n-m+1)p}{2(\delta+k)}\right] + \frac{ps_0}{2(\delta+k)}.$$

Solving the above inequality we have that its solution doesn't depend on r and it holds if $1 \leq m \leq 3$.

If we suppose that m satisfies the constraint given by the second case, i.e. $r < m$, it is necessary to consider two different subcases.

If $m-1 < r$ then $1 \leq r < m < r+1$ and C_i and C_j are given respectively by (22) and (21). Then (25) becomes

$$\frac{L^2}{2\delta n^2} + \frac{p(n-m)}{2\delta(\delta+k)}\left[\frac{L}{n} - \frac{p}{2(\delta+k)}\right] + \frac{ps_0}{2(\delta+k)}$$

$$\leq \frac{p^2}{8\delta(\delta+k)^2} + \frac{p}{2\delta(\delta+k)}\left[L - \frac{(m-1)^2 p}{2(\delta+k)} - \frac{(n-m+1)p}{2(\delta+k)}\right] + \frac{ps_0}{2(\delta+k)}.$$

Solving the above inequality we can conclude that the subcase analysed give us coalitions of size m which are internally stable if $2 \leq m \leq 3$ and $1 \leq r \leq 3$; otherwise, if $r > 3$ there aren't coalitions which are internally stable.

The second subcase is obtained considering, again, that $m > r$ but that also $m-1 > r$; so we have that $m > r+1 > r \geq 1$ and C_i and C_j are given respectively by (22) and (23). Then (25) becomes

$$\frac{L^2}{2\delta n^2} + \frac{p(n-m)}{2\delta(\delta+k)}\left[\frac{L}{n} - \frac{p}{2(\delta+k)}\right] + \frac{ps_0}{2(\delta+k)}$$

$$\leq \frac{p^2}{8\delta(\delta+k)^2} + \frac{p(n-m+1)}{2\delta(\delta+k)}\left[\frac{L}{n} - \frac{p}{2(\delta+k)}\right] + \frac{ps_0}{2(\delta+k)}.$$

Solving the above inequality it is possible to prove that its solution doesn't depend on m and it is always true if and only if $r = 1$. So, we conclude that if $r \neq 1$ there aren't coalitions which are internally stable, if $r = 1$ then any coalition of size m is internally stable.

In order to have the external stability of a coalition of size $m \geq 1$ we need

$$C_i(m+1) \geq C_j(m). \tag{26}$$

If $m = 1$, i.e. if we consider the full non-cooperative abatement, the analysis reduces only to the first case when $r \geq 1$, then C_i and C_j are given respectively by (20) and (21) and (26) becomes $C_i(2) \geq C_j(1)$ which is equivalent to

$$\frac{4p^2}{8\delta(\delta+k)^2} + \frac{p}{2\delta(\delta+k)}\left[L - \frac{4p}{2(\delta+k)} - \frac{(n-2)p}{2(\delta+k)}\right] + \frac{ps_0}{2(\delta+k)}$$

$$\geq \frac{p^2}{8\delta(\delta+k)^2} + \frac{p}{2\delta(\delta+k)}\left[L - \frac{p}{2(\delta+k)} - \frac{(n-1)p}{2(\delta+k)}\right] + \frac{ps_0}{2(\delta+k)}.$$

It is never held, so we conclude that the Nash equilibrium between individual countries is not externally stable and so it is instable.

Now, we suppose that $m > 1$ satisfies the constraint given by the first case, i.e. $r \geq m$, if also $m + 1 \leq r$ then C_i and C_j are given respectively by (20) and (21) and (26) becomes

$$\frac{(m+1)^2p^2}{8\delta(\delta+k)^2} + \frac{p}{2\delta(\delta+k)}\left[L - \frac{(m+1)^2p}{2(\delta+k)} - \frac{(n-m-1)p}{2(\delta+k)}\right] + \frac{ps_0}{2(\delta+k)}$$

$$\geq \frac{p^2}{8\delta(\delta+k)^2} + \frac{p}{2\delta(\delta+k)}\left[L - \frac{m^2p}{2(\delta+k)} - \frac{(n-m)p}{2(\delta+k)}\right] + \frac{ps_0}{2(\delta+k)}$$

which holds $\forall m \geq 2$.

If we suppose that m satisfies the constraint given by the first case, i.e. $r \geq m$ but if $m + 1 > r$ then $r - 1 < m \leq r$, so C_i and C_j are given respectively by (22) and (21) and (26) becomes

$$\frac{L^2}{2\delta n^2} + \frac{p(n-m-1)}{2\delta(\delta+k)}\left[\frac{L}{n} - \frac{p}{2(\delta+k)}\right] + \frac{ps_0}{2(\delta+k)}$$

$$\geq \frac{p^2}{8\delta(\delta+k)^2} + \frac{p}{2\delta(\delta+k)}\left[L - \frac{m^2p}{2(\delta+k)} - \frac{(n-m)p}{2(\delta+k)}\right] + \frac{ps_0}{2(\delta+k)}.$$

Solving it, we obtain that if $2 - \sqrt{2} < r < 2 + \sqrt{2}$ then coalitions of size m, with $2 \leq m \leq 3$, are externally stable, instead, if $r < 2 - \sqrt{2} \vee r > 2 + \sqrt{2}$ then all coalitions of size m with $r - 1 < m < r$ are externally stable.

If we suppose that m satisfies the constraint given by the second case, i.e. $r < m$, then obviously $r < m + 1$ and so C_i and C_j are given respectively by (22) and (23) and (26) becomes

$$\frac{L^2}{2\delta n^2} + \frac{p(n-m-1)}{2\delta(\delta+k)}\left[\frac{L}{n} - \frac{p}{2(\delta+k)}\right] + \frac{ps_0}{2(\delta+k)}$$

$$\geq \frac{p^2}{8\delta(\delta+k)^2} + \frac{p(n-m)}{2\delta(\delta+k)}\left[\frac{L}{n} - \frac{p}{2(\delta+k)}\right] + \frac{ps_0}{2(\delta+k)}$$

which doesn't depend on m and is always true.

Combining internal and external stability we have shown that the requirement of myopic stability is only satisfied for very small coalitions, in particular, only coalitions of size 2 and 3 can be stable, confirming the results obtained in the literature quoted in Sect. 1.

4.2 Farsighted Stability

We want to determine the size of farsighted stable coalitions. We use a recursive argument. Let us assume that a farsighted stable coalition of size $m \geq 1$ exists and we start supposing that m satisfies the constraint given by the first case, i.e. $r \geq m$. In order to have the smallest farsighted stable coalition larger than the coalition of size m, we need to find the smallest integer h such that $1 \leq h \leq n - m$ and

$$C_i(m+h) \leq C_j(m). \tag{27}$$

Before studying the conditions for which (27) is satisfied, we have to characterize the costs of cooperators and of defectors which depend on the relative positions of $m + h$ and r. In fact, if $m + h \leq r$, then C_i and C_j are given, respectively by (20) and (21); if $m + h > r$ then C_i is given by (22) while C_j by (21), again.
If we suppose that $m + h \leq r$, then (27) becomes

$$\frac{(m+h)^2 p^2}{8\delta(\delta+k)^2} + \frac{p}{2\delta(\delta+k)}\left[L - \frac{(m+h)^2 p}{2(\delta+k)} - \frac{(n-m-h)p}{2(\delta+k)}\right] + \frac{ps_0}{2(\delta+k)}$$

$$\leq \frac{p^2}{8\delta(\delta+k)^2} + \frac{p}{2\delta(\delta+k)}\left[L - \frac{m^2 p}{2(\delta+k)} - \frac{(n-m)p}{2(\delta+k)}\right] + \frac{ps_0}{2(\delta+k)}$$

which is satisfied if

$$h \geq \sqrt{2m(m-1)} - (m-1)\,,$$

that is

$$m + h \geq \sqrt{2m(m-1)} + 1.$$

Let $g(m)$ defined as the smallest integer greater than or equal to $\sqrt{2m(m-1)} + 1$, i.e.

$$g(m) = [\sqrt{2m(m-1)} + 1].$$

If $g(m) \leq \min\{[r], n\}$ then the size of the smallest farsighted stable coalition larger than the coalition of size m is $g(m)$.

If we suppose that $m + h > r$, then (27) becomes

$$\frac{L^2}{2\delta n^2} + \frac{p(n-m-h)}{2\delta(\delta+k)}\left[\frac{L}{n} - \frac{p}{2(\delta+k)}\right] + \frac{ps_0}{2(\delta+k)}$$

$$\leq \frac{p^2}{8\delta(\delta+k)^2} + \frac{p}{2\delta(\delta+k)}\left[L - \frac{m^2 p}{2(\delta+k)} - \frac{(n-m)p}{2(\delta+k)}\right] + \frac{ps_0}{2(\delta+k)}$$

which is satisfied if

$$h \leq \frac{1}{2}(r+1) + \frac{m(m-r)}{r-1},$$

and so

$$m+h \leq \frac{2m^2 - 2m - 1 + r^2}{2(r-1)} \equiv \lambda.$$

Let $w(m)$ defined as the smallest integer greater than or equal to λ, i.e.

$$w(m) = [\lambda].$$

If $w(m) \geq [r]+1$ and $[r]+1 \leq n$ then the size of the smallest farsighted stable coalition larger than the coalition of size m is $[r]+1$.

Now, we suppose that $1 \leq r < m$, then C_i and C_j, in (27), are given by (22) and (23) and we have

$$\frac{L^2}{2\delta n^2} + \frac{p(n-m-h)}{2\delta(\delta+k)}\left[\frac{L}{n} - \frac{p}{2(\delta+k)}\right] + \frac{ps_0}{2(\delta+k)}$$

$$\leq \frac{p^2}{8\delta(\delta+k)^2} + \frac{p(n-m)}{2\delta(\delta+k)}\left[\frac{L}{n} - \frac{p}{2(\delta+k)}\right] + \frac{ps_0}{2(\delta+k)}$$

which is satisfied if

$$h \geq \frac{r+1}{2}$$

and so

$$m+h \geq m + \frac{r+1}{2}.$$

Let $z(m)$ defined as the smallest integer greater than or equal to $m + \frac{r+1}{2}$, i.e.

$$z(m) = \left[m + \frac{r+1}{2}\right]$$

then the size of the smallest farsighted stable coalition larger than the coalition of size m is $z(m)$, provided that $z(m) \leq n$.

We propose some numerical examples which show how the size of the farsighted stable coalitions changes, as the value of p varies.

We fix the following values: $n = 100,\ L = 100,\ k = 1,\ \delta = 1,\ s_0 = 0.$ Let $p = 0.01$, then the following coalitions are farsighted stable

m = 2
m = 3
m = 5
m = 8
m = 12
m = 18
m = 26
m = 38
m = 55
m = 79

Let $p = 0.1$, then the following coalitions are farsighted stable

m = 2
m = 3
m = 5
m = 8
m = 12
m = 18
m = 26
m = 38
m = 41
m = 62
m = 83

Let $p = 1$, then the coalitions of size $m = 2$ and $m = 3$ are farsighted stable. Moreover any coalition of size $m = 3t - 1$ with $t = 2, 3, \ldots, 33$ is farsighted stable. The largest farsighted stable coalition is $m = 98$.
Let $p \geq 4$, then any coalition is farsighted stable.

Concluding this section we can say that using the concept of farsighted stability both large and small coalitions can occur. Moreover we observe that the parameter p has an interesting role in the determination of the size of stable coalitions, in fact the numerical simulation shows that if p value increases then the size of coalitions which are farsighted stable increases too. This result underlines how the greater environmental awareness of countries leads to stable coalitions which have greater sizes. In this way it is possible to obtain also the stability of the grand coalition.

5 Feedback Nash Equilibrium

Open loop strategies imply that each player commits himself to his entire course of action at the beginning of the game and will not revise it at any subsequent moment. In this section we abandon this assumption assuming that players use feedback strategies. A feedback strategy consists of a contingency plan that indicates the optimal value of the control variable for each value of the state variable at each point in time. It has the property of being subgame perfect, because after each player's actions have caused the state of the system to evolve from its initial state to a new state, the continuation of the game may be regarded as a subgame of the original game. We can say that in this case each player has committed to a rule which yields the optimal value of the control variable in each moment as a function of the state of the system at that moment. A feedback strategy must satisfy the principle of optimality of dynamic programming.

The Hamilton–Jacobi–Bellman equation for signatories is

$$\delta V_i = \max_{\{a_i\}} \left\{ -\sum_{h=1}^{m} \left(\frac{1}{2} a_h^2 + \frac{1}{2} p\, s \right) + V_i' \left(L - \sum_{h=1}^{m} a_h - \sum_{j=m+1}^{n} a_j - ks \right) \right\}. \tag{28}$$

The Hamilton–Jacobi–Bellman equation for nonsignatories is

$$\delta V_j = \max_{\{a_j\}} \left\{ - \left(\frac{1}{2} a_j^2 + \frac{1}{2} p\, s \right) + V_j' \left(L - \sum_{i=1}^{m} a_i - \sum_{j=m+1}^{n} a_j - ks \right) \right\}, \tag{29}$$

where $V_i(s)$ and $V_j(s)$ denote the optimal control value functions of the coalition and of a nonsignatory associated with the optimization problem (5) and (6), i.e. they denote the minimum present value of the cost flow subject to the dynamic constraint of the accumulated emissions; V_i' and V_j' are the first derivative with respect to the state variable s.

The optimal value of the control variables must satisfy the first order conditions for an interior feedback Nash equilibrium, that is

$$- a_i - V_i' = 0, \qquad i = 1, \dots, m, \tag{30}$$

$$- a_j - V_j' = 0, \qquad j = m + 1, \dots, n. \tag{31}$$

These conditions define the optimal strategies of abatements as functions of accumulated emissions; so, the constraint on the control variables given by (2), (30) and (31) lead to the following conditions on the abatement levels

$$a_i = \begin{cases} 0 & \text{if } -V_i' < 0, \\ -V_i' & \text{if } 0 \leq -V_i' \leq \dfrac{L}{n} \quad i = 1, \ldots, m, \\ \dfrac{L}{n} & \text{if } -V_i' > \dfrac{L}{n}, \end{cases}$$

for a signatory country;

$$a_j = \begin{cases} 0 & \text{if } -V_j' < 0, \\ -V_j' & \text{if } 0 \leq -V_j' \leq \dfrac{L}{n} \quad j = m+1, \ldots, n, \\ \dfrac{L}{n} & \text{if } -V_j' > \dfrac{L}{n}, \end{cases}$$

for a nonsignatory country.

We have analysed all possible combinations between interior and boundary a_i and a_j values.

If we suppose that $0 \leq -V_i' \leq \frac{L}{n}$ and $0 \leq -V_j' \leq \frac{L}{n}$, then $a_i = -V_i'$ and $a_j = -V_j'$. Substituting these abatement level expressions in (28) and in (29), we obtain the following nonlinear differential equations

$$\delta V_i = \frac{m}{2}(V_i')^2 + V_i'(L + (n-m)V_j' - ks) - \frac{1}{2}mps, \tag{32}$$

$$\delta V_j = \left(\frac{2n - 2m - 1}{2}\right)(V_j')^2 + V_j'(L + mV_i' - ks) - \frac{1}{2}ps. \tag{33}$$

In order to compute the solution of the above equations, given the linear quadratic structure of the game, we guess that the optimal value functions are quadratic and consequently the equilibrium strategies are linear in respect to the state variable. Precisely, we postulate quadratic value functions of the form

$$V_i = \frac{1}{2}\alpha_i s^2 + \beta_i s + \mu_i, \tag{34}$$

$$V_j = \frac{1}{2}\alpha_j s^2 + \beta_j s + \mu_j, \tag{35}$$

where α, β, μ are constant parameters of the unknown value functions which are to be determined. Using (34) and (35) to eliminate V_i, V_j, V_i' and V_j' from (32) and from (33) and equating, we yield the following system of algebraic Riccati equations for the coefficients of the value functions

$$\begin{cases} \frac{1}{2}\alpha_i\delta = \dfrac{m}{2}\alpha_i^2 + (n-m)\alpha_i\alpha_j - k\alpha_i, \\ \beta_i\delta = m\alpha_i\beta_i + L\alpha_i + (n-m)\alpha_i\beta_j + (n-m)\beta_i\alpha_j - k\beta_i - \frac{1}{2}mp, \\ \mu_i\delta = \beta_i\left[\dfrac{m}{2}\beta_i + L + (n-m)\beta_j\right], \\ \frac{1}{2}\alpha_j\delta = \left(\dfrac{2n-2m-1}{2}\right)\alpha_j^2 + m\alpha_i\alpha_j - k\alpha_j, \\ \beta_j\delta = (2n-2m-1)\alpha_j\beta_j + L\alpha_j - k\beta_j + m\alpha_j\beta_i + m\alpha_i\beta_j - \frac{1}{2}p, \\ \mu_j\delta = \beta_j\left[\left(\dfrac{2n-2m-1}{2}\right)\beta_j + L + m\beta_i\right]. \end{cases}$$

This system has four solutions, but only one produces value functions satisfying the stability condition. To obtain this condition we substitute the linear strategies

$$a_i = -\alpha_i s - \beta_i, \quad a_j = -\alpha_j s - \beta_j \tag{36}$$

in the dynamical constraint of accumulated emissions. We obtain the following differential equation

$$\dot{s} = [m\alpha_i + (n-m)\alpha_j - k]s + L + m\beta_i + (n-m)\beta_j. \tag{37}$$

The stability condition is

$$\frac{d\dot{s}}{ds} = m\alpha_i + (n-m)\alpha_j - k < 0$$

which is satisfied only by the following solution of the system

$$\alpha_i = \alpha_j = 0, \quad \beta_i = -\frac{mp}{2(k+\delta)}, \quad \beta_j = -\frac{p}{2(k+\delta)},$$

$$\mu_i = -\frac{mp(4kL + 4L\delta - p(m^2 - 2m + 2n))}{8\delta(k+\delta)^2},$$

$$\mu_j = -\frac{p(4kL + 4L\delta - p(2m^2 - 2m + 2n - 1))}{8\delta(k+\delta)^2}.$$

This solution, combined with the constraints $0 \le -V_i' \le \dfrac{L}{n}$ and $0 \le -V_j' \le \dfrac{L}{n}$, gives us the optimal abatement levels:

$$a_i = \frac{mp}{2(\delta + k)} \quad \text{and} \quad a_j = \frac{p}{2(\delta + k)}$$

when the following condition on p is satisfied

$$p \leq \frac{2L(\delta + k)}{mn}$$

which is equivalent to

$$m \leq r .$$

It is possible to conclude that the Feedback Nash equilibrium obtained, coincide with the open loop Nash equilibrium given by case I.

If we suppose that $-V_i' > \frac{L}{n}$ and $0 \leq -V_j' \leq \frac{L}{n}$, then $a_i = \frac{L}{n}$ and $a_j = -V_j'$. Reasoning as above we obtain

$$a_i = \frac{L}{n} \quad \text{and} \quad a_j = \frac{p}{2(\delta + k)}$$

when the following condition on p is satisfied

$$\frac{2L(\delta + k)}{mn} < p \leq \frac{2L(\delta + k)}{n}$$

which is equivalent to

$$1 \leq r < m .$$

Again we conclude that the Feedback Nash equilibrium obtained coincides with the corresponding open loop one given by case II.

If we consider the remaining combinations between a_i anda_j values, it is possible to prove that solutions of the Riccati system don't satisfy the constraints and so they don't represent feedback Nash equilibria.

So, we conclude this analysis, claiming that, in the model proposed, Feedback and Open Loop Nash equilibria are the same.

6 Concluding Remarks

The present paper studies the problem of computing the size of a stable coalition in an International Environmental Agreement.

We studied a differential game in which abatement levels are associated with a stock pollutant. Coalitions formation has been designed as a two stages game in which in the first stage each country decides to join or not a coalition, instead, in the second stage, nonsignatories and signatories determine the optimal paths of the abatements and so, the path of the global emissions.

The model has the objective to reduce pollution at the lowest costs; the cost function of every country is characterized by the presence of a parameter p which gives us the measure of the environmental awareness of countries.

Open loop Nash equilibria and Feedback Nash equilibria have been analysed showing that they carry out to the same solution of the differential game.

Stability conditions, such as myopic stability or farsighted stability have been applied showing that different results about the size of a stable IEA can be obtained. In fact using internal and external stability, which characterize myopic stability, we reached the conclusion that only coalitions of size 2 and 3 are stable, independently of p value. This result confirms the pessimistic conclusion to which some papers quoted in the introduction arrived. Using the concept of farsighted stability we have shown that both large and small stable coalitions can occur. In particular, our numerical simulation shows that the parameter p has an interesting role in the determination of the size of stable coalitions, in fact if p value increases the size of coalitions which are farsighted stable increases too, and this result underlines how the greater environmental awareness of countries leads to stable coalitions which have greater sizes. In this way it is possible to obtain also the grand coalition's stability.

A possible step for a forthcoming paper could be the study of stable coalitions modifying the costs function or relaxing some assumptions of the game's rules.

References

Barrett, S. (1994). Self-enforcing international environmental agreements. *Oxford Economic Paper, 46*, 878–894.

Carraro, C., & Siniscalco, D. (1993). Strategies for the international protection of the environment. *Journal of Public Economics, 52*, 309–328.

Carraro, C., & Siniscalco, D. (1998). International environmental agreement: Incentives and political economy. *European Economic Review, 42*, 561–572.

Chander, P., & Tulkens, H. (1995). A core-theoretic solution for the design of cooperative agreements on transfrontier pollution. *International Tax and Public Finance, 2*, 279–293.

Chwe, M. (1994). Farsighted coalitional stability. *Journal of Economic Theory, 63*, 299–325.

D'Aspremont, C., Jacquemin, J., Gadszeweiz, J., & Weymark, J. (1983). On the stability of collusive price leadership. *Canadian Journal of Economics, 16*, 17–25.

de Zeeuw, A. (2008). Dynamic effects on the stability of international environmental agreements. *Journal of Environmental Economics and Management, 55*, 163–174.

Diamantoudi, E., & Sartzetakis, E. (2002). *International environmental agreements: The role of foresight* (Economics working papers 2002-10). School of Economics and Management, University of Aarhus.

Diamantoudi, E., & Sartzetakis, E. (2006). Stable international environmental agreements: An analytical approach. *Journal of public economic theory, 8*(2), 247–263.

Eyckmans, J. (2001). *On the farsighted stability of the Kyoto Protocol* (ETE working papers series, 2001-03). Energy, Transport and Environment, Katholieke Universiteit Leuven.

Hoel, M. (1992). International environmental conventions: The case of uniform reductions of emissions. *Environmental and Resource Economics, 2*, 141–159.

Rubio, S. J., & Casino, B. (2005). Self-enforcing international environmental agreements with a stock pollutant. *Spanish Economic Review, 7*, 89–109.

Rubio, S. J., & Ulph, A. (2006). Self-enforcing international environmental agreements revisited. *Oxford economic papers, 58*, 233–263.

R&D Cooperation in Real Option Game Analysis

Giovanni Villani

1 Introduction

In recent years, the real option theory has been widely used in evaluating investment decisions in a dynamic environment. The market developments are complex and so the conventional NPV (Net Present Value) rule undertakes the value of a project because this method fails to take into account the market uncertainty, irreversibility of investment and ability to delay entry. The well accepted paradigm in real option theory states the equivalence between investment opportunities of firms and financial contingent claims, allowing for managerial flexibility.

Several models, such as Lee (1997); Shackleton and Wojakowski (2003); Trigeorgis (1991) and so on, are based on the assumption that the option exercise price, and so the investment cost, is fixed. But, particularly for the R&D investments, it is reasonable to consider that the evolution of the investment cost is uncertain. The R&D investment opportunity corresponds to an exchange option, i.e., the swap of an uncertain investment cost for an uncertain gross project value. The most important valuation models of exchange options are given in Armada, Kryzanowsky, and Pereira (2007); Carr (1988, 1995); Margrabe (1978); McDonald and Siegel (1985). In particular way, McDonald and Siegel (1985) value a simple European exchange option while Carr (1988) develops a model to price a compound European exchange option. Both models consider that assets distribute "dividends" that, in real options context, are the opportunity costs if an investment project is postponed (Majd & Pindyck, 1987).

In addition the real option approach, combined with game theory, allows to consider the strategic interactions among real option holders and also the market dynamics.

In this paper we analyze a cooperation between two firms that invest in R&D. Following Dias and Teixeira (2004); Dias (2004); Villani (2008) models, we assume that the R&D investments generate an "information revelation" about their success

G. Villani
Facoltà di Economia - Università di Foggia Via IV Novembre, 1 71100 Foggia,
e-mail: g.villani@unifg.it

G.I. Bischi et al. (eds.), *Nonlinear Dynamics in Economics, Finance and the Social Sciences*, DOI 10.1007/978-3-642-04023-8_6,

and so, by delaying an investment decision, new information can be revealed that might affect the profitability of the R&D projects. By the alliance between two players, we show as the information is wholly revealed and captured by two firms to improve their R&D success probabilities. The mutual information gain implies positive network externalities (as it is shown in Huisman (2001); Kong and Kwok (2007)) which lead more benefits in case of reciprocal R&D success. Therefore, the externalities can involve different entry decisions inducing the cooperation between two firms in order to maximize the partnership return. Accordingly to positive network externalities, we introduce the growth market coefficients depending by the success or failure of two players.

Our model is suitable for joint ventures of car producers, alliance between pharmaceutical and oil companies and other cooperation forms that involve a reduction of R&D risk. For instance, Chi (2000) and Kogut (1991) demonstrate the power of viewing joint ventures as real options to expand in response to future technological and market developments.

The paper is organized as follows. Section 2 reviews the Simple and Compound European exchange option pricing models and Sect. 3 introduces the basic model and also derives the final payoffs of two firms in a noncooperative framework. Section 4 analyses the cooperation between two firms. We show how both players can split the surplus of cooperation. In Sect. 5, we present two numerical examples for the cooperative R&D game. Section 6 concludes.

2 Exchange Options Methodology

In this section we present the final results of the principal models to value European exchange options. McDonald and Siegel (1985)'s model gives the value of a Simple European exchange option (SEEO) to exchange asset D for asset V at time T. Denoting by $s(V, D, T-t)$ the value of SEEO at time t, the final payoff at the option's maturity date T is $s(V, D, 0) = \max[0, V_T - D_T]$. So, assuming that V and D follow a geometric Brownian motion process given by

$$\frac{dV}{V} = (\mu_v - \delta_v)dt + \sigma_v dZ_v, \tag{1}$$

$$\frac{dD}{D} = (\mu_d - \delta_d)dt + \sigma_d dZ_d, \tag{2}$$

$$cov\left(\frac{dV}{V}, \frac{dD}{D}\right) = \rho_{vd}\sigma_v\sigma_d\, dt, \tag{3}$$

where μ_v and μ_d are the equilibrium expected rates of return on two assets, δ_v and δ_d are the corresponding "dividend-yields," σ_v and σ_d are the respective volatilities, Z_v and Z_d are two brownian standard motions with correlation coefficient ρ_{vd}, and considering that the coefficients μ_v, μ_d, δ_v, δ_d, σ_v, σ_d, and ρ_{vd} are nonnegative constants, McDonald and Siegel (1985) show that the value of a SEEO on dividend-paying assets, when the valuation date $t = 0$, is given by

$$s(V, D, T) = Ve^{-\delta_v T} N(d_1(P, T)) - De^{-\delta_d T} N(d_2(P, T)), \tag{4}$$

where

- V and D are the Gross Project Value and Investment Cost, respectively;
- $P = \frac{V}{D}$; $\sigma = \sqrt{\sigma_v^2 - 2\rho_{vd}\sigma_v\sigma_d + \sigma_d^2}$; $\delta = \delta_v - \delta_d$;
- $d_1(P, T) = \frac{\log P + \left(\frac{\sigma^2}{2} - \delta\right)T}{\sigma\sqrt{T}}$; $d_2 = d_1 - \sigma\sqrt{T}$;
- $N(d)$ is the cumulative standard normal distribution.

The current values for V and D are known in (1) and (2). Their future values depend on two components: the first is deterministic corresponding to the drift while the second is a stochastic process with variance increasing with time.

Carr (1988) develops a model to value the Compound European exchange option (CEEO) whose final payoff at maturity date t_1 is $c(s, \varphi D, 0) = \max[0, s - \varphi D]$. The CEEO value, considering the valuation date $t = 0$, is given by

$$\begin{aligned} c(s(V, D, T), \varphi D, t_1) = {} & Ve^{-\delta_v T} N_2\left(d_1\left(\frac{P}{P*}, t_1\right), d_1(P, T); \rho\right) \\ & - De^{-\delta_d T} N_2\left(d_2\left(\frac{P}{P*}, t_1\right), d_2(P, T); \rho\right) \\ & - \varphi De^{-\delta_d t_1} N\left(d_2\left(\frac{P}{P*}, t_1\right)\right), \end{aligned} \tag{5}$$

where

- φ is the exchange ratio of CEEO;
- t_1 is the expiration date of the CEEO;
- T is the expiration date of the SEEO, where $T > t_1$;
- $\tau = T - t_1$ is the time to maturity of the SEEO and $\rho = \sqrt{\frac{t_1}{T}}$;
- $d_1\left(\frac{P}{P*}, t_1\right) = \frac{\log\left(\frac{P}{P*}\right) + \left(-\delta + \frac{\sigma^2}{2}\right)t_1}{\sigma\sqrt{t_1}}$; $d_2\left(\frac{P}{P*}, t_1\right) = d_1\left(\frac{P}{P*}, t_1\right) - \sigma\sqrt{t_1}$;
- P^* is the critical price ratio that solves the following equation:

$$P^* e^{-\delta_v \tau} N(d_1(P^*, \tau)) - e^{-\delta_d \tau} N(d_2(P^*, \tau)) = \varphi;$$

- $N_2(a, b; \rho)$ is the standard bivariate normal distribution function evaluated at a and b with correlation ρ.

3 The Basic Model Game

In our model we consider two firms (A and B) that have the option to realize their R&D investment at initial time t_0 or to delay the decision at time t_1. As it is known, the R&D investment depends on the resolution of several source of uncertainty that may influence the investment decision of each firm. Assuming by q and p the R&D success probability of firms A and B respectively, we introduce two Bernoulli random variates:

$$Y : \begin{cases} 1 & q \\ 0 & 1-q \end{cases} \quad X : \begin{cases} 1 & p \\ 0 & 1-p \end{cases}.$$

The R&D success or failure of one firm generates an information revelation that influences the investment decision of the other firm. So, if firm A's R&D is successful, the firm B's probability p changes in positive information revelation p^+, while p changes in negative information revelation p^- in case of A's failure. Symmetrically, the firm A's R&D success changes in q^+ or in q^- in case of firm B success or failure at time t_0. Using Dias (2004)'s model, it results that:

$$p^+ = Prob[X = 1/Y = 1] = p + \sqrt{\frac{1-q}{q}} \cdot \sqrt{p(1-p)} \cdot \rho(X, Y),$$

$$p^- = Prob[X = 1/Y = 0] = p - \sqrt{\frac{q}{1-q}} \cdot \sqrt{p(1-p)} \cdot \rho(X, Y),$$

$$q^+ = Prob[Y = 1/X = 1] = q + \sqrt{\frac{1-p}{p}} \cdot \sqrt{q(1-q)} \cdot \rho(Y, X),$$

$$q^- = Prob[Y = 1/X = 0] = q - \sqrt{\frac{p}{1-p}} \cdot \sqrt{q(1-q)} \cdot \rho(Y, X),$$

where the correlations $\rho(X, Y)$ and $\rho(Y, X)$ are a measure of information revelation from Y to X and from X to Y, respectively. Obviously, the information revelation is considerable when the investment is not realized in the same time. So, if both players invest simultaneously in R&D or they wait to invest, there is not information revelation and consequently it results that $p = p^+ = p^-$ and $q = q^+ = q^-$. The condition to respect to have $0 \leq p^+ \leq 1$ and $0 \leq p^- \leq 1$ is:

$$0 \leq \rho(X, Y) \leq \min \left\{ \sqrt{\frac{p(1-q)}{q(1-p)}}, \sqrt{\frac{q(1-p)}{p(1-q)}} \right\}. \tag{6}$$

The condition (6) must be respected also for the information revelation process that benefits firm A, namely $\rho(Y, X)$, to have that $0 \leq q^+ \leq 1$ and $0 \leq q^- \leq 1$.

So, with the alliance between A and B, we can assume that information is wholly revealed and we can setting that the cooperative information $\rho_{\max}$ is equal to:

$$\rho_{\max} = \min\left\{\sqrt{\frac{p(1-q)}{q(1-p)}}, \sqrt{\frac{q(1-p)}{p(1-q)}}\right\}. \tag{7}$$

We can observe that, in the case in which both firms have the same success probability $p = q$, it results $\rho_{\max} = 1$ and so $q^+ = 1$ and $p^+ = 1$. This means that, in case of A's R&D success at time t_0, it involves the B's success at time t_1 in the cooperation treatment since the information revelation is fully captured and vice-versa.

Moreover, we assume that R&D investments are characterized by network externalities that induce more benefits in case of reciprocal R&D success. So we denote by

$$K_{0_S 0_S}; \quad K_{0_S 1_S}; \quad K_{1_S 0_S}; \quad K_{1_S 1_S}$$

the growth market coefficients in case of A and B success. The 0 and 1 mean that the R&D investment is realized at time t_0 or t_1 respectively, while the S denotes the success. The first part denotes the operation of considered firm, while the second part is the situation of the other firm. For instance, if A and B invest successfully in R&D at time t_0 and t_1 respectively, firm A takes $K_{0_S 1_S}$ while B obtains $K_{1_S 0_S}$. In the same way, assuming the failure (denoted by F) of the other player and considering that the unsuccess of one firm does not produce network externality, we can write the growth market coefficients for the winning firm as

$$K_{0_S 0_F} = K_{0_S 1_F} \equiv K_{0_S}; \quad K_{1_S 0_F} = K_{1_S 1_F} \equiv K_{1_S};$$

Finally, in case of failure of considered firm, its market coefficient will be equal to zero whether in case of success or failure of other player. Now we can set the relations among the growth market coefficients K using these assumptions:

- *Positive Network Externality*: As it is shown Huisman (2001), the growth market coefficients in case of both R&D success will be bigger than the situation in which only one firm invests successfully and so:

$$K_{SS} > K_S; \tag{8}$$

- *R&D Success Time*: The market coefficient increases if the reciprocal R&D success is realized at time t_0 rather than t_1 because there is more time to benefit both network externalities and R&D innovations. In the situation in which only one firm invests successfully, the market coefficient enlarges if the success is realized at time t_0:

$$K_{0_S 0_S} > K_{1_S 1_S}; \quad K_{0_S} > K_{1_S}; \tag{9}$$

- *First Mover's Advantage*: The firm that realizes with success the R&D investment at time t_0 will receive an higher market coefficient than other player that postpones successfully the project at time t_1:

$$K_{0_S 1_S} > K_{1_S 0_S}. \tag{10}$$

To determine the growth market coefficients K, we assume that they depend by a parameter k involving the R&D innovation and by length of R&D benefits until the expiration time T. For the positive network externality, we take into account two times the one firm market coefficient. So, assuming that the initial time $t_0 = 0$, we have that

$$K_{0_S} = kT, \tag{11}$$

$$K_{0_S 0_S} = 2kT, \tag{12}$$

$$K_{1_S} = k(T - t_1), \tag{13}$$

$$K_{1_S 1_S} = 2k(T - t_1). \tag{14}$$

Finally, to determine $K_{0_S 1_S}$ and $K_{1_S 0_S}$, we assume that

$$K_{0_S 1_S} = 2k(T - t_1) + kt_1, \tag{15}$$

$$K_{1_S 0_S} = 2k(T - t_1) - kt_1. \tag{16}$$

If one firm invests successfully at time t_0 and the other player at time t_1, we have that the first firm takes the network externality starting from time t_1, namely $K_{1_S 1_S}$, plus the first mover's advantage kt_1 until time t_1. Symmetrically, the market coefficient $K_{1_S 0_S}$ for the other firm that postpones its choice will be $K_{1_S 1_S}$ minus kt_1. Finally, to ensure that condition (8) holds, we need to impose that $t_1 < \frac{T}{3}$. This condition is reasonable with the consideration that the information revelation disappears in time and furthermore, if one firm invests at time t_0, then the other firm decision will be made within $t_1 < \frac{T}{3}$ to allow the realization of the development phase in $T - t_1$.

First to start, we state as Leader the pioneer firm (A or B) that invests in R&D at time t_0 earlier than other one, namely the Follower, that postpones the R&D investment decision at time t_1. We denote by R the R&D investment for the development of a new product, V the overall market value deriving by R&D innovations and D is the total investment cost to realize new goods. We consider that the production investment of each firm is proportional to its market share and it can be realized only at time T, that is the time needed to develop the new product.

3.1 *The Leader's Payoff*

We analyze the Leader's payoff assuming that firm A (Leader) invests in R&D at time t_0 while firm B (Follower) decides to wait to invest. So, the Leader spends the investment R at time t_0 and obtains, in case of its R&D success with probability q, the development option. In particular way, if also the Follower's R&D investment is successfully at time t_1, the growth market coefficient will be $K_{0_S 1_S}$ and the Leader holds the development option $s\left(K_{0_S 1_S} V, K_{0_S 1_S} D, T\right)$ that gives the opportunity to invest $K_{0_S 1_S} D$ at time T and to claim a market value equal to $K_{0_S 1_S} V$. So the Leader's payoff is

$$L_A^S(V, D) = -R + qk(2T - t_1)\left(Ve^{-\delta_v T} N(d_1(P, T)) - De^{-\delta_d T} N(d_2(P, T))\right). \tag{17}$$

The probability to have $K_{0_S 1_S}$ depending by the Follower's R&D success that is p^+ since it receives the information revelation from Leader's investment occurred at time t_0. But if the Follower's R&D fails, the Leader's market coefficient, in case of its R&D success, is K_{0_S} and it receives the development option $s\left(K_{0_S} V, K_{0_S} D, T\right)$. So the Leader's payoff is:

$$L_A^F(V, D) = -R + qkT\left(Ve^{-\delta_v T} N(d_1(P, T)) - De^{-\delta_d T} N(d_2(P, T))\right) \tag{18}$$

So, computing the expectation value between (17) and (18), the final Leader's payoff (firm A) at time t_0 is

$$L_A(V, D) = p^+ \cdot L_A^S(V, D) + (1 - p^+) \cdot L_A^F(V, D). \tag{19}$$

Symmetrically, assuming that firm B (Leader) invests at time t_0 while firm A (Follower) decides to postpone its decision, the final Leader's payoff at time t_0 becomes

$$L_B(V, D) = q^+ \cdot L_B^S(V, D) + (1 - q^+) \cdot L_B^F(V, D). \tag{20}$$

3.2 The Follower's Payoff

Now we focus on the Follower's payoff assuming that firm B (Follower) decides to postpone its R&D investment decision at time t_1 and firm A (Leader) invests at time t_0. If the Leader's R&D investment is successfully (with a probability q), then the Follower's probability success becomes p^+ and its growth market coefficient is $K_{1_S 0_S}$. So, after the investment R, the Follower holds with a probability p^+ the development option $s(K_{1_S 0_S} V, K_{1_S 0_S} D, \tau)$ to invest $K_{1_S 0_S} D$ at time T and claims a market value equal to $K_{1_S 0_S} V$. So the Follower's payoff at time t_0 is a CEEO with maturity t_1, exercise price equal to R and the underlying asset is the development option $s(K_{1_S 0_S} V, K_{1_S 0_S} D, \tau)$. According to Carr (1988)'s model, we assume that $R = \varphi D$ is a proportion φ of asset D. Hence, denoting by $c(p^+)$ the CEEO at time t_0, namely:

$$c(p^+) \equiv c(p^+ s(K_{1_S 0_S} V, K_{1_S 0_S} D, \tau), \varphi D, t_1),$$

we can write, using the (5), the value of CEEO with positive information

$$\begin{aligned} c(p^+) = & \; p^+k(2T - 3t_1)Ve^{-\delta_v T} N_2 \left(d_1 \left(\frac{P}{P^*_{upB}}, t_1 \right), d_1 (P, T) ; \rho \right) \\ & - p^+k(2T - 3t_1)De^{-\delta_d T} N_2 \left(d_2 \left(\frac{P}{P^*_{upB}}, t_1 \right), d_2 (P, T) ; \rho \right) \\ & - \varphi De^{-\delta_d t_1} N \left(d_2 \left(\frac{P}{P^*_{upB}}, t_1 \right) \right), \end{aligned} \tag{21}$$

where P^*_{upB} is the critical value that makes the underlying asset of $c(p^+)$ equal to exercise value. Hence P^*_{upB} solves the following equation:

$$p^+ s(K_{1_S 0_S} V, K_{1_S 0_S} D, \tau) = \varphi D,$$

and assuming the asset $K_{1_S 0_S} D$ as numeraire, we can rewrite the above equation as

$$P^*_{upB}\, e^{-\delta_v \tau} N(d_1(P^*_{upB}, \tau)) - e^{-\delta_d \tau} N(d_2(P^*_{upB}, \tau)) = \frac{\varphi}{p^+(2T - 3t_1)}.$$

Alternatively, in case of Leader's failure, the Follower's R&D success probability changes in p^- and its market coefficient is K_{1_S}. Denoting by $c(p^-)$ the CEEO at time t_0 with negative information, we can write, using the (5), the value of CEEO with negative information:

$$\begin{aligned} c(p^-) = & \; p^-k(T - t_1)Ve^{-\delta_v T} N_2 \left(d_1 \left(\frac{P}{P^*_{dwB}}, t_1 \right), d_1 (P, T) ; \rho \right) \\ & - p^-k(T - t_1)De^{-\delta_d T} N_2 \left(d_2 \left(\frac{P}{P^*_{dwB}}, t_1 \right), d_2 (P, T) ; \rho \right) \\ & - \varphi De^{-\delta_d t_1} N \left(d_2 \left(\frac{P}{P^*_{dwB}}, t_1 \right) \right), \end{aligned} \tag{22}$$

where P^*_{dwB} is the critical price that solves the following equation:

$$P^*_{dwB}\, e^{-\delta_v \tau} N(d_1(P^*_{dwB}, \tau)) - e^{-\delta_d \tau} N(d_2(P^*_{dwB}, \tau)) = \frac{\varphi}{p^-k(T - t_1)}$$

The Follower obtains the CEEO $c(p^+)$ in case of Leader's success with a probability q and the CEEO $c(p^-)$ in case of Leader's failure with a probability $(1 - q)$. Hence, the Follower's payoff at time t_0 is the expectation value between (21) and (22):

$$F_B(V, D) = q\,c(p^+) + (1-q)\,c(p^-). \tag{23}$$

Similarly, if we consider that firm B (Leader) invests in R&D at time t_0 and firm A (Follower) decides to wait to invest, we have that

$$F_A(V, D) = p\,c(q^+) + (1-p)\,c(q^-) \tag{24}$$

3.3 The A and B Payoffs in Case of Simultaneous Investment

In this case, we analyze the situation in which both firms invest in R&D at time t_0. We can assume that there is not information revelation since the investment is simultaneous but both players can benefice of network externalities. First of all, we determine the firm's A payoff. Assuming the firm B's R&D success, A receives the development option with a growth market coefficient $K_{0_S 0_S}$ in case of its R&D success. So, after the investment R at time t_0, player A receives the development option $s(K_{0_S 0_S} V, K_{0_S 0_S} D, T)$ with a probability q:

$$S_A^S(V, D) = -R + q2kT\left(Ve^{-\delta_v T} N(d_1(P, T)) - De^{-\delta_d T} N(d_2(P, T))\right). \tag{25}$$

But, assuming the firm B failure, A receives the development option $s(K_{0_S} V, K_{0_S} D, T)$ with a growth market coefficient K_{0_S} in case of its success:

$$S_A^F(V, D) = -R + qkT\left(Ve^{-\delta_v T} N(d_1(P, T)) - De^{-\delta_d T} N(d_2(P, T))\right). \tag{26}$$

So, recalling that firm B's probability success is p, the firm's A payoff in case of simultaneous investment will be the expectation value between (25) and (26):

$$S_A(V, D) = p \cdot S_A^S(V, D) + (1-p) \cdot S_A^F(V, D). \tag{27}$$

Symmetrically, the firm's B payoff will be

$$S_B(V, D) = q \cdot S_B^S(V, D) + (1-q) \cdot S_B^F(V, D). \tag{28}$$

3.4 The A and B Payoffs When Both Firms Wait to Invest

Finally, we suppose that both firms decide to delay their R&D investment decision at time t_1 and we can setting that there is not information revelation. First of all, we analyze the firm A's payoff. Assuming the R&D success of firm B, then the growth market coefficient of player A will be $K_{1_S 1_S}$. So, after the investment R at time t_1, firm A holds with a probability q the development option $s(K_{1_S 1_S} V, K_{1_S 1_S} D, \tau)$

to invest $K_{1_S 1_S} D$ at time T and claims a market value equal to $K_{1_S 1_S} V$. Then the firm's A payoff at time t_0 is a CEEO with maturity t_1, the exercise price equal to R and the underlying asset is the development option $s(K_{1_S 1_S} V, K_{1_S 1_S} D, \tau)$ with a probability q. Thus, according to Carr (1988)'s model and assuming that R is a proportion φ of asset D, the CEEO in case of firm's B success is

$$W_A^S(V, D) = c\left(q \cdot s(K_{1_S 1_S} V, K_{1_S 1_S} D, \tau), \varphi D, t_1\right)$$

and specifically

$$\begin{aligned} W_A^S(V, D) = {} & q2k(T - t_1)Ve^{-\delta_v T} N_2\left(d_1\left(\frac{P}{P_{wsA}^*}, t_1\right), d_1(P, T); \rho\right) \\ & -q2k(T - t_1)De^{-\delta_d T} N_2\left(d_2\left(\frac{P}{P_{wsA}^*}, t_1\right), d_2(P, T); \rho\right) \\ & -\varphi De^{-\delta_d t_1} N\left(d_2\left(\frac{P}{P_{wsA}^*}, t_1\right)\right), \end{aligned} \tag{29}$$

where P_{wsA}^* is the critical value that solves the following equation:

$$P_{wsA}^* e^{-\delta_v \tau} N(d_1(P_{wsA}^*, \tau)) - e^{-\delta_d \tau} N(d_2(P_{wsA}^*, \tau)) = \frac{\varphi}{q2k(T - t_1)}.$$

But, in case of firm's B failure, the firm A growth market coefficient will be K_{1_S}. So, after the investment R at time t_1, firm A obtains with a probability q the development option $s(K_{1_S} V, K_{1_S} D, \tau)$. Thus, using Carr (1988)'s model, the firm' A payoff at time t_0 is a CEEO where the underlying asset is $s(K_{1_S} V, K_{1_S} D, \tau)$ with a probability q and specifically

$$\begin{aligned} W_A^F(V, D) = {} & qk(T - t_1)Ve^{-\delta_v T} N_2\left(d_1\left(\frac{P}{P_{wfA}^*}, t_1\right), d_1(P, T); \rho\right) \\ & -qk(T - t_1)De^{-\delta_d T} N_2\left(d_2\left(\frac{P}{P_{wfA}^*}, t_1\right), d_2(P, T); \rho\right) \\ & -\varphi De^{-\delta_d t_1} N\left(d_2\left(\frac{P}{P_{wfA}^*}, t_1\right)\right), \end{aligned} \tag{30}$$

where, as seen before, P_{wfA}^* is the critical value that solves the following equation:

$$P_{wfA}^* e^{-\delta_v \tau} N(d_1(P_{wfA}^*, \tau)) - e^{-\delta_d \tau} N(d_2(P_{wfA}^*, \tau)) = \frac{\varphi}{qk(T - t_1)}.$$

Hence, recalling that the firm B success is equal to p, we can compute the firm A payoff as the expectation value between (29) and (30):

$$W_A(V,D) = p\, W_A^S(V,D) + (1-p)\, W_A^F(V,D). \tag{31}$$

Similarly, the firm B payoff is

$$W_B(V,D) = q\, W_B^S(V,D) + (1-q)\, W_B^F(V,D). \tag{32}$$

3.5 Noncooperative Critical Market Values

Now, to determine the noncooperative Nash equilibriums denoted by $v(A)$ and $v(B)$, we analyze the relations among the strategic payoffs according to several expected market values V at time t_0 and considering fixed the invest cost D at time t_0. Therefore, we are able to determine the critical market values that delimit the several Nash equilibriums. First of all, we study the relations between the Leader and Waiting strategies considering only the variable V and, to simplify the notation, we do not take into account the dividends to compute the derivatives. We can observe that

- $L_A(0) = L_B(0) = -R; \quad W_A(0) = W_B(0) = 0;$
- $\dfrac{\partial L_A}{\partial V} = qN(d_1(P,T))k[p^+(2T - t_1) + (1-p^+)T];$
- $\dfrac{\partial L_B}{\partial V} = pN(d_1(P,T))k[q^+(2T - t_1) + (1-q^+)T];$
- $$\frac{\partial W_A}{\partial V} = 2pqk(T-t_1)N_2\left(d_1\left(\frac{P}{P^*_{wsA}}, t_1\right), d_1(P,T); \rho\right)$$
$$+ (1-p)qk(T-t_1)N_2\left(d_1\left(\frac{P}{P^*_{wfA}}, t_1\right), d_1(P,T); \rho\right);$$
- $$\frac{\partial W_B}{\partial V} = 2qpk(T-t_1)N_2\left(d_1\left(\frac{P}{P^*_{wsB}}, t_1\right), d_1(P,T); \rho\right)$$
$$+ (1-q)pk(T-t_1)N_2\left(d_1\left(\frac{P}{P^*_{wfB}}, t_1\right), d_1(P,T); \rho\right);$$
- $\dfrac{\partial L_A}{\partial V} > \dfrac{\partial W_A}{\partial V} > 0; \quad \dfrac{\partial L_B}{\partial V} > \dfrac{\partial W_B}{\partial V} > 0;$

as $N(a) > N_2(a,b;\rho)$. Then, the following proposition holds:

Proposition 1. *There exists, for each firm $i = A, B$, a unique critical market value V_i^W that makes $L_i(V_i^W) = W_i(V_i^W)$. Denoting by $V_W^* = \min(V_A^W, V_B^W)$ and $V_Q^* = \max(V_A^W, V_B^W)$, it results that*

$$L_i(V) < W_i(V) \quad \text{for } V < V_W^*;$$
$$L_i(V) > W_i(V) \quad \text{for } V > V_Q^*.$$

Moreover, if A's success probability is higher than B, for $V \in]V_W^*, V_Q^*[$ *it results*

$$L_A(V) > W_A(V); \quad L_B(V) < W_B(V);$$

otherwise, if B's success probability is higher than A, for $V \in]V_W^*, V_Q^*[$ *it results*

$$L_A(V) < W_A(V); \quad L_B(V) > W_B(V).$$

Now we analyze the relation between the Follower and the Simultaneous strategies. Then, we can observe that

- $F_A(0) = F_B(0) = 0; \quad S_A(0) = S_B(0) = -R;$
- $\dfrac{\partial F_A}{\partial V} = pq^+k(2T - 3t_1)N_2\left(d_1\left(\dfrac{P}{P^*_{upA}}, t_1\right), d_1(P,T); \rho\right) + (1-p)q^-k(T - t_1)N_2\left(d_1\left(\dfrac{P}{P^*_{dwA}}, t_1\right), d_1(P,T); \rho\right);$
- $\dfrac{\partial F_B}{\partial V} = qp^+k(2T - 3t_1)N_2\left(d_1\left(\dfrac{P}{P^*_{upB}}, t_1\right), d_1(P,T); \rho\right) + (1-q)p^-k(T - t_1)N_2\left(d_1\left(\dfrac{P}{P^*_{dwB}}, t_1\right), d_1(P,T); \rho\right);$
- $\dfrac{\partial S_A}{\partial V} = qN(d_1(P,T))kT[1+p]; \quad \dfrac{\partial S_B}{\partial V} = pN(d_1(P,T))kT[1+q];$
- $\dfrac{\partial F_i}{\partial V} > 0; \quad \dfrac{\partial S_i}{\partial V} > 0;$

for $i = A, B$. In this case we have that both derivatives are positive but the intersection between Follower and Simultaneous strategies exists if $\frac{\partial S_i}{\partial V} > \frac{\partial F_i}{\partial V}$ for $i = A, B$. So the following proposition holds:

Proposition 2. *If* $\frac{\partial S_A}{\partial V} > \frac{\partial F_A}{\partial V}$, *then there exists a unique critical market value* V_P^* *that makes* $S_A(V_P^*) = F_A(V_P^*)$ *and it results that*

$$\begin{aligned} S_A(V) &< F_A(V) \quad \text{for } V < V_P^*; \\ S_A(V) &> F_A(V) \quad \text{for } V > V_P^*; \end{aligned}$$

otherwise, if $\frac{\partial S_A}{\partial V} \le \frac{\partial F_A}{\partial V}$, *then* $S_A(V) < F_A(V)$ *for every value of* V.

If $\frac{\partial S_B}{\partial V} > \frac{\partial F_B}{\partial V}$, *then there exists a unique critical market value* V_S^* *that makes* $S_B(V_S^*) = F_B(V_S^*)$ *and it results that*

$$\begin{aligned} S_B(V) &< F_B(V) \quad \text{for } V < V_S^*; \\ S_B(V) &> F_B(V) \quad \text{for } V > V_S^*; \end{aligned}$$

otherwise, if $\frac{\partial S_B}{\partial V} \leq \frac{\partial F_B}{\partial V}$*, then* $S_B(V) < F_B(V)$ *for every value of* V*. Moreover, if firm A's success probability is higher than B, then* $V_P^* < V_S^*$ *otherwise* $V_S^* < V_P^*$.

4 The Cooperation Between A and B

In this section we analyze the cooperation between firms A and B that allows to capture the whole information revelation and so to improve the R&D success probabilities. In particular way, we assume that the value achieved by the cooperation can be transferred from one player to the other. We show as the strategic alliance is the joint best response to the noncooperative alternative and so the equilibriums that both firms obtain through the cooperation are Pareto-dominate all the noncooperative ones. As we consider two players, we denote by $C(A \cup B)$ the feasible set for the coalition, namely is the set of outcome which can be obtained by two players acting together. The cooperation value is given by the sum of two firms' payoffs using the whole information revelation $\rho_{\max}$ deriving by their R&D investments. Both players can agree upon several partnership contracts. For instance, A and B can share equitably the surplus of cooperation $C(A \cup B) - (v(A) + v(B))$ using the Shapley values:

$$Sh_A = v(A) + \frac{C(A \cup B) - (v(A) + v(B))}{2}, \tag{33}$$

$$Sh_B = v(B) + \frac{C(A \cup B) - (v(A) + v(B))}{2}. \tag{34}$$

This solution looks natural in the symmetric case $p = q$ in which both firms have the same success probability otherwise, we can assume also asymmetric shares to split the cooperation value according to success probability of each firm:

$$P_A = v(A) + \frac{q}{p+q}\left(C(A \cup B) - (v(A) + v(B))\right), \tag{35}$$

$$P_B = v(B) + \frac{p}{p+q}\left(C(A \cup B) - (v(A) + v(B))\right). \tag{36}$$

We can observe that, if $p = q$, then $Sh_i = P_i$ for $i = A, B$ and the efficiency property is satisfied as $Sh_A + Sh_B = P_A + P_B = C(A \cup B)$. The cooperative information ρ_{max} influences the Leader and Follower payoffs that we denote by $L_i^C(V)$ and $F_i^C(V)$ for $i = A, B$, where C means the cooperative action. The four possible cooperation strategies are:

- Both players decide to wait to invest at time t_0:

$$C(A \cup B) = W_A(V) + W_B(V) \equiv W_C(V)$$

- The firm A invests at time t_0 while the firm B delays its decision at time t_1:

$$C(A \cup B) = L_A^C(V) + F_B^C(V) \equiv LF_C(V)$$

- Symmetrically, B invests at time t_0 and A delays its decision at time t_1:

$$C(A \cup B) = F_A^C(V) + L_B^C(V) \equiv FL_C(V)$$

- Both players decide to invest simultaneously at time t_0:

$$C(A \cup B) = S_A(V) + S_B(V) \equiv S_C(V)$$

4.1 Cooperative Critical Market Values

The aim of two firm acting together is to improve their position compared with no partnership situation and to reach a Pareto optimal solution. To realize this objective, we have to determine the maximum value among the four cooperation strategies according to several expected market values V. Hence we compute the cooperative critical market values that delimit the maximum payoff $C(A \cup B)$. So it results that

- $W_C(0) = 0; \quad S_C(0) = -2R; \quad LF_C(0) = -R; \quad FL_C(0) = -R;$
- $$\begin{aligned}\frac{\partial W_C}{\partial V} &= 2k(T-t_1)pqN_2\left(d_1\left(\frac{P}{P^*_{wsA}}, t_1\right), d_1(P,T); \rho\right)\\ &+ 2k(T-t_1)pqN_2\left(d_1\left(\frac{P}{P^*_{wsB}}, t_1\right), d_1(P,T); \rho\right)\\ &+ k(T-t_1)(1-p)qN_2\left(d_1\left(\frac{P}{P^*_{wfA}}, t_1\right), d_1(P,T); \rho\right)\\ &+ k(T-t_1)(1-q)pN_2\left(d_1\left(\frac{P}{P^*_{wfB}}, t_1\right), d_1(P,T); \rho\right);\end{aligned}$$
- $\dfrac{\partial S_C}{\partial V} = qkTN(d_1(P,T))\,[2p + (1-p)] + pkTN(d_1(P,T))\,[2q + (1-q)];$
- $\dfrac{\partial S_C}{\partial V} > \dfrac{\partial W_C}{\partial V} > 0;$

as $N(a) > N_2(a,b;\rho)$. Now we can remark that, if $q = p$, then it results $LF_C(V) = FL_C(V)$ as $L_A^C(V) = L_B^C(V)$ and $F_A^C(V) = F_B^C(V)$ and so both strategies give the same value. But, if $q > p$, then we have that $LF_C(V) > FL_C(V)$ and, if $q < p$, then $LF_C(V) < FL_C(V)$. So, to determine the maximum value, we consider the cooperation strategy in which the Leader is the firm with the highest success probability. Assuming that $q \geq p$, we take into account the cooperative action LF_C. We have that

- $$\frac{\partial LF_C}{\partial V} = qN(d_1(P,T))k[p^+(T-t_1)+T]$$
$$+ qp^+k(2T-3t_1)N_2\left(d_1\left(\frac{P}{P^*_{upB}},t_1\right),d_1(P,T);\rho\right)$$
$$+ (1-q)p^-k(T-t_1)N_2\left(d_1\left(\frac{P}{P^*_{dwB}},t_1\right),d_1(P,T);\rho\right);$$
- $$\frac{\partial LF_C}{\partial V} > \frac{\partial W_C}{\partial V} > 0;$$

since $P^*_{upi} < P^*_{wsi} < P^*_{wfi}$ for $i = A, B$. So, the following proposition holds:

Proposition 3. *There exists a unique critical market value V_C^* such that $LF_C(V_C^*) = W_C(V_C^*)$ and*

$$LF_C(V) < W_C(V) \quad \textit{for } V < V_C^*;$$
$$LF_C(V) > W_C(V) \quad \textit{for } V > V_C^*.$$

Now we analyze the several cooperative equilibriums that can be occur.

4.1.1 First Case

If $\frac{\partial LF_C}{\partial V} \geq \frac{\partial S_C}{\partial V}$, then there is not intersection between LF_C and S_C. Moreover, the intersection LF_C and W_C occurs before than S_C and W_C. So, in this case, we have to consider only the critical market value V_C^* given by Proposition 3 and we can state that, if $V < V_C^*$, the maximum payoff that both players can obtain by cooperation is $W_C(V)$ otherwise, if $V > V_C^*$, the maximum payoff attainable cooperating is $LF_C(V)$.

In this case, the best strategic cooperation is the waiting policy (W_C) until the expected market value V is below the critical value V_C^* and, if $V > V_C^*$, the optimal strategy is the Leader–Follower one (LF_C) in which the firm with highest success probability realizes the R&D investment at time t_0 and the other player postpones its decision at time t_1.

4.1.2 Second Case

If $\frac{\partial LF_C}{\partial V} < \frac{\partial S_C}{\partial V}$, then there is intersection between the functions LF_C and S_C. So the following proposition holds:

Proposition 4. *If $\frac{\partial LF_C}{\partial V} < \frac{\partial S_C}{\partial V}$, then there exists a unique critical market value V_G^* such that $LF_C(V_G^*) = S_C(V_G^*)$ and it results that*

$$S_C(V) < LF_C(V) \quad \textit{for } V < V_G^*;$$
$$S_C(V) > LF_C(V) \quad \textit{for } V > V_G^*.$$

Moreover, as $\frac{\partial(LF_C-W_C)}{\partial V} \geq \frac{\partial(S_C-LF_C)}{\partial V}$, it results that $V_C^* < V_G^*$. Using the Propositions 3 and 4, we have that the optimal cooperation action is the waiting policy (W_C) when the expected market value V is below V_C^* while, if V is in the range $[V_C^*, V_G^*]$, then the maximum payoff is obtained by the Leader–Follower (LF_C) strategy and finally, if $V > V_G^*$, both firms realize their R&D investment at time t_0.

5 Real Applications

To illustrate the concepts presented, we develop some numerical examples for the cooperative R&D game and, in particular way, we assume that A has a more efficient Know-How than B that implies the highest success probability. So we assume that

- R&D Investment: $R = 250{,}000\,\$$;
- Development Investment: $D = 400{,}000\ \$$;
- Market and Costs Volatility: $\sigma_v = 0.93$; $\sigma_d = 0.23$;
- Proportion of D required for R: $\varphi = \frac{R}{D} = 0.625$
- Correlation between V and D: $\rho_{vd} = 0.15$;
- Dividend-Yields of V and D: $\delta_v = 0.15$; $\delta_d = 0$;
- R&D innovation parameter: $k = 0.30$
- Expiration Time of Simple Option: $T = 3$ years;
- A and B success probability: $q = 0.60$; $p = 0.55$;
- Noncooperative Information Revelation: $\rho(X, Y) = \rho(Y, X) = 0.40$;
- Cooperative Information Revelation: $\rho_{\max} = 0.9026$;

5.1 Numerical Application of First Case

Assuming that the R&D investment decision can be delay at time $t_1 = 0.5$ year, we obtain, using the (11)-(16), the following growth market coefficients:

$$K_{0_S 0_S} = 1.8;\ K_{0_S 1_S} = 1.65;\ K_{1_S 1_S} = 1.50;\ K_{1_S 0_S} = 1.35;$$
$$K_{0_S} = 0.90;\ K_{1_S} = 0.75.$$

As we can show in the Fig. 1a, the $\frac{\partial S_C}{\partial V} < \frac{\partial LF_C}{\partial V}$ and so, using the Proposition 3, we compute the critical market value V_C^* to determine the best cooperation strategy. For our adapted number, it results that $V_C^* = 700{,}037$. So, if $V < 700{,}037$, both players decide to wait to invest and $C(A \cup B) = W_C(V)$ otherwise, if $V > 700{,}037$, the best cooperation strategy is the Leader–Follower one in which firm A invests at time t_0 and B delays its decision at time t_1 and so $C(A \cup B) = LF_C(V)$.

Now, to determine the partnership shares (Sh_A, Sh_B) and (P_A, P_B), we need to compute the noncooperative critical market values V_W^*, V_Q^*, V_P^*, and V_S^* that allow to determine the Nash equilibriums. So, using the Propositions 1 and 2, it results:

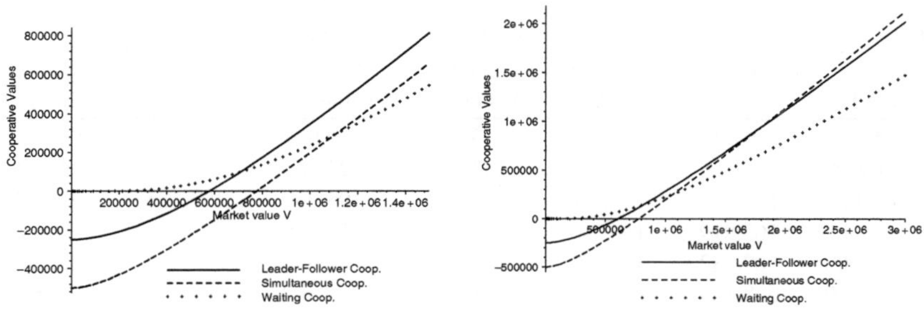

First case: $t_1 = 0.5$ Second case: $t_1 = 0.8$

Fig. 1 Comparison between two cases assuming $t_1 = 0.5$ and $t_1 = 0.8$, respectively

Table 1 Firms A and B cooperative payoffs assuming $k = 0.30$ and $t_1 = 0.5$

Market Value V	Leader–Follower Value LF_C	Follower–Leader Value FL_C	Simultaneous Value S_C	Waiting Value W_C
600,000	17,412	12,486	−146,693	**62,269**
900,000	**257,854**	248,968	107,985	183,345
1,050,000	**390,083**	378,605	241,780	262,199
1,100,000	**435,386**	422,974	287,075	290,593
1,250,000	**574,220**	558,837	424,694	381,060
1,400,000	**716,600**	698,053	564,510	478,170

$$V_W^* = 1,028,380; \quad V_Q^* = 1,066,240; \quad V_P^* = 1,200,470; \quad V_S^* = 1,268,650.$$

If $V < 1,028,380$, the waiting policy (W_A, W_B) is optimal in Nash meaning for both players at time t_0, if $1,028,380 < V < 1,066,240$ and $1,200,470 < V < 1,268,650$, we have one Nash noncooperative equilibrium (L_A, F_B), if $1,066,240 < V < 1,200,470$, then we obtain two Nash equilibriums (L_A, F_B) and (F_A, L_B) and finally, if $V > 1,268,650$, it results one Nash equilibrium (S_A, S_B).

Let us examine the partnership between firms A and B combining the cooperative and noncooperative critical market values. The second and third column of Table 2 contains the noncooperative Nash-equilibriums $v(A)$ and $v(B)$. Moreover, the Table 1 summarizes the cooperative values $C(A \cup B)$ according to four strategic cooperations and, in particular way, the bold type values are the maximum ones deriving by the optimal strategic alliance. Using the (33)-(36), firms A and B can split the cooperative value $C(A \cup B)$ by the Shapley (Sh_A, Sh_B) or the Asymmetric (P_A, P_B) values that are shown in the Table 2. Comparing the cooperative and the noncooperative values, we can observe that the partnership is favorable for both players since each firm improves its payoff deriving from noncooperative Nash equilibrium. So we can state that the couples (Sh_A, Sh_B) and

Table 2 Shapley and asymmetric values assuming $k = 0.30$ and $t_1 = 0.5$

Market Value V	Noncoop. $v(A)$	Noncoop. $v(B)$	Shapley Value Sh_A	Shapley Value Sh_B	Asym. Value P_A	Asym. Value P_B
600,000	33,244	29,024	33,244	29,024	33,244	29,024
900,000	96,736	86,609	133,990	123,863	135,610	122,244
1,050,000	141,889	142,306	194,833	195,250	197,135	192,948
1,100,000	165,819	156,705	222,250	213,136	224,704	210,682
1,100,000	169,582	144,077	230,445	204,940	233,092	202,294
1,250,000	238,525	202,120	305,312	268,907	308,216	266,004
1,400,000	296,958	267,552	373,003	343,597	376,309	340,291

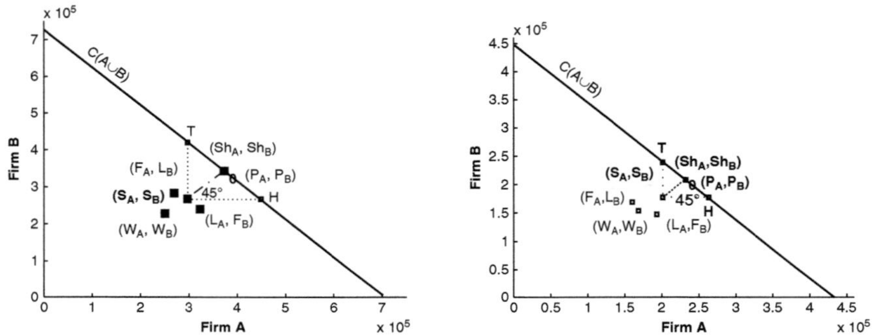

(a) Equilibriums when $V = 1,400,000$ in the first case

(b) Equilibriums when $V = 1,200,000$ in the second case

Fig. 2 A and B equilibriums

(P_A, P_B) are Pareto optimal with respect to $(v(A), v(B))$. Only if $V < 700,037$, and so $V = 600,000$, then the partnership does not add value to each player because the surplus of cooperation $W_C(V) - (W_A(V) + W_B(V))$ is equal to zero.

Finally, the Fig. 2a represents the overall situation assuming $V = 1,400,000$. In particular way, the black line denotes the feasible set of partnership, namely it represents all the combinations to split $C(A \cup B)$. But only the segment T-H is important to analyze because otherwise firms have the incentive to deviate from cooperation. In fact, we can observe that Shapley (Sh_A, Sh_B) and Asymmetric (P_A, P_B) values belong to the segment T-H. Moreover, the Fig. 2b shows the four noncooperative strategies and in particular way the Nash equilibrium (S_A, S_B). We can notice that the segment joins the couples (S_A, S_B) and (Sh_A, Sh_B) has a 45° slope because, by the Shapley value, A and B share equitably the surplus of cooperation $C(A \cup B) - (v(A) + v(B))$.

5.2 Numerical Application of Second Case

If we assume now that $t_1 = 0.8$ year, using (11)-(15), we have that the growth market coefficients are

$$K_{0_S 0_S} = 1.8;\ K_{0_S 1_S} = 1.56;\ K_{1_S 1_S} = 1.32;\ K_{1_S 0_S} = 1.08;$$
$$K_{0_S} = 0.90;\ K_{1_S} = 0.66.$$

As is shown in the Fig. 1b, the $\frac{\partial S_C}{\partial V} > \frac{\partial LF_C}{\partial V}$ and so we have two critical market values: $V_C^* = 815,710$ and $V_G^* = 1,796,130$. Using the Propositions 3 and 4, we are able to state the optimal cooperation strategy. So, if $V < 815,710$, then the best partnership strategy is to wait to invest and $C(A \cup B) = W_C(V)$, if $815,710 < V < 1,796,130$, then both players choose the cooperation form Leader–Follower and hence $C(A \cup B) = LF_C(V)$ and finally, if $V > 1,796,130$, then both firms prefer to invest simultaneously at time t_0 and so $C(A \cup B) = S_C(V)$.

Now, using the Propositions 1 and 2, we compute the four noncooperative critical market values V_W^*, V_Q^*, V_P^*, and V_S^*:

$$V_P^* = 1,019,230;\quad V_S^* = 1,064,060;\quad V_W^* = 1,075,210;\quad V_Q^* = 1,120,840.$$

If $V < 1,064,060$, both players prefer to wait (W_A, W_B) and to defer their R&D decision at time t_1, if $1,064,060 < V < 1,075,210$ we obtain two Nash equilibriums (W_A, W_B) and (S_A, S_B) and finally, if $V > 1,075,210$, then the simultaneous R&D investment (S_A, S_B) at time t_0 is optimal in Nash meaning.

As we have seen in the first case, the second and third column of Table 4 summarizes the noncooperative Nash equilibriums. Moreover, the Table 3 contains the partnership values $C(A \cup B)$ according to four cooperative strategies and, in particular way, the bold type values are the maximum payoffs deriving by best alliance. Both players can split the cooperative value $C(A \cup B)$ by the Shapley (Sh_A, Sh_B) or the Asymmetric (P_A, P_B) values (see (33)–(36)) that are shown in the Table 4. We can observe that, if $V = 600,000$ and more generally $V < 815,710$, then the cooperation does not add any value because the cooperation

Table 3 Firms A and B cooperative payoffs assuming $k = 0.30$ and $t_1 = 0.8$

Market Value V	Leader–Follower Value LF_C	Follower–Leader Value FL_C	Simultaneous Value S_C	Waiting Value W_C
600,000	−2,942	−9,300	−146,693	**70,307**
900,000	**206,691**	195,683	107,985	179,469
1,040,000	**313,413**	299,940	232,760	242,729
1,070,000	**336,830**	322,809	259,860	257,105
1,100,000	**360,419**	345,844	287,075	271,745
1,200,000	**440,190**	423,726	378,553	322,325
1,900,000	1,032,079	1,001,380	**1,041,912**	731,393

Table 4 Shapley and asymmetric values assuming $k = 0.30$ and $t_1 = 0.8$

Market Value V	Noncoop. $v(A)$	Noncoop. $v(B)$	Shapley Value Sh_A	Shapley Value Sh_B	Asym. Value P_A	Asym. Value P_B
600,000	37,169	33,138	37,169	33,138	37,169	33 138
900,000	94,187	85,282	107,798	98,893	108,390	98,301
1,040,000	127,095	115,633	162,437	150,975	163,974	149,439
1,070,000	134,565	122,539	174,428	162,402	176,161	160,669
1,070,000	140,425	119,434	178,910	157,919	180,584	156,246
1,100,000	154,408	132,666	191,080	169,338	192,675	167,744
1,200,000	201,411	177,142	232,229	207,960	233,569	206,621
1,900,000	542,253	499,659	542,253	499,659	542,253	499,659

surplus $W_C(V) - (W_A(V) + W_B(V))$ is equal to zero. So the wait and see policy is optimal also considering the cooperation way between A and B. Even if $V = 1,900,000$ and more generally $V > 1,796,130$, then the cooperative gain $S_C(V) - (S_A(V) + S_B(V))$ is equal to zero. So the simultaneous R&D investment at time t_0 is preferable both in the cooperative strategy and in the noncooperative one. Moreover, the Fig. 2b illustrates the cooperative and noncooperative equilibriums when $V = 1,200,000$.

6 Concluding Remarks

In this paper we have proposed an R&D cooperation between two firms using the real option approach. By the alliance, the information is wholly revealed and captured by two players. Moreover, we have shown that the unique cooperation strategy that allows to increase the information revelation with respect to the noncooperative situation is the Leader-Follower strategy in which one firm realizes the R&D investment at time t_0 while the other one delays its decision at time t_1. In particular way, as the mutual information gain implies positive network externalities, we have shown that the Leader role is assumed by the firm with the highest success probability.

Finally, computing the noncooperative and the cooperative critical market values, we are able to determine the range game in which is optimal each partnership strategy. Moreover, using the Shapley values (Sh_A, Sh_B), both firms split equitably the cooperation surplus but they can agree upon several partnership contracts such as the Asymmetric shares (P_A, P_B) based on different success probability.

References

Armada, M. R., Kryzanowsky, L., & Pereira, P. J. (2007) A modified finite-lived American exchange option methodology applied to real options valuation. *Global Finance Journal, 17*(3), 419–438.

Carr, P. (1988) The valuation of sequential exchange opportunities. *The Journal of Finance, 43*(5), 1235–1256.

Carr, P. (1995) The valuation of American exchange options with application to real options. In L. Trigeorgis (Ed.) *Real options in capital investment: models, strategies and applications*. Westport Connecticut, London: Praeger.

Chi, T. (2000) Option to acquire or divest a joint venture. *Strategic Management Journal, 21*(6), 665–688.

Dias, M. A. G. & Teixeira, J. P. (2004) Continuous-time option games part 2: oligopoly, war of attrition and bargaining under uncertainty. *Working Paper, PUC-Rio, Proceeding at 8th Annual International Conference on Real Options, Montreal, June 2004.*

Dias, M. A. G. (2004) Real options, learning measures, and Bernoulli revelation processes. *Working paper, Puc-Rio, Proceeding at 8th Annual International Conference on Real Options, Paris, June 2005.*

Lee, M. H. (1997) Valuing finite-maturity investment-timing options. *Financial Management, 26*(2), 58–66.

Majd, S. & Pindyck, R. S. (1987) Time to build, option value and investment decisions. *Journal of Financial Economics, 18*(1), 7–27.

Huisman, K. J. M. (2001) *Technology investment: a game theoretic real options approach*. Boston: Kluwer.

Margrabe, W. (1978) The value of an exchange option to exchange one asset for another. *The Journal of Finance, 33*(1), 177–186.

McDonald, R. L. & Siegel, D. R. (1985) Investment and the valuation of firms when there is an option to shut down. *International Economic Review, 28*(2), 331–349.

Kogut, B. (1991) Joint ventures and option to expand and acquire. *Management Science, 37*(1), 19–33.

Kong, J. J. & Kwok, Y. K. (2007) Real options in strategic investment games between two asymmetric firms. *European Journal of Operational Research, 181*(2), 967–985.

Shackleton, M. & Wojakowski, R. (2003) The expected return and exercise time of merton-style real options. *Journal of Business Finance & Accounting, 29*(3–4), 541–555.

Trigeorgis, L. (1991) Anticipated competitive entry and early preemptive investment in deferrable projects. *Journal of Economics and Business, 43*(2), 143–156.

Villani, G. (2008) An R&D investment game under uncertainty in real option analysis. *Computational Economics, 32*(1–2), 199–219.

Unifying Cournot and Stackelberg Action in a Dynamic Setting

Tonü Puu

1 Introduction

1.1 Background

Heinrich von Stackelberg, like Harold Hotelling, was one of those scientists of the early twentieth Century, who put down many seeds for new original ways to look at old problems in theoretical economics, often pointing at paradoxical issues with no obvious solution. Stackelberg's contribution to duopoly (von Stackelberg, 1934, 1938), one Century after Cournot's initiation (Cournot, 1838), is probably his most well known contribution.

Even if Stackelberg departs from Cournot in his leader/follower dichotomy, in contrast to Cournot, there is no clue to any dynamization of the model. It is all static equilibrium theory. One competitor can learn and take account of the Cournot reaction function of the other and then maximize profits. If the latter indeed follows the proper reaction function, everything is fine, there is a Stackelberg equilibrium. The other competitor can do the same, and again, if the first then follows its reaction function, everything is fine again, there is another Stackelberg equilibrium. They can also both adhere to their reaction functions, and then one is back to Cournot's original case. However, if both competitors attempt leadership at once, then there is trouble; both are disappointed as expectations show up wrong.

What will then happen? Will one of the competitors, or both, resign leadership, and the system go to the Cournot or one of the Stackelberg equilibria? Stackelberg did not give any clue at all to this.

A consequence of this is also that it is difficult to see how the Cournot and Stackelberg models at all fit together. Suppose one sets up a dynamic Cournot model, with, as it is popular to say now, "naive" expectations (as if not all hitherto modelled expectations were naive). Further, suppose one of the competitors chooses Stackelberg leadership action, and the other responds according to the Cournot reaction function. In the Cournot setting, even if the other competitor responds as

T. Puu
Department of Economics, Umea University 90187 UMEA, Sweden,
e-mail: Tonu.Puu@econ.umu.se

G.I. Bischi et al. (eds.), *Nonlinear Dynamics in Economics, Finance and the Social Sciences*, DOI 10.1007/978-3-642-04023-8_7,

expected, the first would in the next run of the Cournot game always find Cournot action better than adhering to the leadership attempt. The Stackelberg equilibria can therefore simply not be considered as fixed points of the Cournot iterative map. And, as for the reverse, Stackelberg theory is just static.

Current literature on Stackelberg duopoly focuses traditional static issues such as existence of equilibrium, or compares equilibrium profits to those from the Cournot case. (See Flam, Mallozzi, and Morgan (2002); Dastidar (2004); Colombo and Labrecciosa (2008).) When dynamics is addressed the Stackelberg case is put in an adaptive system, as is often done with the Cournot case (see Richard (1980)), but the present author has not seen any attempt to integrate the traditional Cournot and Stackelberg cases so that they turn up as fixed points of the same dynamic system.

1.2 Agenda

We would like to formulate a single more general map in which both the Cournot and the Stackelberg equilibria can exist as fixed points. To this end we suggest a rule for flipping between Cournot and Stackelberg action on the basis of expected profits. The idea is to compare expected next period profits under Cournot action to profits resulting from Stackelberg leadership. But, as already mentioned, (naively expected) Cournot profits always exceed Stackelberg leadership profits, so we propose to locate the balance point between current Cournot and Stackelberg profits, incorporating a scaling parameter. Such a rule is contestable, but it is at least a start.

By this, we can formulate a single model which contains both the Cournot and Stackelberg solutions as fixed points, even including the point where both competitors go on trying to remain leaders. The model may display multistability, and it is also a specimen of heterogenous agent models. Normally, agent heterogeneity is taken to mean that the agents stick to the same kind of behavior decided once and for all. In the proposed model they can and do shift behavior, depending on the unfolding of the process itself. By the way, Stackelberg's original model must be one of the earliest heterogenous agent models ever proposed before the term itself came in use.

Having this agenda we choose a simple setup, used repeatedly by the present author: A smooth iso-elastic demand function combined with constant marginal costs for the competitors (see Puu (1991, 2004, 2000)). Arguments can be given both for and against the iso-elastic case. To the *advantage* of the iso-elastic case speaks that it results for individual consumers from Cobb-Douglas utility functions, and further aggregates to the same kind of iso-elastic aggregate market demand function. The obvious alternative, linear demand functions, have discontinuity points, and they aggregate to a kinked train of linear segments resulting in a Robinson type of marginal revenue that jumps up and down. Such models with kinked linear demand functions may themselves offer prospects for interesting models (see Puu and Sushko (2002)), though they would complicate things unnecessarily in the present setting. An arguably minor *disadvantage* of the iso-elastic is that it is not suitable for the discussion of collusion or monopoly.

2 Model Setup

2.1 Cournot Action

2.1.1 Reaction Functions

Assume the inverse demand function

$$p = \frac{1}{x + y}, \tag{1}$$

where p denotes market price and x, y denote the outputs of the duopolists. Given the competitors have constant marginal costs, denoted a, b respectively, the profits are

$$U = \frac{x}{x + y} - ax, \tag{2}$$

$$V = \frac{y}{x + y} - by. \tag{3}$$

Putting the derivatives $\partial U/\partial x = 0$ and $\partial V/\partial y = 0$, and solving for x, y, one obtains

$$x' = \sqrt{\frac{y}{a}} - y, \tag{4}$$

$$y' = \sqrt{\frac{x}{b}} - x, \tag{5}$$

as the reaction functions. The dash, as usual, represents the next iterate, i.e., the "best reply" of one competitor given the observed supply of the other.

2.1.2 Constraints

Obviously, (4) returns a negative reply x' if $y > 1/a$, and (5) a negative reply y' if $x > 1/b$. To avoid this, we put $x' = 0$ whenever $y > 1/a$, and $y' = 0$ whenever $x > 1/b$. This means reformulating (4)–(5) as follows

$$x' = \begin{cases} \sqrt{\frac{y}{a}} - y, & y \leq \frac{1}{a} \\ 0, & y > \frac{1}{a} \end{cases}, \tag{6}$$

$$y' = \begin{cases} \sqrt{\frac{x}{b}} - x, & x \leq \frac{1}{b} \\ 0, & x > \frac{1}{b} \end{cases}. \tag{7}$$

2.1.3 Equilibrium

Putting $x' = x,\ y' = y$, one can solve for the coordinates of the Cournot equilibrium point

$$x = \frac{b}{(a+b)^2}, \tag{8}$$

$$y = \frac{a}{(a+b)^2}. \tag{9}$$

Substituting back from (8)–(9) in (2)–(3), one gets the profits of the competitors in the Cournot equilibrium point

$$U = \frac{b^2}{(a+b)^2}, \tag{10}$$

$$V = \frac{a^2}{(a+b)^2}. \tag{11}$$

It is obvious that the firm with lower unit costs obtains the higher profit.

2.1.4 Profits

Note that (10)–(11) are the profits in the Cournot equilibrium point. During the Cournot iteration process (4)–(5) profits can also be considerably higher than in equilibrium. As we will see below, the Stackelberg leadership profits are always higher than the Cournot equilibrium profits, but in the Cournot process temporary profits can exceed the leadership profits. To calculate temporary profits, just substitute from (4) into (2), and from (5) into (3), and obtain

$$U = \left(1 - \sqrt{ay}\right)^2, \tag{12}$$

$$V = \left(1 - \sqrt{bx}\right)^2. \tag{13}$$

Note that these may always seem to be nonnegative. But if $y > 1/a$ or $x > 1/b$, i.e., the constraints for the first branches of (6)–(7) are violated, then this is due to the fact that negative costs dominate over negative revenues, which in terms of subject matter is nonsense. Anyhow, we already restricted the map (4)–(5) to (6)–(7), so we need not be further concerned.

2.1.5 Stability

Of course, the Cournot equilibrium (8)–(9) can be stable or unstable. To check, calculate the derivatives of (4)–(5)

$$\frac{\partial x'}{\partial y} = \frac{1}{2}\frac{1}{\sqrt{ay}} - 1, \tag{14}$$

$$\frac{\partial y'}{\partial x} = \frac{1}{2}\frac{1}{\sqrt{bx}} - 1. \tag{15}$$

In the Cournot point, substitute from (8)–(9) to obtain

$$\frac{\partial x'}{\partial y} = \frac{1}{2}\frac{b}{a} - \frac{1}{2},$$

$$\frac{\partial y'}{\partial x} = \frac{1}{2}\frac{a}{b} - \frac{1}{2}.$$

Note that both these Cournot equilibrium derivatives have an infimum value $-\frac{1}{2}$; hence neither can become smaller than -1, but either can become larger than 1, if $\frac{b}{a} > 3$ or $\frac{a}{b} > 3$. Note that the derivatives always have opposite signs. To check stability, compose the Jacobian

$$J = \begin{bmatrix} 0 & \frac{b-a}{2a} \\ \frac{a-b}{2b} & 0 \end{bmatrix}. \tag{16}$$

Iff

$$\left| \frac{(a-b)^2}{4ab} \right| < 1 \tag{17}$$

holds, then the Cournot equilibrium is stable, otherwise not. The condition is equivalent to $\frac{b}{a} \in \left[3 - 2\sqrt{2}, 3 + 2\sqrt{2}\right]$. In previous publications (see Puu (1991, 2004, 2000)) it was shown that, once the equilibrium point loses stability, there follows a period doubling cascade of bifurcations to chaos.

2.2 *Stackelberg Action*

2.2.1 First Equilibrium

According to Stackelberg's idea, the competitor controlling x can take the reaction function (5) of the other for given, and substitute it in its own proper profit function (2), to obtain

$$U = \sqrt{bx} - ax.$$

Putting $dU/dx = 0$, and solving, one gets

$$x = \frac{b}{4a^2}. \tag{18}$$

The corresponding value of y, provided that firm really adheres to its Cournot reaction function, is obtained through substituting (18) in (5) and equals

$$y = \frac{2a - b}{4a^2}. \tag{19}$$

Using (18)–(19) in (3), the Stackelberg leadership profit can be easily calculated

$$U = \frac{b}{4a}. \tag{20}$$

2.2.2 Second Equilibrium

Similarly, the second firm can try leadership, substituting (4) in (3), and maximizing

$$V = \sqrt{ay} - by,$$

to obtain

$$y = \frac{a}{4b^2}. \tag{21}$$

The corresponding Cournot response of the first firm would then be

$$x = \frac{2b - a}{4b^2}, \tag{22}$$

and the Stackelberg leadership profit

$$V = \frac{a}{4b}. \tag{23}$$

It is easy to check that Stackelberg leadership profits always exceed the respective Cournot equilibrium profits, i.e., (20) is higher than (10), and (23) higher than (11).

2.2.3 Stability

Even if the Stackelberg model is static, we could consider stability as only the leader keeps to constant leadership action, (18) or (21), whereas the follower reacts according to the Cournot reaction function, (5) or (4) respectively. We can hence consider the stability of these partial reactions, as it is obvious that tiny disturbances of the reacting competitor might make the system diverge from the Stackelberg equilibria. So, if the first firm is a Stackelberg leader, then substituting from (18) in (15) yields

$$\frac{\partial x'}{\partial y} = \frac{a}{b} - 1, \tag{24}$$

which can never become less than -1, but can exceed 1. Hence, for stability we require

$$a < 2b. \tag{25}$$

Likewise we can substitute from (21) in (14), obtaining

$$\frac{\partial y'}{\partial x} = \frac{b}{a} - 1, \tag{26}$$

and the stability condition

$$a < 2b \tag{27}$$

for the second Stackelberg equilibrium.

Note that the fulfilment of these stability conditions (25) and (27) also guarantee nonnegativity of the followership reactions according to (19) and (22). Further, note that the Stackelberg stability conditions (25) and (27) are stronger than the condition for Cournot stability (17).

2.2.4 Problems

These equilibrium solutions are consistent, provided one competitor in each case agrees to act as a Cournot follower. If not, there is a problem that was noted by Stackelberg.

But there is more to it because Cournot's theory is dynamic, Stackelberg's static. One may ask: Could the Stackelberg equilibria (18)–(19) and (21)–(22) be fitted into the Cournot dynamical system (4)–(5) as fixed points of the map?

The answer is no! If one competitor chooses to try leadership according to (18), and the other indeed follows according to (19), then, in the next move, retaining leadership by the first firm is no longer optimal under the dynamic Cournot map. The leader could in fact obtain a higher profit through returning to Cournot action; after all Cournot best reply was defined that way. The Stackelberg equilibria hence cannot be fixed points of the dynamic Cournot map; it has to be modified in order to accommodate these additional fixed points.

3 A Proposed Map

3.1 Profit Considerations

Above in (6)–(7), the output positivity constrained Cournot map was given. We now want to amend this map through considering possible jumping to Stackelberg leadership. It is close at hand then to compare current expected Cournot action profit, as given in (12)–(13), to Stackelberg leadership profit according to (20) and (23). We assume that there is a certain proportionality constant k such that the firms keep to Cournot action as long as expected current profits are not less than Stackelberg leadership profits multiplied by this k, i.e.,

$$\left(1-\sqrt{ay}\right)^2 \geq k\frac{b}{4a}, \tag{28}$$

$$\left(1-\sqrt{bx}\right)^2 \geq k\frac{a}{4b}. \tag{29}$$

The value of the parameter k indicates how adventurous the competitors are at attempting a jump to leadership action.

3.2 The Map

We can now specify the map resulting from these considerations with three branches for each actor.

$$x' = \begin{cases} \sqrt{\frac{y}{a}} - y, & y \leq \frac{1}{a} \;\&\; \left(1-\sqrt{ay}\right)^2 \geq k\frac{b}{4a} \\ \frac{b}{4a^2}, & y \leq \frac{1}{a} \;\&\; \left(1-\sqrt{ay}\right)^2 < k\frac{b}{4a} \\ 0, & y > \frac{1}{a}, \end{cases} \tag{30}$$

$$y' = \begin{cases} \sqrt{\frac{x}{b}} - x, & x \leq \frac{1}{b} \;\&\; \left(1-\sqrt{bx}\right)^2 \geq k\frac{a}{4b} \\ \frac{a}{4b^2}, & x \leq \frac{1}{b} \;\&\; \left(1-\sqrt{bx}\right)^2 < k\frac{a}{4b} \\ 0, & x > \frac{1}{b}. \end{cases} \tag{31}$$

This map can accommodate the Cournot equilibrium as well as both Stackelberg equilibria.

4 Fixed Points

4.1 Cournot Equilibrium

4.1.1 Existence

In order to confirm that the Cournot equilibrium point is indeed located in its proper region of application we have to check (28) and (29) for the Cournot point (8)–(9). Substitutions result in just one single condition

$$k \leq \frac{4ab}{(a+b)^2}, \tag{32}$$

which depends on k and the ratio b/a. The Cournot equilibrium (8)–(9) exists in its region of definition iff (32) holds. As the right hand side attains a unitary maximum value for $a = b$, we conclude that for (32) to hold it is then necessary that $k \leq 1$.

4.1.2 Stability

The stability conditions for Cournot equilibrium were already stated above,

$$\frac{b}{a} \in \left[3 - 2\sqrt{2}, 3 + 2\sqrt{2}\right].$$

As Cournot equilibria according to (32) can be defined for $b/a \in [0, \infty)$ the requirement on stability constrains the existence region.

4.2 Stackelberg Equilibria

4.2.1 Existence

First Stackelberg Equilibrium

In a similar way we can deal with the existence problem for the Stackelberg fixed points. Assume first that the firm supplying x is the leader, the firm supplying y the follower. Then the coordinates of the fixed point are (18)–(19), whereas the conditions for the branch definition are (28), sign reversed, and (29). Substituting from (18)–(19), we get two conditions,

$$k > \left(2 - \sqrt{2 - \frac{b}{a}}\right)^2 \frac{a}{b},$$

$$k \leq \left(2 - \frac{b}{a}\right)^2 \frac{b}{a}.$$

Again note that they only depend on the parameter k and on the ratio b/a.

Second Stackelberg Equilibrium

Reversing the roles of the firms, consider the fixed point (21)–(22), which must fulfil the branch conditions (28), and (29), sign reversed. Again, substitution gives

$$k \leq \left(2 - \frac{a}{b}\right)^2 \frac{a}{b},$$

$$k > \left(2 - \sqrt{2 - \frac{a}{b}}\right)^2 \frac{b}{a}.$$

In parameter space $b/a, k$ (displayed in Fig. 1), the Stackelberg equilibria are located in their proper branch ranges in the two small lens shaped areas. Note that

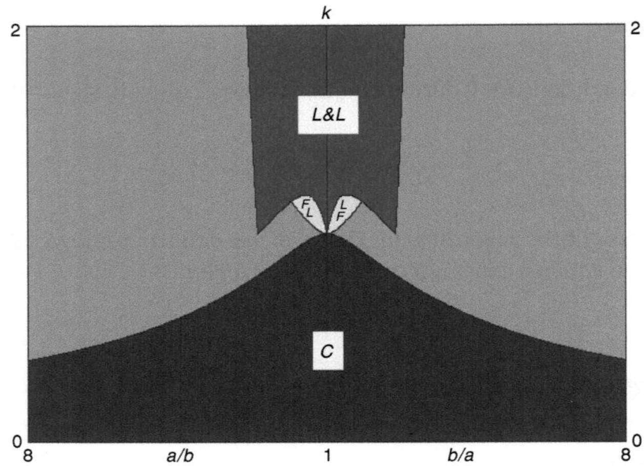

Fig. 1 Existence regions in parameter space for Cournot point (*lower hump*), the regular Stackelberg leader/follower pairs (*small lens shaped areas*), and simultaneous persistence on Stackelberg leadership (*upper region with cuspoid bottomline*)

they are disjoint, and separated by a vertical line at $b/a = 1$. Hence, for each parameter combination, only one of the Stackelberg equilibria exists as a fixed point (the one where the firm with lower unit cost is the leader).

In the same picture we also display the lower region of existent Cournot fixed points. Note that it is disjoint from both Stackelberg equilibrium areas.

Persistence on Leadership by both Competitors

Finally, under the proposed map, there can also exist a fixed point where both firms keep to Stackelberg action, choosing (18) and (21) respectively. It is a fixed point of the map provided (28) and (29), both signs reversed, hold, which upon substitutions from (18) and (21) become

$$k > \left(2 - \frac{a}{b}\right)^2 \frac{a}{b},$$
$$k > \left(2 - \frac{b}{a}\right)^2 \frac{b}{a}.$$

The region in Fig. 1 of the parameter plane with the saw-toothed lower boundary represents the cases where both firms keep to Stackelberg leadership, though it is not any normal Stackelberg equilibrium. (In addition we have two more fixed points, where any one firm persists at its Stackelberg leadership, and supplies so much that the other firm drops out. Their regions of definition are not represented in the picture.)

As for the region outside the depicted existence regions for the fixed points, it is not possible to say from the preceding simple analysis which kinds of attractors may emerge there. Numerical experiment indicates periodic orbits.

Note the way Fig. 1 is constructed. The system is completely symmetric with respect to $\frac{a}{b}$ and $\frac{b}{a}$. Therefore, actually two diagrams, in $\frac{b}{a}, k$-space and in $\frac{a}{b}, k$-space respectively were put back to back, with $\frac{b}{a}$ increasing to the right, and $\frac{a}{b}$ increasing to the left and $\frac{b}{a} = \frac{a}{b} = 1$ as the common dividing line. Had we chosen just one of the ratios as the horizontal coordinate, then one half of the diagram would have been squeezed in the interval $(0, 1]$ whereas the other would extend over $[1, \infty)$. This distortion would make it impossible to see the symmetry.

4.2.2 Stability

We already concluded that the regular Stackelberg equilibria are stable if

$$\frac{1}{2} < b/a < 2.$$

Accordingly, the Stackelberg equilibria, unlike the Cournot equilibria, are always stable when they exist.

5 Numerical Study

5.1 Bifurcation Diagram

5.1.1 Fixed Points

In Fig. 2, the bifurcation diagram, resulting from numerical experiment performed to detect periodic orbits of the system (30)–(31) when starting from a point close to the Cournot equilibrium point, is shown. To improve resolution, only the right half of the diagram as compared to Fig. 1 is displayed. Obviously most of the plane contains areas representing period 1 orbits, i.e., attracting fixed points. These are of four different types (six in the format of Fig. 1), most of which can also be seen in Fig. 1 that represented existence regions for fixed points.

1. At the bottom of the picture the Cournot attracting equilibrium area is shown. It looks quite as in Fig. 1, though it yields periodic orbits where the Cournot equilibrium turns unstable, at $b/a = 3 + 2\sqrt{2}$, as we saw.
2. There is also the tiny lens shaped attracting Stackelberg area for the leader/follower pair, also quite as in Fig. 1.
3. Likewise there is the region with the cuspoid bottom line, representing the non-standard Stackelberg case where both competitors insist on sticking to Stackelberg leadership (which Stackelberg in the equilibrium format considered

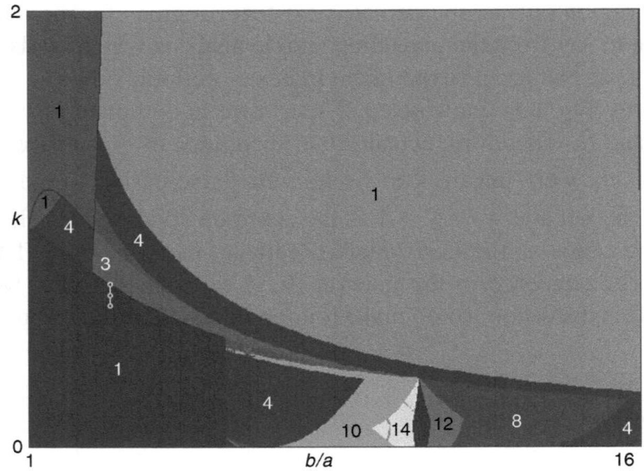

Fig. 2 Observable are regions of periodic orbits. There are fixed points (period 1) of four different types, further orbits of periods 3, 4, 8, 10, 12, and 14. Enlargements may display more detail. The *black area* indicates higher periodicity, quasiperiodicity, or chaos

unresolvable). True, both competitors get disappointed, as the assumption that the other acts as a follower does not hold, but under the defined map, neither could find expected current profits from Cournot action more profitable. Also this region has a corresponding one in Fig. 1.

4. There is further the region in the upper right part of Fig. 2, marked with a period 1 label, which was not shown if Fig. 1. It represents the case where one competitor sticks to Stackelberg leadership, whereas the other finds no better action than to suppress production altogether. It might seem to be a kind of monopoly. However, as mentioned in the introduction, the model, in particular the iso-elastic demand function, is not suitable to treat either monopoly or collusion. The reason is that under iso-elastic demand market revenue is a constant. Hence a monopolist (or a pair of collusive duopolists) could retrieve the whole revenue as profit without incurring any production costs if they produce nothing and sell this nothing at an infinite price. Such a solution is purely mathematical and has no meaning in terms of economics; it is just a shortcoming of one assumption.

However, it can hardly be regarded as a major defect that the model is unsuitable for the analysis of market situations which in the real world as a rule are forbidden by law. Further, the *dynamic* model proposed can *never* end up at monopoly or collusion, because the origin where the reaction functions (4) and (5) intersect (the collusion state) is as unstable as anything can be, both derivatives according to (14) and (15) being infinite at $x = y = 0$. There is just a little snag in the numerics; the computer interprets zero as an exact number, so a computation may stick to the origin despite of its instability. To avoid this, the exact zero in the definition of (30)–(31) is replaced by a small number $\varepsilon = 10^{-6}$ in the experiment.

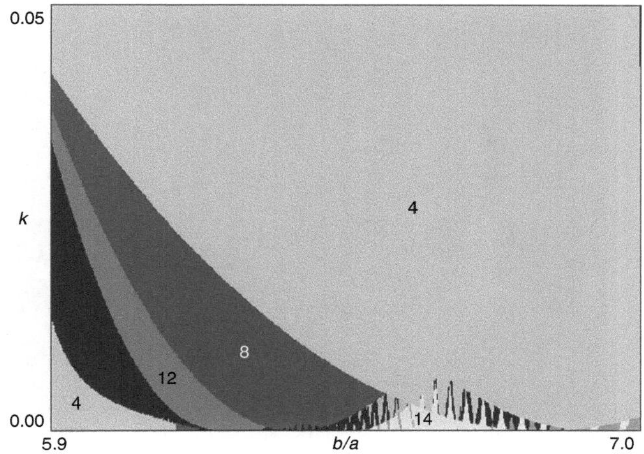

Fig. 3 Blowup picture of part of Fig. 2 to show more of periodic orbits and overlaps. The *black regions* indicate higher periodicities or more complex dynamics

5.1.2 Periodic Orbits

As so much of the area in Fig. 2 is taken up by fixed point regions, little is left for other attracting orbits. We can see periodic orbits, labelled 3, 4, 8, 10, 12, and 14. This is at the chosen part of parameter plane and the resolution of the picture, as no higher periods were checked by the computer. The black area may contain higher periodicities, quasiperiodicity, or even chaos.

There seems to be some irregularity in the lower part of Fig. 2 in the interval $a/b \in [6, 7]$, so we take a close up picture of it in Fig. 3. Notable are the two humps of a spiky appearance. They indicate coexistence of attractors; for some parameter combinations the chosen initial condition leads to one periodicity, for a nearby combination to another. As we will see below, the model is proficient in such coexisting attractors.

There are many potentially interesting bifurcation scenarios to pursue. We will just pick one, with $a/b = 3$ fixed, and $k \in [0.65, 0.75]$, indicated by the tiny line segment in Fig. 2. In Figs. 4–7, the phase diagram is displayed for $k = 0.65$, $k = 0.7$, $k = 0.745$, and $k = 0.75$.

5.2 Attractors and Basins

5.2.1 Coexistence

The phase diagrams show the two Cournot reaction curves, along with the lines representing Stackelberg leadership action, further the different attractors, and their

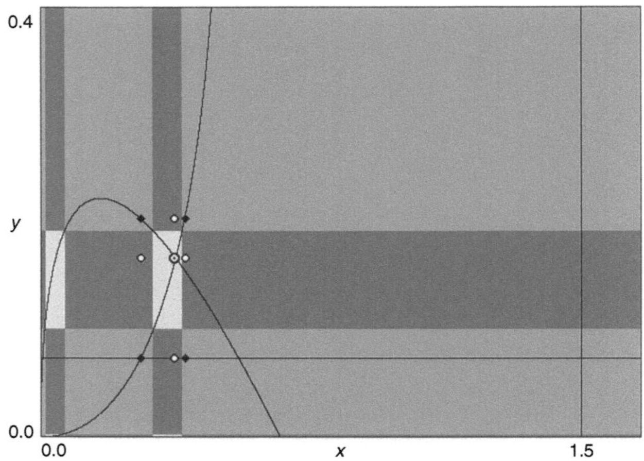

Fig. 4 Coexistent Cournot equilibrium and six-period SIM cycle at $a/b = 3, k = 0.65$

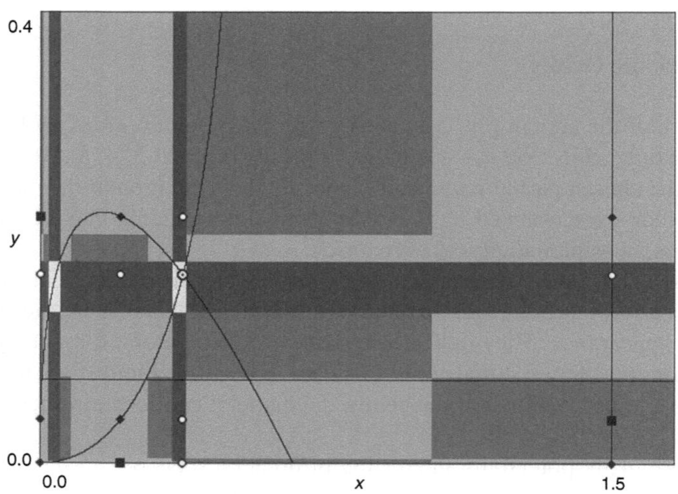

Fig. 5 Coexistence of Cournot equilibrium three-period cycle, and two six-period cycles, one SIM and one SEQ, at $a/b = 3, k = 0.7$

basins of attraction. Note that the periodic orbits are located on the intersection points of a grid of three or four horizontal and vertical lines, hence producing nine or sixteen attractor points. These are distributed between orbits of different periodicities, such as $1 + 3 + 6 + 6 = 16$, or $3 + 6 = 9$, or $1 + 4 + 4 = 9$.

First, take a look at Fig. 5, where $b/a = 3, k = 0.7$. As the fixed Cournot point is destabilized only at $k = 0.75$, it remains stable. Its basin is the tiny rectangle around the intersection of the Cournot reaction curves, and a few other tiny rectangles in

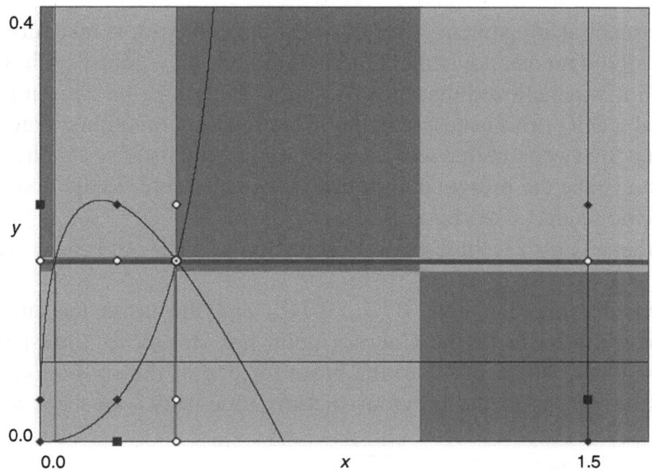

Fig. 6 Same as previous picture at $a/b = 3, k = 0.745$, though two basins are about to disappear

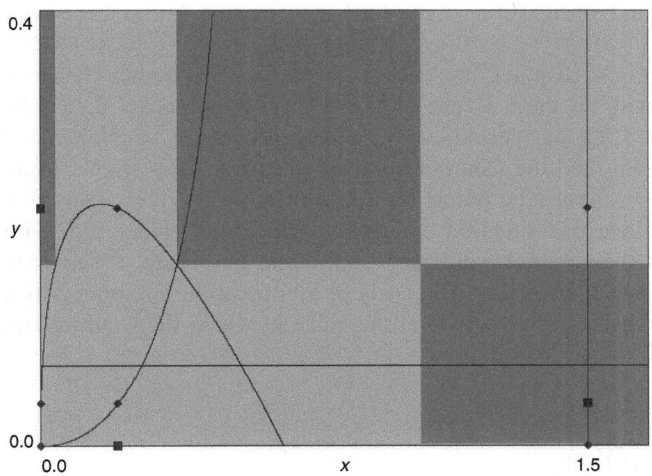

Fig. 7 The subcritical bifurcation, only the three-period cycle, and the six-period SEQ cycle remain at $a/b = 3, k = 0.75$

the same shade. The cross, in the middle of which the main basin for the Cournot point lies, also contains the points of a six-period orbit.

In addition there is another six-period orbit whose points lie on the Cournot reaction function (including the Stackelberg lines). Some researchers call such orbits "Markov perfect equilibria" (MPE).In the present author's opinion this is misleading, as it is a periodic orbit and no fixed point. Earlier literature in stead distinguished between "sequential adjustment" (SEQ), where the competitors take turns in adjustment, and "simultaneous adjustment" (SIM), where they adjust both at the same time; though it was believed that the adjustment type must be chosen beforehand. The points of a SEQ orbit necessarily lie on the reaction functions. Later it became obvious that a SIM system, that was more general, could itself settle on a SEQ orbit. Despite this mistake the present author finds it better to refer to the first six-orbit as SIM, and to the second six-orbit as SEQ.

Finally, there is also a three-period orbit. So, $1 + 3 + 6 + 6 = 16$, quite as suggested.

Next, consider Fig. 6, where $k = 0.745$, and the cross has almost disappeared. Hence the basin of the Cournot point has shrunk to almost nothing, as has the cross itself, which provides the basin for one of the six-period orbits. The same orbits remain as in the previous picture, though two of them are about to disappear.

This indeed happens at $k = 0.75$, the subcritical bifurcation in Fig. 7, where both the Cournot equilibrium, and one of the six-period cycles have disappeared. What remains are two attractors; the three-period orbit, and the six-period SEQ cycle. In all, there are now $3 + 6 = 9$ grid points in the intersections of three (horizontal and vertical) lines. For further increasing k, the three- and six-cycles remain.

Finally, Fig. 4 displays the case $k = 0.65$. The higher Stackelberg line is not yet visited, so there is one SIM cycle, now of period 4, within the cross, and another SEQ four-period cycle, with points on the reaction functions, as its companion. Further, the Cournot point is, of course, also stable. In all, there are $1 + 4 + 4 = 9$ periodic points on a grid of three by three lines. For k increasing from below, this situation emerges at about $k \in [0.61, 0.62]$. For lower k, there is just the Cournot equilibrium point and one basin. It seems that the two four-period orbits arise simultaneously at a bifurcation for some critical k, which notably is considerably lower than the value at which the Cournot equilibrium is destabilized.

5.2.2 Subcriticality

The fact that Cournot point, contained in a small basin, remains along with other attractors indicates that the bifurcation from Cournot point to periodic orbits is subcritical, i.e., the fixed point is not just destabilized and replaced by another attractor as in the supercritical bifurcation. It rather disappears through its basin contracting and eventually vanishing, so that other attractors that already coexisted with it remain the only ones. The scenarios connected with such subcritical bifurcations often display quite complicated bifurcation structures in the boundary regions as shown in Agliari, Gardini, and Puu (2005a,b). Further, subcritical bifurcations are "hard" compared to the "soft" produced by supercritical bifurcations, and lead to dramatic changes in the system when they occur.

5.2.3 Profits

It may be of interest to check average profits over the various coexistent cycles. Again take Fig. 5 for a start.

The Cournot point profits are then

$$U = 0.562,$$
$$V = 0.062,$$

as can be calculated from (10)–(11) when $a = 0.5, b = 1.5$. The firm facing three times higher marginal cost hence receives slightly more than one tenth of the profits of the other firm. According to (20), the first firm could even obtain a profit $U = 2.25$ if it becomes a Stackelberg leader in equilibrium, but this situation is not sustainable under the assumed parameter combination.

For the three-cycle, profits become

$$U = 0.377,$$
$$V = 0.240,$$

for the SEQ six-cycle

$$U = 0.446,$$
$$V = 0.172,$$

and for the SIM six-cycle

$$U = 0.461,$$
$$V = 0.160.$$

It hence always pays for the firm facing the higher marginal cost to try to break out from a Cournot equilibrium to a cyclic solution, as this alters the profit shares to its advantage; especially in the three-period cycle, where the profits of the duopolists have the same order of magnitude.

For $k = 0.745$ or $k = 0.75$, Figs. 6 and 7, the situation remains the same, the profit entries are not even changed at the chosen number of significant digits displayed, though in the latter case the Cournot equilibrium and the SIM six-cycle no longer exist.

As for the case $k = 0.65$, in Fig. 4, the facts are changed; the Cournot profits remain the same, but for the four-period SEQ orbit we get

$$U = 0.601,$$

$$V = 0.074,$$

and for the four-period SIM orbit

$$U = 0.579,$$

$$V = 0.070.$$

Thus, the competitor facing the higher production cost can not earn much from periodic orbits, the profits remain about the same as in Cournot equilibrium.

5.2.4 Rational Expectations

The question now arises if the agents could learn the periodicity they produce. This is easiest if the periodicity is low and the coexistent attractors have the same period. We could choose the case displayed in Fig. 4, with just two four-period orbits. But things become complicated enough, so check the facts at a parameter point $b/a = 6, k = 0.4$ within the bigger four-period tongue in Fig. 2.

There then exists just one four-period cycle of SEQ type, and a unique basin. The four orbit points are:

(A) The competitor controlling y having chosen Stackelberg leadership, the one controlling x responds with Cournot action.
(B) The competitor controlling y chooses Cournot action as it is better under the map than keeping to Stackelberg action
(C) The competitor controlling x again responds with Cournot action
(D) The competitor controlling y now finds it better to return to Stackelberg action. The profits from this four-cycle are now

$$U = 0.725,$$

$$V = 0.030,$$

which is not much different from the Cournot profits

$$U = 0.735,$$
$$V = 0.020,$$

but the point is no longer stable.

Given the simple regularity, suppose the agents learn the periodicity, and react, not to the competitor's action one period back, but four periods back. What would be the outcome?

If one only considers the dynamic every fourth period, the outcome is exactly as before. But, in the three intervening periods, the system will settle to an independent cycle of the same type. The result is a composition of four four-cycles successively displaced in time, and, even if we just have one original cycle, it is impossible to make the composition break down to a four-cycle. The outcome is a 16-period cycle. To see the point, consider a single four-cycle, whose points are denoted $ABCD$.

Try to arrange a four cycle. In the first line of the table there are three blanks between the entries. To produce $ABCD$ again, choose $BCDA$ for the first blanks, $CDAB$ for the second, and $DABC$ for the third, thus completing the sequence. (Each new choice of entries is indicated in bold face.)

$$\begin{array}{cccccccccccccccc}
\mathbf{A} & - & - & - & \mathbf{B} & - & - & - & \mathbf{C} & - & - & - & \mathbf{D} & - & - & - \\
A & \mathbf{B} & - & - & B & \mathbf{C} & - & - & C & \mathbf{D} & - & - & D & \mathbf{A} & - & - \\
A & B & \mathbf{C} & - & B & C & \mathbf{D} & - & C & D & \mathbf{A} & - & D & A & \mathbf{B} & - \\
A & B & C & \mathbf{D} & B & C & D & \mathbf{A} & C & D & A & \mathbf{B} & D & A & B & \mathbf{C}
\end{array}$$

So what is the outcome? The resulting sequence in the bottom row indeed starts with $ABCD$, but it is *not* repeated, because then follows $BCDA$, $CDAB$, and $DABC$. Only the full 16-sequence of the bottom line is repeated.

So, if the agents learn the periodicity of 4, they actually produce a periodicity of 16. The reader can try to arrange the sequences differently, but the outcome is always the same; a 16-period cycle. The first four entries in the sequence of 16 determine all the following, according to the order established in the original four-cycle, and as we have four choices for each entry, there would now seem to exist $4^4 = 256$ different cycles.

However, only 16 are different. The reason is as follows. Each recurrent sequence of 16 entries contains 16 different four-sequences of entries, as we can start with any one of the entries. (If we start with one of the three last entries, we just have to add one, two, or three of the entries in the beginning of the full sequence.) But, as any four subsequent entries in the sequence determine the whole 16-period cycle, any of the 256 sequences belongs to a group of 16 identical orbits (only chosen with different starting points). Hence, the total number of distinct orbits is 16.

Note that average profits in the 16-period cycle remain the same as in the four-period cycle, because the same points $ABCD$ are visited with the same relative frequency as before though under different permutations.

To sum up; starting with one four-period cycle and one basin, learning produced 16 different 16-period cycles, and, of course, 16 different basins, though. we cannot produce any picture of the basins in the phase diagram as it is eight-dimensional. And so it goes on, learning and adjusting to the 16 periodicity produces a host of different 256-period cycles. The conclusion is that the idea of "rational expectations" is untenable even in the simplest case of only one original orbit of low order. The only periodicity that is possible to learn is a fixed point. For a general argument, see Puu (2006).

6 Discussion

The purpose of this article was to define a dynamic duopoly system, as simple as possible, but able to contain both the Cournot and the Stackelberg points as particular equilibria or fixed points. For this reason a traditional Cournot type of stepwise adjustment was assumed, but it was amended through assuming that the competitors might jump to Stackelberg leadership action if they were too disappointed by currently expected Cournot profits. Depending on the two parameters of the simple system, the unit production cost ratio and the coefficient for jumping to attempts at leadership, different kinds of attractors could be detected.

There could exist a stable Cournot equilibrium state or various Stackelberg equilibria, and the system could also go to periodic solutions, such as 3, 4, 6, 8, 10, 12, or 14. In certain parameter regions there could even be attractors of higher complexity. A typical feature was that periodic attractors coexisted, with each other, and with fixed points, and that the bifurcations were subcritical.

Stable Stackelberg points could exist for traditional leader/follower pairs (depending on the unit cost ratio), but the system could also stick to both competitors persisting in being Stackelberg leaders. It could also happen that a Stackelberg leader managed to drive the competitor out of business.

In certain situations of multistabilty, the periodic orbits provided for considerably better profit situations for the competitor facing high productions costs than would the Cournot equilibrium.

The question was also raised if the competitors could learn the actual periodicity and react accordingly with an appropriate delay. It was shown that even in the simplest cases of just one attractor of low periodicity this was impossible, because such learning and adaptation would inevitably alter the resulting periodicity. In a superficial sense this is similar to the uncertainty principle in physics; in the present case learning and adapting to a periodicity the competitors produce is bound to change the periodicity itself. Rational expectations are hence impossible, except for the single case of a fixed point.

The proposed integration of Cournot and Stackelberg, admittedly, is very simple, and should be amended by other strategic action rules that tell how the competitors break out from situations that are neither Cournot points nor traditional Stackelberg leader/follower pairs. But it seems important to start trying to integrate Cournot and Stackelberg in one dynamic model.

References

Agliari, A., Gardini, L., & Puu, T. (2005) Global bifurcations in duopoly when the Cournot point is destabilized through a subcritical Neimark bifurcation. *Int Game Theory Rev, 8*, 1–20.

Agliari, A., Gardini, L., & Puu, T. (2005) Some global bifurcations related to the apperance of closed invariant curves. *Math Comput Simul, 68*, 201–219.

Cournot, A. (1838) Récherces sur les principes mathématiques de la théorie des richesses. Paris

Colombo, L. & Labrecciosa, P. (2008) When Stackelberg and Cournot equilibria coincide. *B E Journal of Theoretical Economics, 8*, Article #1.

Dastidar KG (2004) On Stackelberg games in a homogeneous product market. *European Economics Review, 48*, 549–562.

Flam, S. D., Mallozzi, L., & Morgan, J. (2002) A new look for Stackelberg-Cournot equilibria in oligopolistic markets. *Economic Theory, 20*, 183–188.

Puu, T. (1991, 2004) Chaos in duopoly pricing. *Chaos, Solitons & Fractals, 1*, 573–581, republished in: Rosser, J. B. (Ed.) *Complexity in economics: the international library of critical writings in economics* (Vol. 174). Cheltenham: Edward Elgar.

Puu, T. (2000) *Attractors, bifurcations, & chaos – nonlinear phenomena in economics*. Berlin: Springer.

Puu, T. (2006) Rational expectations and the Cournot-Theocharis problem. *Discrete Dynamics in Nature and Society, 32103*, 1–11.

Puu, T., Sushko, I. (Eds.) (2002) *Oligopoly dynamics*. Berlin: Springer.
Richard, J. A. (1980) Stability properties in a dynamic Stackelberg Oligopoly Model. *Economic Record, 56*, 253–256.
von Stackelberg, H. (1934) *Marktform und Gleichgewicht*. Berlin: Springer.
von Stackelberg, H. (1938) Probleme der unvollkommenen Konkurrenz. *Weltwirtschaftliches Archive, 48*, 95.

Issues on Strategy-Switching Dynamics

Weihong Huang

1 Introduction

In many economic dynamic models, economic agents rationally and constantly switch to more profitable strategies in response to the outcomes as well as the environments.[1] In most cases with finite agents, the payoff for adopting an optimal strategy by each agent depends not only on the strategies space but also on the frequency with which different strategies are adopted. A strategy that is relatively superior to the others in a particular distribution can turn inferior in other distributions. The nonexistence of a strategy, that is superior to all other strategies for all distributions, forces rational agents to switch their strategies now and then in response to changing frequency. In other words, economic agents "migrate" constantly among the different strategy groups.

In such a set-up, the traditional concept of static equilibrium frequency (at which agents in different strategy groups have identical benefits) is not applicable in determining the final outcome of strategy-switching dynamics, for the reason that certain variation may improve some agents' payoffs and hence induce the "migration trend". Such "migration" continues until none of them have incentive to change further, a state of dynamic equilibrium, at which the profits across different strategy groups may differ significantly.

[1] The term of "strategy" adopted in this article is more general than the common usage in game theory and most researches in industrial economics. It refers to the different options of *dynamic behavioral principle* in contrast to the different choices of action variables. In quantity-competed oligopolistic model, it is not the particular level of quantity but the different ways that determine the quantity, such as "Cournot best-response", "Stackelberg leader"and "price taking" that form different strategies. In other words, we regard different "reactions curves" as different "strategies", while game theory interprets different points along a particular reaction curve (such as "the best-response curve") as different "strategies".

W. Huang
Nanyang Technological University, Nanyang Avenue, Singapore 639798,
e-mail: awhhuang@ntu.edu.sg

G.I. Bischi et al. (eds.), *Nonlinear Dynamics in Economics, Finance and the Social Sciences*, DOI 10.1007/978-3-642-04023-8_8,

The aim of current research is to explore the new equilibrium and stability concepts for the dynamic competition model involving frequency-dependent payoffs and to evaluate the dynamic strategies from the perspective that is consistent with the normative economic theory.

2 Illustration of Strategy-Switching Dynamics

To illustrate, consider a simple situation in which N players are allowed to take two strategies: $\mathcal{S} = \{X, Y\}$, with n players taking strategy X (belonging to strategy group X) and remaining $N - n$ players taking strategy Y (belonging to strategy group Y).

We shall refer the distribution of players in the different strategy groups, denoted by $\bar{\mathbf{n}} = (n, N - n)$ with $0 \leq n \leq N$, as a *state* of the switching dynamics. Given the state $\bar{\mathbf{n}}$, we denote $\pi^x(\bar{\mathbf{n}})$ and $\pi^y(\bar{\mathbf{n}})$ as the payoffs for the relevant players, respectively.

All players are allowed to join or switch to each strategy group freely. The following two assumptions are essential to get a vivid picture of switching dynamics.

Assumption (Intra-group ordering). *For each group, the members follow a certain order in deciding whether to adjust his strategy (migrate to another), one member at a time.*

Assumption (One-step limited foresight). *All players are* bounded-rational *in the sense that they can only predict the outcome of their own action (whether to switch or which strategy to switch) and make the decision based on such limited foresight.*

At first, we shall see that, so long as one of π^x and π^y varies with $\bar{\mathbf{n}}$, the traditional concept of static equilibrium frequency $\bar{\mathbf{n}}^* = (n^*, N - n^*)$ at which $\pi^x(\bar{\mathbf{n}}^*) = \pi^y(\bar{\mathbf{n}}^*)$ is not always consistent with the final outcome.

Example 1. Let $N = 6$ and consider three different payoffs structures.

Table 1a depicts a typical situation in which *increasing the size of strategy group X benefits all agents.* There exists a unique static equilibrium $(3, 3)$ at which we have $\pi^x(3, 3) = \pi^y(3, 3)$. However, it is not dynamically stable since the agents in group Y has incentive to defect to group X due to $\pi^x(4, 2) > \pi^y(3, 3)$. In fact, we have $\pi^x(n, 6 - n) > \pi^y(n - 1, 6 - n + 1)$ for all $1 \leq n < 6$, thus starting with any $n < 3$, the static equilibrium can be reached but cannot be sustained due to one-way migration trend:

$$(0, 6) \to (1, 5) \to (2, 4) \to (3, 3) \to (4, 2) \to (5, 1) \to \boxed{(6, 0)}\,.$$

The monotonicity characteristics of the above one-way migration makes $(6, 0)$ be the final outcome, which will be referred to as a dynamical equilibrium.

Table 1b reflects another typical situation of the so-called "size advantage", i.e., *the strategy group with more players profits more than the one with less players*. The

Table 1a The unique static equilibrium is reachable but dynamically unstable

	(0, 6)	(1, 5)	(2, 4)	(3, 3)	(4, 2)	(5, 1)	(6, 0)
$\pi^x(n, 6-n)$	–	60	70	80	90	100	110
		↗	↗	↗	↗	↗	↗
$\pi^y(n, 6-n)$	50	65	75	80	85	95	–

Table 1b A unique static equilibrium is both dynamically unstable and unreachable

	(0, 6)	(1, 5)	(2, 4)	(3, 3)	(4, 2)	(5, 1)	(6, 0)
$\pi^x(n, 6-n)$	–	60	70	80	90	100	110
		↙	↙	↙	↗	↗	↗
$\pi^y(n, 6-n)$	110	100	90	80	70	60	–

Table 1c Static equilibrium is dynamically stable and reachable

	(0, 6)	(1, 5)	(2, 4)	(3, 3)	(4, 2)	(5, 1)	(6, 0)
$\pi^x(n, 6-n)$	–	75	85	90	80	70	60
		↗	↗	↗	↙	↙	↙
$\pi^y(n, 6-n)$	60	70	80	90	85	75	–

distribution $(3, 3)$ is again a static equilibrium. It is dynamically unstable because the players in group X has incentive to defect to group Y due to $\pi^y(2, 4) > \pi^x(3, 3)$ while the players in group Y has incentive to defect to group X for the reason of $\pi^x(4, 2) > \pi^y(3, 3)$. If the defections do not occur simultaneously, we shall observe an one-way migration route until all firms adopt Strategyr X:

$$(3,3) \rightarrow (4,2) \rightarrow (5,1) \rightarrow \boxed{(6,0)}$$

or an one-way migration pattern until all firms adopt Strategy Y:

$$(3,3) \rightarrow (2,4) \rightarrow (1,5) \rightarrow \boxed{(0,6)}\,.$$

Monotonicity exhibited in the above two migration routes implies that $(3, 3)$ is not reachable unless it happens to be the initial distribution. In contrast, the two extremes, $(6, 0)$ and $(0, 6)$ are both dynamically stable states.

As shown in Table 1c, the static equilibrium may also coincide with the dynamic equilibrium, as shown by the payoff structure that is associated with the so-called "size disadvantage", that is, *the strategy with less players makes higher payoffs than the other*[2] (where migrations occur so long as the groups sizes are not equal).

$$(0,6) \rightarrow (1,5) \rightarrow (2,4) \rightarrow \boxed{(3,3)} \leftarrow (4,2) \leftarrow (5,1) \leftarrow (6,0)\,.$$

[2] Such situations reflect the over-exploit of commons.

Table 2a A unique dynamic equilibrium exists at one extreme

	(0, 6)	(1, 5)	(2, 4)	(3, 3)	(4, 2)	(5, 1)	(6, 0)
$\pi^x(n, 6-n)$	–	60	70	80	90	100	110
		↗	↗	↗	↗	↗	↗
$\pi^y(n, 6-n)$	50	60	70	80	90	100	–

Table 2b Both extremes are dynamic equilibria

	(0, 6)	(1, 5)	(2, 4)	(3, 3)	(4, 2)	(5, 1)	(6, 0)
$\pi^x(n, 6-n)$	–	60	50	40	50	60	70
		↙	↙	↙	↗	↗	↗
$\pi^y(n, 6-n)$	70	60	50	40	50	60	–

Table 2c The even distribution is the unique dynamic equilibrium

	(0, 6)	(1, 5)	(2, 4)	(3, 3)	(4, 2)	(5, 1)	(6, 0)
$\pi^x(n, 6-n)$	–	60	70	80	70	60	50
		↗	↗	↗	↙	↙	↙
$\pi^y(n, 6-n)$	50	60	70	80	70	60	–

Example 2 (All bimodal distributions are static equilibria).

The inconsistency of the static equilibrium concept with the outcome of strategy-switching dynamics even exists when two strategies have identical payoffs for all bimodal distributions, that is, $\pi^x\ (n, 6-n) = \pi^y(n, 6-n)$ for $1 < n < N$. As illustrated in Tables 2a–c, all the possibilities shown in Example 1 can be reproduced.

Secondly, we shall see that, even when one strategy "statically" dominates the other, say, $\pi^y(n, N-n) > \pi^x(n, N-n)$ for all n, all the above possibilities can still appear. Moreover, there exist possibilities in which none of the players is willing to adopt the very strategy.

Example 3 ("Static dominant strategy" is unfavorable).

In Table 3a, even when $\pi^y\ (n, N-n) > \pi^x(n, N-n)$ for all n, the players in Group Y have the incentive to migrate to Group X such that not a single player is willing to take static dominant strategy in the final outcome.

In contrast, Table 3b demonstrates the possibilities of multi-equilibria. Although $\pi^y\ (n, 6-n) > \pi^x(n, 6-n)$ for all $n \neq 0$, 6, all distributions are stable in the dynamic sense because players in Group X are indifferent before and after strategy-switching.

The above examples call for meaningful concepts relevant to the strategy-switching dynamics, which leads us to the next section.

Table 3a Static inferior strategy is dynamically preferred for all distributions

	(0, 6)	(1, 5)	(2, 4)	(3, 3)	(4, 2)	(5, 1)	(6, 0)
$\pi^x(n, 6-n)$	–	60	70	80	90	100	110
	↗	↗	↗	↗	↗	↗	
$\pi^y(n, 6-n)$	55	65	75	85	95	105	–

Table 3b Static inferior strategy is dynamically indifferent for all distributions

	(0, 6)	(1, 5)	(2, 4)	(3, 3)	(4, 2)	(5, 1)	(6, 0)
$\pi^x(n, 6-n)$	–	60	70	80	90	100	110
	/	/	/	/	/	/	
$\pi^y(n, 6-n)$	60	70	80	90	100	110	–

3 Dynamic Stability and Dynamic Dominance

Examples shown in Sect. 2 demonstrated the importance to distinguish the final dynamic equilibrium of a strategy-switching regime from its traditional counterpart as well as distinguish the static dominance concept from the dynamic dominance concept for the strategies involved. This section intends to provide the relevant definitions.

3.1 Dynamical Stability of Switching-Dynamics

In general, for a dynamic model involving finite population (N) and finite dynamic strategy space $\mathcal{S} = \{s_j\}_{j=1}^{l}$, we are able to associate a state $\mathbf{n} \doteq (n_1, n_2, \ldots, n_l)$ with $\sum_{j=1}^{l} n_j = N$ and a payoff $\pi^i(\mathbf{n})$ for the firms who adopt strategy i when $n_i > 0$, for $i = 1, 2, \ldots, l$.

Definition 1. A *state* $\bar{\mathbf{n}}$ is said to be *statically stable* (or equivalently *static equilibrium*) if all strategic groups enjoy identical payoffs, i.e., for all $\bar{n}_i > 0$, $\bar{n}_j > 0$ and $i \neq j$, we have $\pi^i(\bar{\mathbf{n}}) = \pi^j(\bar{\mathbf{n}})$.

Definition 2. A *state* $\bar{\mathbf{n}}$ is said to be *dynamically stable* (or equivalently *dynamic equilibrium*) if

$$\pi^j\left(\bar{n}_1, \ldots, \bar{n}_i - 1, \ldots, \bar{n}_j + 1, \ldots, \bar{n}_l\right) < \pi^i\left(\bar{n}_1, \ldots, \bar{n}_i, \ldots, \bar{n}_j, \ldots, \bar{n}_l\right), \text{ if } \bar{n}_i > 1, \tag{1}$$

$$\pi^i\left(\bar{n}_1, \ldots, \bar{n}_i + 1, \ldots, \bar{n}_j - 1, \ldots, \bar{n}_l\right) < \pi^j\left(\bar{n}_1, \ldots, \bar{n}_i, \ldots, \bar{n}_j, \ldots, \bar{n}_l\right), \text{ if } \bar{n}_j > 1, \tag{2}$$

for all $i \neq j$, $i, j \in \{1, 2, \ldots, l\}$.

Remark 1. Condition (1) implies the internal stability for Group i: *no firm in Group i has incentive to defect to any other group.* Condition (2) indicates the external stability for Group i because *no firm in any other group has incentive to join Group* i for all $i \neq j, i, j \in \{1, 2, \ldots, l\}$.

For the two-strategy competition presented in Example 1.(b), both $(6, 0)$ and $(0, 6)$ are stable distributions. However, in between these two stable distributions, there exists an intermediate distribution $(3, 3)$, at which there is mutual attraction between both groups. A member of Group Y finds its more profitable to migrate to the Group X while a member of Group X also finds it beneficial to join Group Y (or regret to betray from the latter). If the migrations to each group occur *simultaneously*, then the migration will last forever so that the distribution $(3, 3)$ remains invariant. Such an equilibrium-alike phenomenon stimulates us to introduce the concept of transiently stable distribution.

Definition 3. A *state* $\bar{\mathbf{n}}$ is said to be *transient-stable* (or equivalently *transient equilibrium*) if there exists some pair of (i^*, j^*) such that $\bar{n}_{i^*} > 1, \bar{n}_{j^*} > 1$ and

$$\pi^{j^*}\left(\bar{n}_1, \ldots, \bar{n}_{i^*} - 1, \ldots, \bar{n}_{j^*} + 1, \ldots, \bar{n}_l\right) > \pi^{i^*}\left(\bar{n}_1, \ldots, \bar{n}_{i^*}, \ldots, \bar{n}_{j^*}, \ldots, \bar{n}_l\right),$$
$$\pi^{i^*}\left(\bar{n}_1, \ldots, \bar{n}_{i^*} + 1, \ldots, \bar{n}_{j^*} - 1, \ldots, \bar{n}_l\right) > \pi^{j^*}\left(\bar{n}_1, \ldots, \bar{n}_{i^*}, \ldots, \bar{n}_{j^*}, \ldots, \bar{n}_l\right),$$

while inequalities (1) and (2) are satisfied for other irrelevant indices.

Remark 2. A *state* $\bar{\mathbf{n}}$ is said to be *transient-stable* if there exists incentive for migration among some strategy groups and no incentive among the other groups. Unlike the dynamically stable distribution, a transiently stable distribution is dynamically unstable because any bias from it will direct the migration towards one of the nearby dynamically stable distributions.

For a finite strategy space $\mathcal{S} = \{s_j\}_{j=1}^{l}$, the dynamical-stability of a state $\bar{\mathbf{n}}$ requires $\binom{l}{2}$ times of checking inequalities (1) and (2). Apparently, some of the checking are overlapping. Therefore, q more efficient measure is resorted to simplify the comparison and checking as well as to enable us to analyze how players migrate from one group to the other to pursue higher payoff.

Definition 4. *Marginal benefit of strategy-switching:* by which we mean the posterior increment in absolute payoff if a member of group j migrates to group i:

$$\begin{aligned}\delta_{ij}\left(n_i, n_j\right) &\doteq \pi^i\left(n_1, \ldots, n_i, \ldots n_j, \ldots n_l\right) \\ &\quad -\pi^j\left(n_1, \ldots, n_i - 1, \ldots n_j + 1, \ldots n_l\right), n_i \geq 1.\end{aligned} \tag{3}$$

The definition (3) suggests that

$$\delta_{ij}\left(n_i, n_j\right) = -\delta_{ji}\left(n_j + 1, n_i - 1\right) \quad \text{for any } j \neq i,$$

which implies that we only need to define the number of marginal benefit of strategy-switching for those $j > i$. Then the set of $\binom{l}{2}$ marginal benefits of "defections" serve well for the purpose of simplifying the analysis of migration effects. This is because, assuming that all n_s, $s \neq i, j$, are being fixed, δ_{ij} characterizes vividly the trend of strategy-switching between Group i and Group j. Starting with any strategic distribution (n_i, n_j), an incentive exists for a member of Group j to migrate to Group i if $\delta_{ij}(n_i, n_j) > 0$.[3] The migration ceases at (n_i, n_j) in both directions when $\delta_{ij}(\bar{n}_i, \bar{n}_j) > 0$ and $\delta_{ij}(\bar{n}_i + 1, \bar{n}_j - 1) < 0$, which are nothing but (1) and (2). Therefore, just by examining the signs of $\delta_{ij}(\bar{n}_i, \bar{n}_j)$ for different $(\bar{n}_i, \bar{n}_j)$ combinations, we are able to see how the migration proceeds between Group i and Group j (that is, how the states transit among them).

Example 4. For a two-strategies competition, say, $\mathcal{S} = \{X, Y\}$, for any given fixed number of players N, we have $n_y = N - n_x$. By denoting $\delta_N(n_x) \doteq \delta_{xy}(n_x, N - n_x)$, the marginal benefit of switching from strategy x (Group X) to strategy y (Group Y) degenerates into a mathematical function with single variable n_x, should N be fixed. Since it is the sign of $\delta_N(n)$ that determines the direction of strategy-switching, the monotonicity of $\delta_N(n)$ characterized by the derivative $\delta'_N(n)$ offers the sufficient condition to attain the relevant stable distribution. In addition, $\delta_N(1)$ and $\delta_N(N)$ are two important indicators for the trend of strategy-switching. Based on the signs of these two indicators and the condition that the sign of $\delta_N(n)$ is a monotonically function of n, we are able to characterize some of the most typical types of stable distributions. These are summarized in Table 4 and illustrated in Fig. 2.

A state $(\tilde{n}, N - \tilde{n})$ is transiently stable if $\delta_N(\tilde{n}) \leq 0$ and $\delta_N(\tilde{n} + 1) > 0$. A transiently stable distribution always occurs in between two *extremal* stable distributions $(N, 0)$ and $(0, N)$, as illustrated by Pattern III of Fig. 1.

3.2 Stable Cluster

The relationships between the analytical property of payoffs and the possible patterns of migration dynamics are by no mean straightforward. To see this, we proceed

Table 4 Typical patterns of stable distributions

Pattern	Sign pattern of $\delta_N(n)$	Stable distribution(s)	Sufficientconditions
I	$(-, -, \cdots, -, -, \cdots, -, -)$	$(0, N)$	$\delta'_N(n) < 0, \delta_N(1) < 0$ $\delta'_N(n) > 0, \delta_N(N) < 0$
II	$(+, +, \cdots +, +, \cdots, +, +)$	$(N, 0)$	$\delta'_N(n) > 0, \delta_N(1) > 0$ $\delta'_N(n) < 0, \delta_N(N) > 0$
III	$(-, -, \cdots -, +, \cdots, +, +)$	$(0, N)$ and $(N, 0)$	$\delta'_N(n) > 0, \delta_N(1) < 0, \delta_N(N) > 0$
IV	$(+, +, \cdots +, -, \cdots, -, -)$	$(n^*, N - n^*)$	$\delta'_N(n) < 0, \delta_N(1) > 0, \delta_N(N) < 0$

[3] Additionally, we may assume that a member of Group j has incentive to migrate to Group i even when $\delta_{ij}(n_i, n_j) = 0$ but $\pi^i(n_1, \ldots, n_i, \ldots n_j, \ldots n_l) > \pi^j(n_1, \ldots, n_i, \ldots n_j, \ldots n_l)$. In other words, the migration can be proceeded even if it results in relative profitability without sacrificing the absolute profitability.

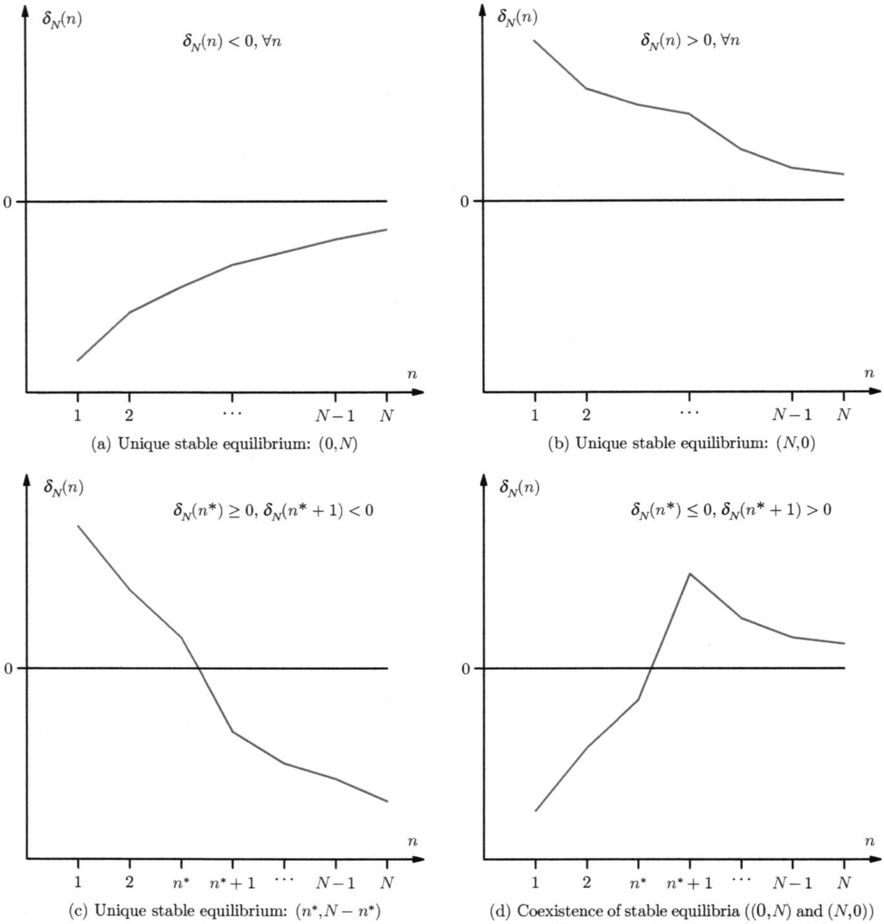

(a) Unique stable equilibrium: $(0,N)$

(b) Unique stable equilibrium: $(N,0)$

(c) Unique stable equilibrium: $(n^*,N-n^*)$

(d) Coexistence of stable equilibria $((0,N)$ and $(N,0))$ with a transient equilibrium $(k*,N-k*)$

Fig. 1 Typical patterns of stable distributions

to examine the simplest case where all payoffs are monotonic to n.[4] Even if $\pi^i(n)$, $i = x, y$, are both monotonic, the marginal benefit of switching $\delta_N(n)$ is monotonic only if $(d\pi^x/dn)(d\pi^y/dn) < 0$, under which, the stable distribution patterns are limited to four possible cases illustrated in Fig. 1. However, if $(d\pi^x/dn)(d\pi^y/dn) > 0$, more possible patterns can occur, as illustrated by the following two examples for the cases in which $d\pi^x/dn < 0$ and $d\pi^y/dn < 0$.

[4] A payoff function $\pi^i(n)$, $i = x, y$, is said to be monotonic with respect to n if $d\pi^i(n)/dn$ does not change sign for all n.

Table 5a Coexistence of $(N, 0)$ and $(n^*, N - n^*)$

	(0, 6)		(1, 5)		(2, 4)		(3, 3)		(4, 2)		(5, 1)		(6, 0)
$\pi^x(n, 6-n)$	–		150		135		125		120		115		110
		↗		↙		↙		↙		↙		↗	
$\pi^y(n, 6-n)$	145		140		135		125		120		105		–
$\delta_N(n)$	–		5		−5		−10		−5		−5		5

Table 5b Coexistence of $(0, N)$ and $(n^*, N - n^*)$

	(0, 6)		(1, 5)		(2, 4)		(3, 3)		(4, 2)		(5, 1)		(6, 0)
$\pi^x(n, 6-n)$	–		145		140		135		125		115		105
		↙		↗		↗		↗		/		↙	
$\pi^y(n, 6-n)$	150		135		125		120		115		110		–
$\delta_N(n)$	–		−5		5		10		5		0		−5

Table 5c Multiple stable states

	(0, 6)		(1, 5)		(2, 4)		(3, 3)		(4, 2)		(5, 1)		(6, 0)
$\pi^x(n, 6-n)$	–		85		80		65		62		55		50
		↙		↗		↙		↗		↙		↗	
$\pi^y(n, 6-n)$	90		75		70		60		58		45		–
$\delta_N(n)$	–		−5		5		−5		2		−3		10

Example 5 (Coexistence of two stable states). There may coexist two stable states such as $(N, 0) \cup (n^*, N - n^*)$ or $(n^*, N - n^*) \cup (0, N)$ as illustrated in Tables 5a and 5b, respectively. If the derivative $\delta'_N(n)$ behaves irregularly, then multiple bimodal equilibria can occur, as seen in Tables 3b and 5c.

The possibilities of coexistence of three or more dynamically stable states would in turn suggest the coexistence of multiple transiently stable states. The following proposition follows from intuition straightforwardly.

Proposition 1. *For a two-strategy competition, between every two dynamically stable states* $(n_1^*, N - n_1^*)$ *and* $(n_2^*, N - n_2^*)$, *with* $n_2^* > n_1^* + 1$, *there exists at least one transiently stable state.*

For a two-strategies competition, there always exists a stable state. Starting with any unstable distribution, the strategy-switching behavior proceeds if necessary until a local stable state is reached. In the terminology of dynamic theory, the strategy-switching dynamics is monotonic converging, either to a local equilibrium, or to a global equilibrium. Additionally, a stable cluster of states can also appear as shown in the next few examples.

For a two-strategies competition, two neighborhood states $(n, N - n)$ and $(n - 1, N - n + 1)$ are said to be *indifferent* if $d_N(n) = 0$. Indifferent states form a cluster, which is locally stable if all the states outside of this cluster will transit to it.

Table 6a A stable cluster formed by indifferent states

	(0, 6)		(1, 5)		(2, 4)		(3, 3)		(4, 2)		(5, 1)		(6, 0)
$\pi^x(n, 6-n)$	–		70		65		55		45		40		35
		↗		↗		/		/		↙		↙	
$\pi^y(n, 6-n)$	65		60		55		45		55		60		–

Table 6b An unstable cluster formed by indifferent states

	(0, 6)		(1, 5)		(2, 4)		(3, 3)		(4, 2)		(5, 1)		(6, 0)
$\pi^x(n, 6-n)$	–		40		45		50		55		60		65
		↙		↙		/		/		↗		↗	
$\pi^y(n, 6-n)$	45		48		50		55		57		60		-

Table 6c An unstable cluster formed by indifferent states

	(0, 6)		(1, 5)		(2, 4)		(3, 3)		(4, 2)		(5, 1)		(6, 0)
$\pi^x(n, 6-n)$	–		40		45		50		55		60		65
		↗		↗		/		/		↗		↗	
$\pi^y(n, 6-n)$	35		40		50		55		57		60		–

Table 7a A stable cluster formed by transient state

	(0, 6)		(1, 5)		(2, 4)		(3, 3)		(4, 2)		(5, 1)		(6, 0)
$\pi^x(n, 6-n)$	–		40		50		40		50		40		35
		↗		↗		↙		↗		↙		↙	
$\pi^y(n, 6-n)$	35		45		50		40		45		40		–

Table 7b A stable cluster formed by transient state and indifferent state

	(0, 6)		(1, 5)		(2, 4)		(3, 3)		(4, 2)		(5, 1)		(6, 0)
$\pi^x(n, 6-n)$	–		40		50		40		50		45		35
		↗		↗		↙		↗		/		↙	
$\pi^y(n, 6-n)$	35		45		50		40		45		40		–

Example 6 (Stable and unstable cluster resulted from indifferent states). The payoff structure shown in Table 6a presents a case of stable cluster. An extreme case occurs when all states form a stable cluster, as shown in Table 3b.

A cluster can also be unstable if at least one of the states can transit to the states outside the cluster, as illustrated in Tables 6b,c where the states $(2, 4)$, $(3, 3)$ and $(4, 2)$ form a cluster.

Example 7 (Stable cluster resulted from transient equilibrium). A stable cluster can also consist of a transient equilibrium and two nearby stable states, as shown in Table 7a.

A mixed type of stable cluster that consists of indifferent transition and transient equilibrium can be easily constructed, as shown in Table 7b.

The possibility for stable cluster stimulates us to generalize the concept of dynamic stability for a single state to the one for a stable cluster. When more than three strategies are involved, the strategy-switching dynamics turns out to be much more complex than the two-strategies competition. With some initial distribution, the players may migrate cyclically. If all possible distributions result in a cyclical evolutionary pattern, no stable distribution can exist. Instead of defining the stability of the neighborhood of a cluster directly, a more general concept of stability to cover both cyclical path and neighborhood cluster is provided as the following:

Definition 5. *Directly linked distributions:* two distributions $\mathbf{n}^{(k)} \doteq (\bar{n}_1^{(k)}, \bar{n}_2^{(k)}, \ldots, \bar{n}_l^{(k)})$, $k = 1, 2$ are said to be directly linked if there exist two indices i and j such that $\bar{n}_i^{(1)} = \bar{n}_i^{(2)} + 1$ and $\bar{n}_j^{(2)} = \bar{n}_j^{(1)} + 1$ and $\bar{n}_s^{(1)} = \bar{n}_s^{(2)}$ for all $s \neq i, j$. In other words, two distributions are directly linked to each other if one can reach the other by a single migration activity.

Definition 6. A collection of distributions $\{\mathbf{n}^{(k)}\}$ is said to form a *cluster* if any one of them is directly linked to at least one other member of the collection.

Remark 3. If we visualize a distribution $\mathbf{n}^{(k)}$ as a node in a multi-dimensional lattice (network) and connect any two directly linked distributions either with an one-way arrow (to indicate possible one-way transit) or a two-way arrow (to indicate a possible two-way transit, which occurs when $\delta_{ij}\left(n_i, n_j\right) = 0$), then starting with any initial node in a neighborhood, all other nodes can be reached with at least one arrow.

Definition 7. A *cluster of states* $\mathbf{N}^* \doteq \{\bar{\mathbf{n}}_k\}$ is said to be a *dynamically stable attractor* if:

1. *External stability*: all distributions linked directly to any one member of $\mathbf{N}^*$will transit to it.
2. *Internal stability*: the distributions within the collection do not transit to those that do not belong to the collection.

The above definition is particularly important for the strategy-switching dynamics involving more than two-strategies, which will not be explored in the current presentation (see Huang, 2009). But it is essential to point out that it is exactly when more than two strategies are involved that the analysis of dynamics becomes interesting: the stable attractor can be either a cyclical path or a chaotic cluster.

3.3 Dominance of Dynamic Strategy

As a complement to the concepts for the static "strategy" defined in game theory, we conclude this section with an analogous one for the "dynamic strategy" under the framework of strategy-switching dynamics.

Definition 8. A strategy $s_j \in \mathcal{S}$ is said to be *dynamically feasible* if there exists at least a dynamically stable state $\bar{\mathbf{n}}$ with $\bar{n}_j \geq 1$.

Definition 9. A strategy $s_j \in \mathcal{S}$ is said to be *dynamically stable* if there exists a *unique* dynamically stable state $\bar{\mathbf{n}}$ such that $\bar{n}_j \geq 1$.

Definition 10. A strategy $s_j \in \mathcal{S}$ is said to be *dynamically infeasible* if for all dynamically stable state $\bar{\mathbf{n}}$, we have $\bar{n}_j = 0$.

Remark 4. When there are more than two strategies involved, there may not exist any dynamically stable distribution. Equivalently to say, there may exist possibilities in which all strategies are infeasible.

Definition 11. An economically stable strategy $s_i \in \mathcal{S}$ is said to be *dynamically dominant* if for all $j \neq i$, we have

$$\delta_{ij}\left(n_i, n_j\right) > 0 \qquad \text{for all } 1 \leq n_i \leq N_{ij},$$

where $N_{ij} = N - \sum_{s \neq i,j} n_s$.

Remark 5. To distinguish from the "static" concept of strategy dominance in the standard game theory, we may regard the economic-dominance as a dynamic dominance concept.

If the strategy s_i is *dynamically dominant*, then $(0 \cdots 0, N, 0 \cdots 0)$ must be the unique dynamically stable distribution. The converse, however, is always true for two-strategy competition but not when the strategy space consists more than two strategies.

The following facts, however, can be proved straightforward.

Proposition 2.

1. *There exists at most one dynamically dominant strategy.*
2. *If a strategy is dynamically dominant, all other strategies must be dynamically infeasible.*

4 Economic Applications: Dominance of Price-Taking Strategy

The ad-hoc examples provided in previous sections serve only as illustrations. To appreciate the newly defined concepts and their implications, it is necessary at this point to apply these concepts and the relevant framework to a concrete economic problem. We choose the traditional quantity-competed oligopoly model to serve such a purpose for the reason that the distinction between the static dominance and dynamic dominance can be best illustrated.

4.1 Static Dominance of Price-Taking Strategy

Consider an oligopoly market, in which N firms produce a homogeneous product with quantity q_t^i, $i = 1, 2, \ldots, N$, at period t. The market inverse demand for the product is given by $p_t = D(Q_t)$, where $D' \leq 0$. It is assumed that *the actual market price adjusts to the demand so as to clear the market at every period*, that is, $Q_t = \sum_{i=1}^{N} q_t^i$. The payoffs of firm i is given by

$$\pi^i \left(Q_t, q_t^i\right) = D\left(Q_t\right) q_t^i - C_i \left(q_t^i\right).$$

Definition 12. By *the price-taking strategy* we mean any firm adjusts its output $\bar{q}^i$ dynamically in response to market price so that[5]

$$D\left(Q_t\right) = C_i' \left(\bar{q}_t^i\right). \tag{4}$$

Then we have the following beautiful result on the (static) dominance of price-taking strategy (Huang, 2008):[6]

Theorem 1. *If $D' < 0$ and $C_i'' > 0$ are satisfied, for all j such that $C_j = C_i$, we have*

$$\pi^i (Q_t, \bar{q}_t^i) \geq \pi^j (Q_t, q_t^j), \tag{5}$$

where $Q_t = \bar{q}_t^i + q_t^j + \sum_{l \neq i,j} q_t^l$ and the equality holds if and only if $q_t^j = \bar{q}_t^i$. In other words, a firm adopting price-taking strategy has the relative profitability over any other firm who has identical cost but produces at different output level.

The conclusions drawn in Theorem 1 is generic in the sense that it is robust to the changes (during the evolutionary process) in the market environments such as the market demand, entry and exit of oligopolistic firms, advances in some or overall technology level.

The driving force behind the long run outcome in the evolutionary literature is the relative payoff (Schaffer, 1989), rather than the absolute payoff assumed in most fields of economics, including standard game theory. From this regard, the price-taking strategy defined above is definitely evolutionary stable. However, according to the normative economic theory, it is the absolute profit, not the relative profit, that an economic agent should pursue. Although a firm can enjoy a relative payoff advantage over the other firm by adopting a certain evolutionary stable strategy, it may find it worthwhile to give up such advantage by switching to the other "inferior" strategies to gain extra increment in the payoff. Extreme cases may arise in which

[5] It needs to emphasize that this definition is different from the price-taking strategy defined in evolutionary game theoretic literature, where the output level of competitive equilibrium is defined as the price-taking strategy.

[6] The relative profitability of the price-taking strategy was first formally proposed in Huang (2002) where the proof was mistakenly omitted in the editorial process.

none of the agents adopts an evolutionarily stable strategy. These points can be made clearer with a symmetric oligopolistic model.

Assume that N oligopolistic firms have identical cost and are divided into two strategic groups, X and Y. Group X consists of n *price-takers* whose output x is determined by

$$D(Q) = C'(x). \tag{6}$$

The rest $m = N - n$ firms in Group Y are *Cournot optimizers* who adopt individually the conventional Cournot best-response strategy (C). Denote these firms' average output as y, then y is implicitly determined by[7]

$$D(Q) + yD'(Q) = C'(y), \tag{7}$$

where $Q \triangleq nx + my$.

Apparently, the equilibrium profits at an equilibrium $(\bar{x}, \bar{y})$ established by (6) and (7) for each firm in these two strategy groups, denoted as π^x and π^y, respectively, depend on (n, m), the *state*:

$$\pi^x(n,m) = \bar{x} \cdot D(n\bar{x} + m\bar{y}) - C(\bar{x}),$$
$$\pi^y(n,m) = \bar{y} \cdot D(n\bar{x} + m\bar{y}) - C(\bar{y}).$$

The *extremal* distributions $(N, 0)$ and $(0, N)$ thus correspond to the *competitive equilibrium* and the *Cournot equilibrium*, respectively.

Denote

$$\Delta\pi^{xy}(n,m) \triangleq \pi^x(n,m) - \pi^y(n,m) \tag{8}$$

as the *profit difference* for $n \cdot m > 0$. It follows from (5) that $\Delta\pi^{xy}(n,m) > 0$ for all $n \cdot m > 0$, providing $C'' > 0$ and $D' < 0$. In words, regardless of how N firms are distributed between the two groups, any member of Group X (if is not empty) makes higher profit than the firms in Group Y. If firms are after the relative payoffs, all members of Group Y will defect to become price-takers so that *competitive equilibrium* $(N, 0)$ ends up as the unique dynamic stable equilibrium distribution. On the other hand, if firms pursue absolute profits, at any distribution of $(n, N - n)$, $1 \leq n \leq N$, there exists economic incentive for any member of Group X to join Group Y if

$$\delta_N(n) \triangleq \pi^x(n, N - n) - \pi^y(n - 1, N - n + 1) < 0.$$

This is the situation where a price-taker is willing to sacrifice its relative profitability in exchange for the absolute profitability by migrating back to Group Y. In consequence, an exact opposite outcome may occur, that is, all firms may abandon the evolutionary stable strategy and switch to Cournot best-response strategy so that

[7] Here we assume that firms in Group Y do not collude together, which is equivalent to assume that each and every firm in Group Y knows nothing about the group size $N - n$.

$(0, N)$ becomes the unique stable distribution. Alternatively speaking, the Cournot best-response strategy turns out to be dynamically dominant, although the price-taking strategy is static-dominant. Such extreme case can be easily constructed, as shown in the next subsection.

4.2 *The Price-Taking Strategy Can Be Dynamically Infeasible*

Let the market demand be iso-elastic: $D(Q) = 1/Q$, and the cost function be quadratic (linear marginal cost): $C(q) = cq^2/2$. (6) and (7) turn out to be

$$\frac{1}{nx + (N-n)y} = cx,$$
$$\frac{1}{nx + (N-n)y} - \frac{y}{(nx + (N-n)y)^2} = cy,$$

from which an equilibrium $(\bar{x}, \bar{y})$ is solved as

$$\bar{x} = \frac{\sqrt{g(n) - N + m}}{\sqrt{2cn}}, \qquad \bar{y} = \frac{n(N + 1 - g(n))}{(N-n)\sqrt{2cn(g(n) - N + 1)}},$$

where $g(n) \triangleq \sqrt{(N-1)^2 + 4n}$. The equilibrium profits for each firm of two groups are respectively

$$\pi^x(n, N-n) = \frac{1}{4n}(g(n) - N + 1),$$
$$\pi^y(n, N-n) = \frac{(N+1)^2 + (N-n)(N-1) - (2N+1-n)g(n))}{4(N-n)^2}.$$

It can be easily checked that $\Delta\pi^{xy}(n, N-n) > 0$ for all $1 \leq n \leq N-1$, i.e., the price-takers make more profit than the Cournot optimizers for all distributions.

It turns out that $\delta_N(n) < 0$ for all $1 \leq n \leq N$, which suggests that $(0, N)$ is the unique stable distribution and hence *the price-taking strategy is completely abandoned by all firms*. Following our earlier definitions, the Cournot best-response strategy is dynamically dominant although the price-taking strategy is static dominant.

However, the beauty of price-taking strategy should not be undermined just by this particular example. The dynamic dominance of a particular strategy relies strongly on the market demand, the cost structure, and even on the number of firms involved, as illustrated in the next section.

4.3 The Price-Taking Strategy Can Also Be Dynamically Dominant

Let the market demand be linear:[8] $D(Q) = 1 - Q$, and the cost function remains to be $C(q) = cq^2/2$. Equations (6) and (7) now become

$$\begin{aligned} 1 - nx - (N-n)y &= cx, \\ 1 - nx - (N-n)y - y &= cy, \end{aligned}$$

which yield $(\bar{x}, \bar{y}) = ((1+c)/(n + c(N+1+c)), c/(n + c(N+1+c)))$ so that

$$\begin{aligned} \pi^x(n, N-n) &= \frac{c(1+c)^2}{2(n + c(N+1+c))^2}, \\ \pi^y(n, N-n) &= \frac{c^2(2+c)}{2(n + c(N+1+c))^2}. \end{aligned}$$

Again it can be easily checked that $\Delta\pi^{xy}(n, N-n) > 0$ for all $1 \le n \le N-1$, that is, the price-taking strategy is static dominant. However, the sign of marginal benefit of switching $\delta_N(n)$ now depends on the size of the industry.

Denote $\lceil\cdot\rceil$ and $\lfloor\cdot\rfloor$ as the ceiling function and floor functions, respectively, which returns the smallest (largest) integer greater (less) than or equal to a non-negative number. Let $N_l \doteq \max\{2, \left\lfloor 1 + \sqrt{c(c+2)} \right\rfloor\}$ and $N_u \doteq \left\lceil 1 + (c+1)\sqrt{1 + 2/c} \right\rceil$, then the relationship between the market size and the dominance of price-taking strategy can be seen in Fig. 2 and summarized in the following proposition.

Proposition 3. *When $D(Q) = 1 - Q$ and $C(q) = cq^2/2$, the outcome of the strategy-switching dynamics depends strictly on the number of N:*

Case (I) when $N \le N_l$, the Cournot best-response strategy is dynamically dominant (equivalently, $(0, N)$ is the unique stable distribution).

Case (II) when $N_l < N < N_u$, there is no dynamically dominant strategy so that both strategies are economically feasible (equivalently, two stable distributions $(N, 0)$ and $(0, N)$ coexist with a transiently stable distribution $(\tilde{n}, N - \tilde{n})$).

Case (III) when $N \ge N_u$, the price-taking strategy is dynamically dominant (equivalently, $(N, 0)$ is the unique stable distribution).

We are able to conclude that:

(1) Unless c is very large ($c \gg 1$), condition $N \le N_l$ is rarely satisfied for N greater than 2. On the other hand, regardless of c, condition $N > N_u$ is easily satisfied when N is sufficiently large.

[8] The linear demand is conventionally assumed to take the form of $D(Q) = a - bQ$ (see Fisher, 1961 and Rothschild, 1990). It can be verified that, with appropriate change of quantity variable Q as well as redefinition of parameters, the effects of parameters a and b can be summarized into the cost parameter c.

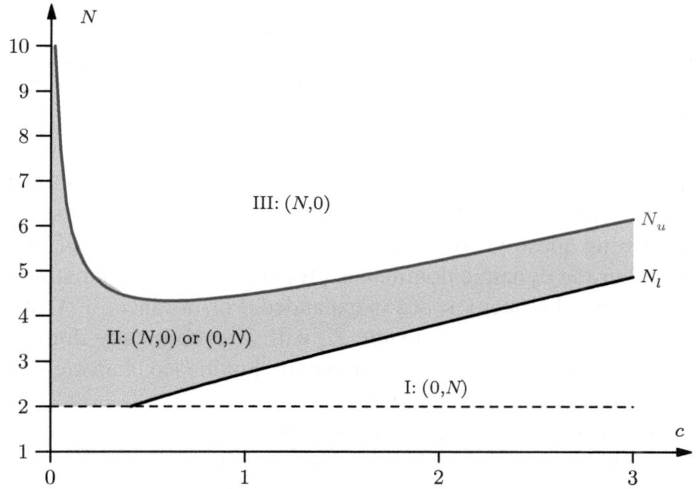

Fig. 2 Illustrations of N_l and N_u

(2) The difference between N_l and N_u decreases with increasing c and approaches a limit that equals to unity. When $c \geq 2$, that is, c is relatively large, the difference between N_l and N_u peaks at unity, which suggests that Case II do not occur. Only when c is relative small ($c \ll 1$) that the difference between N_l and N_u is large enough to generate Case II.
(3) When $c \ll 1$, a single new firm may make a huge difference as it may turn the stable distribution from $(0, N)$ to $(N + 1, 0)$.
(4) Regardless of c, increasing N increases the dynamic dominance of the price-taking strategy.

5 Conclusions

The strategy-switching dynamics in the dynamic competition models where the agents are after the absolute profitability, with the bounded rationality, and the payoffs are frequency-dependent are studied. The relevant equilibrium and stability concepts have been defined for the dynamic strategies from a perspective that is consistent with the normative economic theory, and complement the ones defined for the static strategies from the evolutionary game-theoretical perspective. Based on such a framework, the appreciation of price-taking strategy can be reexamined.

Compared to the model with two strategies, the strategy-switching dynamics for multi-strategies are more difficult to analyze. This is because, for the two-strategy competition, the actual realization of switching path is consistent with the "incentive"of switching because only one target is considered. On the contrary, when the size of the strategy space is more than two, there exists the possibility that a member of a strategy group has incentive to migrate to several other strategy groups. If payoffs are identical for some of these target groups, the actual migration path is not unique, even after an additional assumption that the agents are all after higher payoff is imposed. Such indeterminacy of strategy-switching results in the indeterminacy of the next state along a migration path. Moreover, if such phenomenon occurs within a stable cluster, a chaotic-like behavior can be observed along the trajectory of migration. These issues will be explored further in the subsequent study.

Many interesting questions remain unanswered from the aspect of "strategy". It is unclear whether the dynamic dominance property of a particular strategy can be affected if the original strategy space is expanded. For instance, if X dynamically dominates Y, Y dynamically dominates Z, will X dynamically dominate Z? In other words, the answer to whether the dynamic dominance characteristics can be preserved transitively remains unknown.

Even for the symmetric oligopoly model with heterogeneous strategies as presented in Sect. 4, it is unclear when the price-taking strategy can be dynamic dominant. Future research will include identifying the necessary and/or sufficient conditions for the dynamic dominance of price-taking strategy.

Acknowledgements The research is partly supported by the research grant RG68/06 from the Ministry of Education of Singapore. I would like to thank an anonymous referee, Barkley Rosser Jr., Carl Chiarella, Laura Gardini, Gian-Italo Bischi, Volker Bohm and the participants of MDEF 2008 for positive comments and suggestions.

References

Fisher, F. M. (1961). The stability of the cournot oligopoly solution: The effects of the speeds of adjusting and increasing marginal costs. *Review of Economic Studies, 28*, 125–135.

Huang, W. (2002). On the incentive for price-taking behavior. *Management Decision, 40*, 682–692.

Huang, W. (2008). *On the relative profitability of dynamic walrasian strategies* (EGC working paper). Nanyang Technological University.

Huang, W. (2009). *On the strategy-switching dynamics* (EGC working paper). Nanyang Technological University.

Rothschild, R. (1990). On the incentives for horizontal merger in an industry with dominant firms. *European Journal of Political Economy, 6*, 473–485.

Schaffer, M. E. (1989). Are profit-maximizers the best survivors? *Journal of Economic Behavior and Organization, 12*, 29–45.

R&D Public Expenditure, Knowledge Spillovers and Agglomeration: Comparative Statics and Dynamics

Pasquale Commendatore, Ingrid Kubin, and Carmelo Petraglia

1 Introduction

Since the introduction of the influential core-periphery (CP) model by Krugman (1991), New Economic Geography (NEG) models have provided a natural framework for non-linear dynamic analysis.[1] Moreover, as shown by the comprehensive picture of policy implications of the NEG paradigm provided by Baldwin et al. (2003), a well established finding is that policy changes have non-linear effects on industrial location, in general.

Why economic activities tend to cluster in space is an old question. Often referring to Marshall's famous classification, regional and urban economists analysed the agglomerative effects of better access to public goods in central locations, of knowledge spillovers between firms and of labour market pooling (see for a recent survey Duranton and Puga 2004). Instead, the NEG focuses on the trade costs, increasing returns at the firm level and factor mobility, and determines endogenously the spatial distribution of (monopolistically competitive) firms by the interplay of agglomeration and dispersion forces (see Venables 2008).

In Krugman's CP model, mobile workers spend their incomes locally and the spatial distribution of industrialized activities is driven by three effects. The "market-access" effect (i.e. the tendency of imperfectly competitive firms to locate in the large market and export to small markets) and the "cost-of-living" effect (i.e. goods are cheaper in regions with higher concentration of industrial firms) encourage agglomeration. The "market-crowding" effect (i.e. the tendency of imperfectly

[1] Noticeable previous contributions on non-linear dynamics in regional economics are those by Mees (1975), Puu (1981) and White (1985). Mees (1975) demonstrated that slow improvements in transportation and communication can result in the catastrophic agglomeration of population in cities. In 1981, Puu established the possibility of catastrophic changes in the structure of intra-regional trade flow patterns. In 1985, White examined the onset and character of chaotic behaviour in a multisector multicentre simulation model.

P. Commendatore (✉)
Università di Napoli, Dipartimento di Teoria Economica e Applicazioni Via Rodinò, 22 80138 Napoli, Italy,
e-mail: commenda@unina.it

G.I. Bischi et al. (eds.), *Nonlinear Dynamics in Economics, Finance and the Social Sciences*, DOI 10.1007/978-3-642-04023-8_9,

competitive firms to locate in regions with less competitors) favors dispersion. Combining the "market-access" effect and the "cost-of-living" effect with migration creates the potential for cumulative causality and self-reinforcing agglomeration processes (Baldwin et al. 2003, p. 10). As a result, complete agglomeration in one region may be a stable equilibrium. Within the NEG, the CP model impresses with the richness of delivered results; however, due to cumulative causality, it is difficult to manipulate analytically and most results are obtained via numerical simulation.

The Footlose Capital (FC) variant of the CP model, originally proposed by Martin and Rogers (1995), assumes that the mobile factor (capital) repatriates all of its earnings to its region of origin. Such an assumption cuts off "cumulative causality" thus rendering the analysis much more tractable.

The use of NEG models for policy analysis is a quite new (and still controversial) achievement (see Neary 2001; and also Sachs and McCord 2008).[2] In two recent papers we contribute to this literature by introducing into a FC model public goods as another potentially agglomerative force;[3] in particular, we explicitly modeled productivity enhancing public expenditure and decomposed its overall effect on industrial location into two components. First, the productivity effect: an increase in the provision of public services in one region lowers labour input requirements and leads firms to relocate there. Second, the demand effect: the increase in taxation required to finance this provision and the consequent contraction in private expenditure for manufactured goods favour dispersion via a change in the relative market size. In Commendatore et al. (2008b) we adopt a two-region FC model with endogenous capital and focus our analysis on the equilibrium outcomes. In Commendatore et al. (2009), we consider a much simpler analytical framework – a FC model without an investment sector – and fully characterize the dynamic process underlying capital movements (including an explicit existence and local stability analysis of the emerging fixed points).

Our present contribution departs from Commendatore et al. (2008b, 2009) in three main aspects. First, we restrict our attention to R&D public expenditure and assume that the productivity enhancing effect induced by public policy takes place via a reduction of fixed costs in the manufacturing sector. Consequently, the standard equalisation between number of firms and capital units does not apply. Second, we are able to consider how public policy impact on the spatial distribution of capital, also altering both the localisation and the number of firms. Third, we relax the assumption of pure local effects of productive public expenditure, studying the spatial spillovers effects of public policy.

[2] Commendatore et al. (2008a) show that the policy propositions derived in Baldwin et al. (2003) are not robust with respect to the temporal specification of the economic geography model employed (continuous vs discrete time).

[3] Note that we do not incorporate direct knowledge spillovers between firms (another of the potentially agglomerative forces mentioned above). Bischi and Lamantia (2002) and Bischi et al. (2003a, 2003b) analyse in an industrial organization modelling framework the complex clustering dynamics generated by such direct knowledge spillovers. Extending our model in this direction is left for further research.

Assuming that knowledge creation is financed by the public sector, we will explore the location effect of knowledge creation and diffusion across regions. Publicly financed local universities and research centres (or the government itself) undertake R&D activities which have a positive productivity effect on firms via a reduction of fixed costs. We will pursue the analysis for both the cases of global and of local knowledge spillovers. In the first scenario, we will assume that once new ideas have been generated they can freely circulate across regions and firms can benefit to the same extent from publicly financed R&D activities, regardless of their location (perfectly global spillovers). In the second case, we will maintain that knowledge generated by R&D activities undertaken in one region is beneficial to both local and foreign firms to a different extent though (perfectly/partially local spillovers).

We will disentangle the demand and productivity effects induced by changes in public expenditure on R&D and their impact on industrial location equilibria. This will be done under the alternative assumptions of global and (partially/perfectly) local knowledge spillovers. Both local and global dynamics of the model will be studied. In particular, we will deliver a local dynamic analysis on the impact of trade freeness on the long-term regional allocation of capital and an analysis of the complex structure of the basins of attraction of the boundary fixed points.

The remainder of the paper unfolds as follows. Section 2 introduces the assumptions of the model. Section 3 reports the derivation of the short-run equilibrium. Section 4 introduces the complete dynamic model. In Sects. 5 and 6 we study the impact of public policy on industrial location in the cases of global and local knowledge spillovers respectively, focussing on the local dynamics properties. Section 7 deals with global dynamics properties. Section 8 concludes.

2 Assumptions

Our analytical framework is based on the FC model (Martin and Rogers 1995). The economy is composed of two regions, labelled 1 and 2. Each region has a perfectly competitive agricultural sector and, potentially, a manufacturing sector. There are two productive factors, labour and capital, equally distributed among L households, which are in turn symmetrically distributed across regions (that is, $L/2$ households reside in each region). Each household supplies inelastically one unit of labour, thus L represents also total labour supply. Households move freely within a region but are not prepared to migrate across regions and spend all their earnings in the region where they live. Physical capital, however, is allowed to move between regions. We denote by K the overall number of capital units.

A local government may provide a public good which reduces manufacturing firms (fixed) costs. For simplicity we assume that the provision of the public good does not impact on consumers' utility. We interpret such public good as R&D activities undertaken by universities or by other institutions financed by the government.

The representative household's utility function is[4]

$$U = C_A^{1-\mu} C_M^{\mu} \tag{1}$$

where C_A and C_M correspond to the consumption of the homogeneous agricultural good and of a composite of manufactured goods:

$$C_M = \sum_{i=1}^{n} c_i^{\frac{\sigma-1}{\sigma}} \tag{2}$$

where c_i is the consumption of good i, n is the total number of manufactured goods and $\sigma > 1$ is the constant elasticity of substitution; the lower σ, the greater the consumers' taste for variety. The exponents in the utility function $1 - \mu$ and μ indicate, respectively, the invariant shares of disposable income devoted to the agricultural and manufactured goods, with $0 < \mu < 1$.

One unit of L is required to produce one unit of the homogeneous agricultural good. We also assume that the so-called "non-full specialisation condition" holds; i.e. the agricultural sector in each region is not large enough to satisfy the demand of the overall economy and both regions produce the agricultural good (Baldwin et al. 2003, p. 72).

Each local public sector finances R&D activities aiming to increase productivity in the regional manufacturing sector. For one unit of such a public good, labelled C_{AG}, one unit of the agricultural good is used:

$$H = C_{AG} \tag{3}$$

R&D public expenditure is financed by income taxation under a balanced government budget constraint.

Manufacturing is modelled as a Dixit-Stiglitz monopolistically competitive sector. Increasing returns prevail: each manufacturer requires a fixed input of α units of capital to operate and has a constant labour requirement β for each unit of production.

We assume that α depends (negatively) upon R&D activities financed by the local public sector. Moreover, fixed cost reduction may also result from the use of new ideas spilled over from the other region.[5] More formally, we assume that the provision of public goods affects the capital input requirement according to the

[4] In this section, we differentiate notation between regions 1 and 2 only when necessary. Moreover, we introduce the notation concerning the time sequence only in the next section.

[5] For simplicity, we assume that the provision of public goods only affects the fixed input. For the case in which the provision of productivity enhancing public goods affects the variable input requirement β, see Commendatore et al. (2008b, 2009). In a more general framework, one could assume that public goods affect both fixed and variable factor productivity (see, for instance, the Footloose Entrepreneur model by Brakman et al. 2008).

following relationship:

$$\alpha_r = f(H_r, H_s) = \frac{1}{1 + A(H_r + \theta H_s)} \tag{4}$$

where $r, s = 1, 2, r \neq s, A > 0$ and θ measures the ability of firms located in region r to absorb knowledge created in region s. Hence, the term $(H_r + \theta H_s)$ defines "global knowledge" potentially available to firms located in region r and the parameter $0 \leq \theta \leq 1$ measures the degree to which R&D activities undertaken in region s affect capital productivity in region r via knowledge diffusion from the former to the latter.[6] Imposing $\theta = 1$ implies that knowledge spillovers are perfectly global, i.e. the impact of knowledge on productivity is independent of where knowledge is originated and firms can benefit to the same extent from public expenditure on R&D regardless of their location. When $0 \leq \theta < 1$, instead, a firm's capacity to take advantage of knowledge is affected by its location (the extreme case $\theta = 0$ representing "perfectly local spillovers", with no knowledge transfer from one region to the other).[7]

Given consumers' preference for variety and increasing returns, a firm will always produce a variety different from those produced by other firms. Furthermore, denoting by λ the share of capital located and used in region 1 and considering that α_r units of capital are required for each manufacturing firm in region r ($r = 1, 2$), the number of varieties produced in region r corresponds to

$$n_1 = \frac{\lambda K}{\alpha_1} \text{ and } n_2 = \frac{(1-\lambda)K}{\alpha_2} \tag{5}$$

The total number of firms/varieties is

$$n = n_1 + n_2 = \frac{\alpha_2 \lambda + \alpha_1 (1-\lambda)}{\alpha_1 \alpha_2} K \tag{6}$$

Equations (4), (5) and (6) imply that the total number of firms and, for $0 < \lambda < 1$, the number of firms located in each region depend on the local provision of public goods.

---

6 For a similar concept, applied to "technological spillovers", see Baldwin and Martin (2004).

7 The driving forces of knowledge diffusion across borders are not studied here. A possible extension to our model, that we leave for future work, would be to endogenize the degree of international knowledge spillovers, with θ being a function of some concept of distance. The empirical evidence provided by Jaffe et al. (1993), for instance, would suggest imposing a linear (negative) relationship between θ and geographical distance (knowledge spillovers are mainly of the local type). On the other hand, using the concept of "cognitive" distance (Nooteboom 2000) one should impose a non-linear relationship between the former and θ in order to have effective knowledge spillovers, as agents (or regions) exchanging knowledge should have a sufficient level of common knowledge in order to communicate (i.e. a sufficiently small cognitive distance) and a sufficient level of heterogenous knowledge (i.e. sufficiently large cognitive distance) in order to have non-redundant transfers of knowledge. Furthermore, increasing values of θ could capture a higher "technological" distance dividing firms located in the two regions, implying that knowledge becomes more and more specific to local production as such a distance increases.

Finally, it is assumed that the agricultural good is traded costlessly, while transport costs for manufacturers take an iceberg form (Samuelson 1954): when 1 unit is shipped only a fraction $1/T$ arrives at the destination, where $T \geq 1$. Following Baldwin et al. (2003), we introduce the "trade freeness" parameter $\phi \equiv T^{1-\sigma}$; where $0 < \phi \leq 1$, with $\phi = 1$ corresponding to no trade cost ($T = 1$) and with $\phi \to 0$ corresponding to trade cost becoming prohibitive ($T \to \infty$).

3 Short-run General Equilibrium

In this section, we explicitly introduce time by characterising a short-run general equilibrium. In period t, a short-run general equilibrium is contingent on the given spatial allocation of capital, λ_t. As λ_t changes through time, the short-run equilibrium changes accordingly. In the agricultural market, short-run equilibrium is instantaneously established. With perfect competition agricultural profits are zero and the equilibrium nominal wage of workers in period t is equal to the price of the agricultural good. Moreover, in the absence of transport costs, the agricultural price is identical in both regions; it follows that nominal wages are also equalised across regions. We take this wage/agricultural price as the *numéraire*. Under the assumption of identical behaviour, each firm sets the same local (mill) price p using the Dixit-Stiglitz pricing rule. Given that the wage is 1, the local price of every variety is:

$$p = \frac{\sigma}{\sigma - 1}\beta \tag{7}$$

Taking into account transport costs, the effective price paid by consumers for one unit of a variety produced in the other region is pT.

Short-run general equilibrium in period t requires that each manufacturer meets the demand for its variety.[8] For a variety produced in region r:

$$x_{r,t} = d_{r,t} \tag{8}$$

where $x_{r,t}$ is the output for each manufacturing firms in region r and $d_{r,t}$ is the demand for that firm's variety. From (7), the short-run return to a unit of capital in region r is:

$$\pi_{r,t} = \frac{p_r x_{r,t} - \beta x_{r,t}}{\alpha_r} = \frac{p_r x_{r,t}}{\alpha_r \sigma} = \frac{\beta}{\alpha_r(\sigma - 1)} x_{r,t} \tag{9}$$

Consumers face regional manufacturing price indices given by:

$$P_{1,t} = (n_{1,t} p^{1-\sigma} + n_{2,t} p^{1-\sigma} T^{1-\sigma})^{\frac{1}{1-\sigma}} = \left[\frac{\lambda_t}{\alpha_1} + \frac{(1-\lambda_t)\phi}{\alpha_2}\right]^{\frac{1}{1-\sigma}} K^{\frac{1}{1-\sigma}} p$$

[8] As a result of Walras' Law, equilibrium in all product markets implies equilibrium in the regional labour markets.

$$P_{2,t} = (n_{1,t}p^{1-\sigma}T^{1-\sigma}+n_{2,t}p^{1-\sigma})^{\frac{1}{1-\sigma}} = \left[\frac{\lambda_t\phi}{\alpha_1}+\frac{(1-\lambda_t)}{\alpha_2}\right]^{\frac{1}{1-\sigma}} K^{\frac{1}{1-\sigma}} p \quad (10)$$

Consumption per variety in each region is:

$$\begin{aligned} d_{1,t} &= (M_1 P_{1,t}^{\sigma-1} + M_2 P_{2,t}^{\sigma-1}\phi)p^{-\sigma} \\ d_{2,t} &= (M_1 P_{1,t}^{\sigma-1}\phi + M_2 P_{2,t}^{\sigma-1})p^{-\sigma} \end{aligned} \quad (11)$$

M_r denotes the expenditure on manufactured goods in region r; $M = M_1 + M_2$ defines the world expenditure on manufactures and $s_E = M_1/M$ its regional split. As we will see below, M_r, M and s_E are independent of λ_t. From (8), (10) and (11), it follows:

$$\begin{aligned} x_{1,t} = d_{1,t} &= \left[\frac{s_E}{\frac{\lambda_t}{\alpha_1}+\frac{(1-\lambda_t)\phi}{\alpha_2}} + \frac{(1-s_E)\phi}{\frac{\lambda_t\phi}{\alpha_1}+\frac{(1-\lambda_t)}{\alpha_2}}\right]\frac{1}{p}\frac{M}{K} \\ x_{2,t} = d_{2,t} &= \left[\frac{s_E\phi}{\frac{\lambda_t}{\alpha_1}+\frac{(1-\lambda_t)\phi}{\alpha_2}} + \frac{1-s_E}{\frac{\lambda_t\phi}{\alpha_1}+\frac{(1-\lambda_t)}{\alpha_2}}\right]\frac{1}{p}\frac{M}{K} \end{aligned} \quad (12)$$

Therefore, from (9), the short-run equilibrium return to a unit of capital in region r is:

$$\begin{aligned} \pi_{1,t} &= \left[\frac{s_E a}{\lambda_t a + (1-\lambda_t)\phi} + \frac{(1-s_E)\phi a}{\lambda_t a\phi + (1-\lambda_t)}\right]\frac{1}{\sigma}\frac{M}{K} \\ \pi_{2,t} &= \left[\frac{s_E\phi}{\lambda_t a + (1-\lambda_t)\phi} + \frac{1-s_E}{\lambda_t a\phi + (1-\lambda_t)}\right]\frac{1}{\sigma}\frac{M}{K} \end{aligned} \quad (13)$$

where $a \equiv \alpha_2 / \alpha_1$ in order to simplify the notation. Moreover, regional and world capital incomes, $\Pi_{r,t}$ and Π respectively, are given by

$$\Pi_{1,t} = \lambda_t K\pi_{1,t} \quad \Pi_{2,t} = (1-\lambda_t)K\pi_{2,t} \quad \Pi = \frac{M}{\sigma} \quad (14)$$

(for the latter use (13)) and world gross income is given by $Y = L + \frac{M}{\sigma}$.

Denoting by H_r the level of public goods provided in region r, the overall provision of public goods corresponds to $\Sigma H = H_1 + H_2$. Moreover, considering that one unit of the public good has a unitary cost (the agricultural commodity is the *numéraire*), the regional government balance budget constraint implies that the tax burden TB_r for region r is

$$TB_r = H_r \quad (15)$$

Regional expenditure on manufactured goods is therefore given as

$$M_r = \mu\left(\frac{L+\Pi}{2} - TB_r\right) = \mu\left(\frac{L+\frac{M}{\sigma}}{2} - H_r\right) \quad (16)$$

and the overall expenditure on manufactured goods is $M = M_1 + M_2 = \mu\big(L + \frac{M}{\sigma} - \Sigma H\big)$. Therefore,

$$M = \frac{\mu\sigma}{\sigma - \mu}(L - \Sigma H) \tag{17}$$

and its regional split is

$$s_E = \frac{1}{2}\left(1 - \frac{\sigma - \mu}{\sigma}\frac{H_1 - H_2}{L - \Sigma H}\right) \tag{18}$$

where, to have $0 < s_E < 1$ it must be $H_1^{\min} < H_1 < H_1^{\max}$, where $H_1^{\max,\min} = H_2 \pm \frac{\sigma}{\sigma - \mu}(L - \Sigma H)$.

From (18), it follows

$$s_E \leq (>)\frac{1}{2} \qquad \textit{for} \qquad H_1 \geq (<)H_2 \tag{19}$$

That is, given the balanced government budget constraint, if public expenditure and the corresponding provision of public goods is larger in region 1 than in region 2, then the after-tax expenditure share for manufactured goods will be smaller in region 1 than in region 2.

Finally, (13), (17) and (18) give the short-run equilibrium regional returns to one unit of capital, $\pi_{r,t}$. The relative profitability of capital

$$\frac{\pi_{1,t}}{\pi_{2,t}} = R(\lambda_t) = a\frac{s_E[\lambda_t a\phi + (1 - \lambda_t)] + (1 - s_E)\phi[\lambda_t a + (1 - \lambda_t)\phi]}{s_E\phi[\lambda_t a\phi + (1 - \lambda_t)] + (1 - s_E)[\lambda_t a + (1 - \lambda_t)\phi]} \tag{20}$$

is crucial for the subsequent dynamic analysis.[9]

4 Capital Movements and the Complete Dynamic Model

In a FC model, the representative capitalist does not move herself but allocates the physical capital she owns between the regions and repatriates the income. Her incentive to move capital from one region to the other is based on *real net* capital income, taking as given the level of publicly provided goods. Since we assumed taxation according to the residence principle, capital income is taxed and spent in the home region of the capital owners and the relevant tax rate and price index for calculating *real net* capital income are the ones at home, irrespective of the regional capital allocation. Therefore, location choices based on *real net* income and on *nominal gross* income are equivalent.

[9] Note that, for a constant s_E, $R(\lambda_t)$ has a negative slope, that is, $\partial R(\lambda_t) / \partial \lambda_t < 0$. This is known as "competition" effect: a higher λ_t increases the competition in region 1 and therefore reduces relative profitability.

The central dynamic process follows the "replicator dynamics", widely used in evolutionary economics and evolutionary game theory (see e.g. Weibull 1997; see also Fujita et al. 1999). At the transition between period t and period $t+1$, the representative capitalist modifies the share of physical capital in region 1 – and, consequently, in region 2 – in response to a discrepancy in period t between the rate of profit in region 1 and the average rate of profit, given by $\lambda_t \pi_{1,t} + (1-\lambda_t)\pi_{2,t}$:

$$\frac{F(\lambda_t) - \lambda_t}{\lambda_t} = \gamma \frac{\pi_{1,t} - [\lambda_t \pi_{1,t} + (1-\lambda_t)\pi_{2,t}]}{\lambda_t \pi_{1,t} + (1-\lambda_t)\pi_{2,t}} \tag{21}$$

We refer to $\gamma > 0$ as the "speed" at which the representative capitalist alters the share of capital in region 1 in response to economic incentives. Equation (21) can be transformed into a law of motion depending upon the ratio in regional profitability, $R(\lambda_t)$:[10]

$$F(\lambda_t) = \lambda_t + \gamma \lambda_t (1-\lambda_t) \frac{R(\lambda_t) - 1}{\lambda_t R(\lambda_t) + (1-\lambda_t)} \tag{22}$$

Taking into account the constraint $0 \leq \lambda_{t+1} \leq 1$, the piecewise smooth one-dimensional map whereby λ_{t+1} is determined by λ_t is:

$$\lambda_{t+1} = Z(\lambda_t) = \begin{cases} 0 & if \quad F(\lambda_t) < 0 \\ F(\lambda_t) & if \ 0 \leq F(\lambda_t) \leq 1 \\ 1 & if \quad F(\lambda_t) > 1 \end{cases} \tag{23}$$

where λ_t in [0,1] implies that λ_{t+1} is in [0,1]. Fixed points for the dynamic system, which correspond to long-run equilibria, are defined by $Z(\lambda) = \lambda$. Core-periphery equilibria, i.e. $\lambda^{CP(0)} = 0$ or $\lambda^{CP(1)} = 1$, are boundary fixed points of the map (23).

A central question of the NEG concerns critical values for trade freeness (or for any other parameter) at which agglomeration in either region is sustainable. The so-called sustain points give conditions under which "the advantages created by such a concentration, should it somehow come into existence, [are] sufficient to maintain it" (Fujita et al. 1999, p. 9). Sustain points therefore specify conditions at which the boundary equilibria $\lambda^{CP(i)}$ (where $i = 0, 1$) become (at least locally) stable. These sustain point values are defined by $F'(\lambda^{CP(i)}) = 1$, with the latter indicating the derivative of the first return map (22). The latter condition can be reduced to $R(\lambda^{CP(i)}) = 1$ and solved for

$$\phi_{1,2}^{S(0)} = \frac{1 \pm \sqrt{1 - 4a^2 s_E (1-s_E)}}{2a(1-s_E)} \qquad \phi_{1,2}^{S(1)} = \frac{a \pm \sqrt{a^2 - 4 s_E (1-s_E)}}{2 s_E} \tag{24}$$

where $\phi^{S(i)}$ indicates the sustain point for $\lambda^{CP(i)}$. Tables 1 and 2 summarize the properties of the sustain values.

[10] Note that – from an analytic perspective – this specification is a good approximation to the discrete time counterpart of the process assumed by Puga (1998) in his core-periphery model.

Table 1 Properties of sustain values for the boundary equilibrium $\lambda^{CP(1)}$

Properties of $\phi_{1,2}^{S(1)}$		
$1 < a$	$\phi_{1,2}^{S(1)}$ are both real and $0 < \phi_2^{S(1)} < 1 < \phi_1^{S(1)}$ holds	
$2\sqrt{s_E(1-s_E)} < a < 1$	$\phi_{1,2}^{S(1)}$ are both real $s_E < 0.5$: $0 < \phi_2^{S(1)} < \phi_1^{S(1)} < 1$	$s_E > 0.5$: $1 < \phi_2^{S(1)} < \phi_1^{S(1)}$
$a < 2\sqrt{s_E(1-s_E)}$	No real $\phi_{1,2}^{S(1)}$ exists	

Table 2 Properties of sustain values for the boundary equilibrium $\lambda^{CP(0)}$

Properties of $\phi_{1,2}^{S(0)}$		
$1 < \frac{1}{2\sqrt{s_E(1-s_E)}} < a$	No real $\phi_{1,2}^{S(0)}$ exists	
$1 < a < \frac{1}{2\sqrt{s_E(1-s_E)}}$	$\phi_{1,2}^{S(0)}$ are both real $s_E < 0.5$: $0 < \phi_2^{S(0)} < \phi_1^{S(0)} < 1$	$s_E > 0.5$: $1 < \phi_2^{S(0)} < \phi_1^{S(0)}$
$a < 1$	$\phi_{1,2}^{S(0)}$ are both real and $0 < \phi_2^{S(0)} < 1 < \phi_1^{S(0)}$ holds	

Note that for $a = 1$ the sustain points reduce to $\phi^{S(0)} = \frac{s_E}{1-s_E}$ and $\phi^{S(1)} = \frac{1-s_E}{s_E}$. Moreover, for $s_E = \frac{1}{2}$, it holds that $\phi^{S(0)} = \phi^{S(1)} = 1$.

In addition to the boundary fixed points, an interior fixed point is given by

$$\lambda^* = \frac{1}{2} + \frac{a(1-\phi)(1+\phi)}{(a-\phi)(1-\phi a)}\left[s_E - \frac{1}{2}\frac{(a+\phi)(1-a\phi)}{(1-\phi)(1+\phi)a}\right] \tag{25}$$

Note that in order to have $0 < \lambda^* < 1$, the condition $\frac{\phi}{a}\frac{1-\phi a}{1-\phi^2} < s_E < \frac{1-\phi a}{1-\phi^2}$ must hold.

A second central question of the NEG concerns crucial values for the trade freeness (or for any other parameter) at which an (interior) equilibrium without spatial concentration "breaks up". This so-called break point gives conditions under which "small differences among locations [will] snowball into larger differences over time, so that the symmetry between identical locations will spontaneously break" (Fujita et al. 1999, p. 9). That is, it gives conditions under which an interior fixed point λ^* becomes (at least locally) unstable and the dynamics is attracted to one of the boundary equilibria. Analytically, the break point is defined by $F'(\lambda^*) = 1$. In our model, the break point ϕ^B arises when the interior fixed point coincides with one of the boundary fixed points and it is equal to the corresponding sustain point. At this value of the trade freeness a transcritical bifurcation occurs (see Alligood et al. 1997). Two fixed points (that is, the interior fixed point and one of the boundary fixed points) cross each other (with the interior fixed point leaving the admissible interval) and exchange stability (with the interior fixed point losing and one of the boundary fixed points gaining local stability).

In our model, the interior fixed point can also lose stability when the derivative of the map (22) evaluated at λ^* crosses – 1. If by varying one parameter (or more parameters simultaneously) the equality $F'(\lambda^*) = -1$ is violated, an attracting period two-cycle emerges through a flip bifurcation. The condition $F'(\lambda^*) = -1$ allows determining critical parameter values. More specifically, a flip bifurcation occurs, if the parameters satisfy the following condition:

$$\frac{\gamma-2}{\gamma} a s_E (1-s_E) \frac{(1-\phi^2)^2}{\phi} = s_E(a^2-1)(1-\phi^2) + (1-a\phi)^2 \qquad (26)$$

This equation cannot be solved explicitly. However, focusing on the trade freeness parameter we state the following proposition:

Proposition 1. *For $\gamma > 2$, (26) implicitly defines a unique bifurcation value ϕ^{bif} for the trade freeness parameter, with $0 < \phi^{bif} < 1$.*

For a proof, see the Appendix.

5 Perfectly Global Spillovers

In this section we explore the case of perfectly global spillovers. That is, knowledge diffuses from one region to the other without impediments and firms "absorptive capacity" is independent of where knowledge is originated. In terms of our notation, perfectly global spillovers correspond to the assumption $\theta = 1$, from which it follows $a = 1$.[11]

5.1 Comparative Statics

When spillovers are perfectly global, the interior equilibrium becomes

$$\lambda^* = \frac{1}{2} + \frac{1+\phi}{1-\phi}\left(s_E - \frac{1}{2}\right)$$

where $0 < \lambda^* < 1$ for $\frac{\phi}{1+\phi} < s_E < \frac{1}{1+\phi}$ and $\lambda^* \leq (>) \frac{1}{2}$ for $s_E \leq (>) \frac{1}{2}$.

Hence, for a given degree of trade freeness (i.e. economic integration), the equilibrium share of capital located in region 1 depends on the relative market size, i.e. on the expenditure share for manufactured goods. The lower (higher) the expenditure share in region 1, the lower (higher) the share of capital located in the same region.

Concerning the effect of ϕ on λ^*, it is positive, negative or nil depending on region 1's relative market size. That is, noticing that $\frac{\partial \lambda^*}{\partial \phi} = \frac{2}{(1-\phi)^2}\left(s_E - \frac{1}{2}\right)$, we have $\frac{\partial \lambda^*}{\partial \phi} \leq (>) 0$ for $s_E \leq (>) \frac{1}{2}$.

[11] Notice that with this assumption the model's structure becomes analytically equivalent to the one used in Commendatore and Kubin (2006).

Hence, a higher degree of trade freeness (i.e. an increase in ϕ) leads capital to relocate in the region with the relatively larger market. For instance, when region 2 concentrates a higher share of expenditure in manufactured goods as compared to region 1, the equilibrium value of the share of capital located and used in the former region will be positively affected by higher economic integration. That is, in the NEG terminology, lowering trade costs magnifies the market-access effect.

Turning to policy analysis, (15) and (18) imply that, given public expenditure in region 2, the relative market size of region 1, s_E, depends on the provision of public goods in the same region, H_1, and on the corresponding taxation. We call demand effect the impact of the provision of public goods in region 1 on λ^* via a change in the relative market size (see Commendatore et al. 2008, 2009). For the case of perfectly global spillovers, public expenditure in R&D can affect λ^* only through this channel:

$$\frac{\partial \lambda^*}{\partial H_1} = \frac{1+\phi}{1-\phi} \frac{\partial s_E}{\partial H_1}$$

where $\frac{\partial s_E}{\partial H_1} = -\frac{(\sigma-\mu)(L-2H_2)}{2\sigma(L-H)^2}$ is negative for $H_1 - H_2 > -(L - \Sigma H)$ and *a fortiori* for $H_1 > H_2$.

That is, when knowledge freely circulates across regions and firms in the two regions benefit to the same extent from new ideas, increasing public expenditure on R&D in region 1 leads to a reduction in the share of capital located in the same region. This occurs because the relative market size of region 1 shrinks due to higher taxation. In the terminology used in Commendatore et al. (2008b, 2009), higher productive public expenditure in one region only affects the regional distribution of capital via the demand effect.

On the other hand, the impact of public expenditure on R&D activities affects the regional distribution of firms through the combination of two effects. Indeed, as suggested by expression (5), H_1 may affect the equilibrium number of firms located in region 1, n_1^*, via the impact on λ^*, due only to the demand effect, and also via a change in the capital input coefficient. Intuitively, the equilibrium number of firms located in the region where public expenditure increases depends on the extent to which the share of local capital drops due to the demand effect as well as on the extent to which the productivity of local capital improves. The overall effect is given by

$$\frac{\partial n_1^*}{\partial H_1} = \frac{n_1^*}{H_1}\left(\frac{\partial \lambda^*}{\partial H_1}\frac{H_1}{\lambda^*} - \frac{\partial \alpha_1}{\partial H_1}\frac{H_1}{\alpha_1}\right) \tag{27}$$

From the above expression, it is possible to deduce the following: for $H_1 - H_2 < -(L - \Sigma H)$, the effect of public expenditure in region 1 on n_1^* is positive and the spatial distribution of firms changes in favour of region 1. On the other hand, for $H_1 - H_2 > -(L - H)$, the effect on the equilibrium number of firms located in region 1 will be positive (negative or nil) if the effect of H_1 on λ^* is not larger than (smaller than or equal to) its effect on α_1 (in percentage terms). That is, following an expansion in public expenditure in region 1, the number of firms located in the same region will increase only if the positive effect of more productive local capital will prevail over the negative effect of a smaller relative local market.

Given the assumption of $\theta = 1$, which implies $\alpha_1 = \alpha_2 = \alpha$, the opposite holds for n_2^*; whereas the effect on the total number of firms n^*, is always positive, $\frac{\partial n^*}{\partial H_1} = -\frac{n^*}{\alpha}\frac{\partial \alpha}{\partial H_1} > 0$.

Turning to the core-periphery equilibria, when manufacturing is agglomerated in one region, then the number of firms in that region is equal to K/α and zero in the other region. The total number of firms always increases with H_1.

5.2 Local Bifurcation Analysis

Next, we explore the local stability properties of the map (23) for the case of perfectly global spillovers. Given the assumption $\theta = 1$, we simplify (26) as follows

$$\frac{\gamma - 2}{\gamma} s_E (1 - s_E) = \frac{\phi}{(1 + \phi)^2} \tag{28}$$

As stated above (see also Commendatore and Kubin 2006), this equation allows to identify a unique bifurcation value ϕ^{bif} for the trade freeness parameter within the range $0 < \phi < 1$. It is easy to verify that starting from the symmetric case, $H_1 = H_2$ ($s_E = 1/2$), ϕ^{bif} is reduced by increasing H_1 (or H_2).

Figures 1a,b present bifurcation diagrams showing the impact of trade freeness, ϕ, on the long-run allocation of capital λ_t for global spillovers and for (a) $H_1 = H_2 = 0$ and (b) $H_1 = 0.05$ and $H_2 = 0$. When $H_1 = H_2 = 0$, regions 1 and 2 are symmetric, i.e. the market for manufactured goods is equally split between the regions, i.e. $s_E = 1/2$. The map $Z(\lambda_t)$ exhibits symmetric properties.[12] Figure 1a shows that the interior equilibrium, $\lambda^* = \frac{1}{2}$, is stable over the range $\phi^{bif} < \phi < 1$. At $\phi = 1$, the interior fixed point loses stability via a transcritical bifurcation. Since for our model break and sustain points are equal, corresponding to $\phi^{S(0)} = \phi^{S(1)} = 1$, it is also true that the two boundary equilibria $\lambda^{CP(0)}$ and $\lambda^{CP(1)}$ are always (locally) unstable for $0 < \phi < 1$.[13] For the symmetric case, the bifurcation value of trade freeness only depends on the speed at which capital owners react at the economic incentive, with ϕ^{bif} increasing in γ. Below ϕ^{bif} a period doubling bifurcation route to chaos takes place. The time path exhibits locally attracting cycles of any period or even chaotic cycles (not all of them visible in Fig. 1: symmetric counterparts co-existing with a different basin of attraction) with an ever increasing volatility of the regional shares of capital. Below ϕ^A the volatility that results for relatively high trade costs leads to the concentration of all manufacturing activity in one of the regions. Given the mobility hypothesis specified in expression (23), the share of capital does not longer change once one of the boundary values

---

[12] For a detailed account of the symmetric properties of a FC model with no government sector, see Commendatore et al. (2007).

[13] Since they coincide with the sustain points, we do not mark the break points in the Figures.

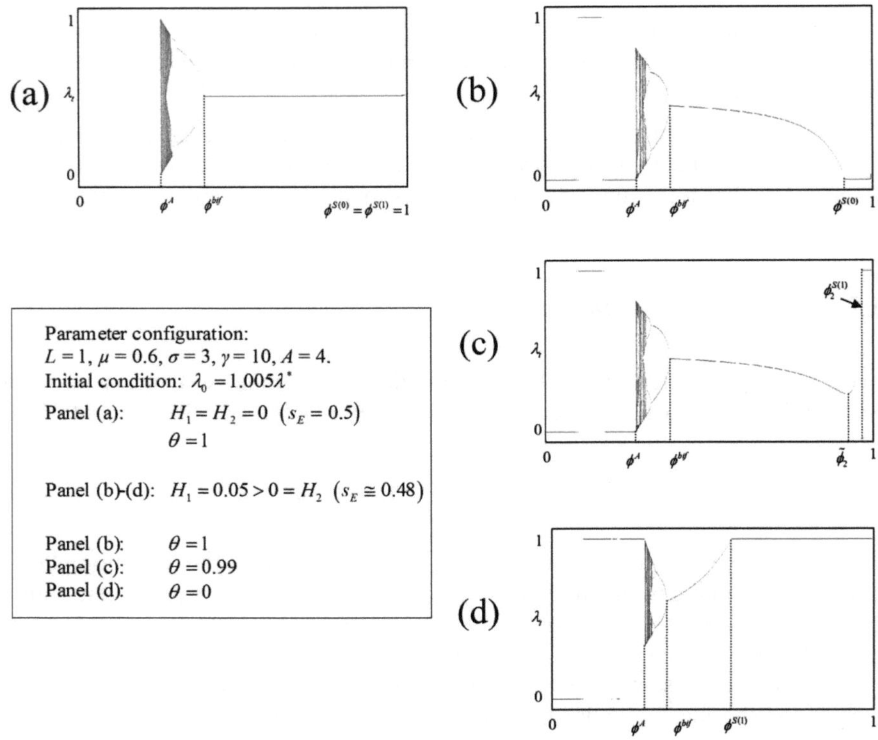

Fig. 1 Bifurcation on trade freeness with different values of R&D public expenditure and knowledge spillovers

0 or 1 is reached. A core-periphery outcome emerges even though both boundary fixed points are locally unstable.

When $H_1 > H_2$ (as assumed for Fig. 1b) the regions are asymmetric, with region 1 characterized by a smaller relative market size, $s_E < 1/2$. Comparing Fig. 1a with Fig. 1b, the following differences emerge. First, from the properties of the sustain points (see above), it follows:

$$\phi^{S(0)} = \frac{s_E}{1 - s_E} < 1 < \phi^{S(1)} = \frac{1 - s_E}{s_E}$$

The boundary equilibrium $\lambda^{CP(1)}$ is (locally) unstable for $0 < \phi \leq 1$. Moreover, in an economy highly integrated (i.e. low transport costs or high trade freeness), specifically for $\phi^{S(0)} < \phi \leq 1$, no interior fixed point exists in the interval (0, 1) and the boundary equilibrium $\lambda^{CP(0)}$ is (locally) stable. Second, trade freeness affects the value of the interior equilibrium. As ϕ falls below the sustain point value $\phi^{S(0)}$ a transcritical bifurcation occurs, $\lambda^{CP(0)}$ loses stability; the interior fixed point enters the interval (0, 1) and becomes locally stable for $\phi^{bif} < \phi < \phi^{S(0)}$. In this interval

λ^* increases as ϕ declines. The larger market size favors capital location in region 2, this effect being reduced as trade becomes less free. Finally, also the dynamic properties holding in the interval $0 < \phi < \phi^{bif}$ are altered: the flip bifurcation value ϕ^{bif} is smaller and fluctuations in the regional shares of capital are narrower. Below ϕ^A, due to the asymmetry of the map $Z(\lambda_t)$, the boundary equilibrium $\lambda^{CP(0)}$ emerges more often.

6 Local Spillovers

In this section, we study the case when knowledge produced in public research laboratories does not diffuse uniformly across space. This can occur because of its specificity to local production. Also, other impediments could be responsible for a lower absorptive capacity of foreign firms as compared to the one of firms located in the region where new ideas are originated. When spillovers are (at least partially) local we have $0 \leq \theta < 1$, implying $a \geq (<) 1$ for $H_1 \geq (<) H_2$.

6.1 Comparative Static Analysis

With local spillovers, the effects of trade freeness on the interior equilibrium, λ^*, given by (25), are summarised in the following proposition:

Proposition 2. *A) Let* $0 < s_E < \frac{1}{1+a^2} < \frac{1}{2}$, *then there exists a value of trade freeness,* $\tilde{\phi}_2$, *such that* λ^* *reaches a minimum at* $\phi = \tilde{\phi}_2$, *that is* $\lambda^*_{\min} = \lambda^*(\tilde{\phi}_2)$, *with* $0 < \tilde{\phi}_2 = \frac{a(1-2s_E)-(a^2-1)\sqrt{s_E(1-s_E)}}{a^2(1-s_E)-s_E} < \phi_2^{S(1)} < 1$ *and* $\lambda^*_{\min} < 1$. *Moreover,* $\lambda^*_{\min} > 0$ *if* $s_E < \frac{a-\sqrt{a-1}}{2a} < \frac{1}{1+a^2}$ *holds. B) Let* $s_E > \frac{1}{1+a^2} > \frac{1}{2}$, *then there exists a value of trade freeness,* $\tilde{\phi}_2$, *such that* λ^* *reaches a maximum at* $\phi = \tilde{\phi}_2$ *that is,* $\lambda^*_{\max} = \lambda(\tilde{\phi}_2)$, *with* $0 < \tilde{\phi}_2 < \phi_2^{S(0)} < 1$ *and* $\lambda^*_{\max} > 0$. *Moreover,* $\lambda^*_{\max} < 1$ *if* $s_E > \frac{1+\sqrt{1-a^2}}{2} > \frac{1}{1+a^2}$ *holds. For a proof, see the Appendix.*

Turning to policy analysis, the provision of public services in region 1 affects λ^* as follows:

$$\frac{\partial \lambda^*}{\partial H_1} = \frac{\phi}{(a-\phi)^2}\left[1 + \frac{(1-\phi^2)(a^2-1)}{(1-a\phi)^2} s_E\right]\frac{\partial a}{\partial H_1} + a\frac{1-\phi^2}{(a-\phi)(1-a\phi)}\frac{\partial s_E}{\partial H_1} \quad (29)$$

According to this expression, when spillovers are localised, it is possible to identify two effects that an increase in H_1 could exert on the location of the manufacturing sector: the demand effect and the productivity effect.

The demand effect is expressed by the second term on the right-hand side of (29) and, as before, its direction depends on the sign of $\partial s_E / \partial H_1$. The productivity effect is expressed by the first term on the right-hand side in (29) and is equal to zero for the case of perfectly global spillovers. According to the productivity effect, the provision of public goods in region 1 affects λ^* via its impact on capital productivity in region 1. Since the term in the square brackets is positive for $0 \leq s_E \leq 1$ and $\frac{\partial a}{\partial H_1} > 0$, the productivity effect is positive on λ^*. As stated in Commendatore et al. (2008b, 2009), the overall effect of H_1 on λ^* can be non monotonic (with at least an initial range of values of ϕ for which λ^* is increasing), depending on the relative strength of demand and productivity effects.

Taking into account (27), when the demand effect is larger than the productivity effect, $\partial\lambda^*/\partial H_1 < 0$, the results concerning the sign of the effect of H_1 on the number of firms located in region 1 (and in region 2) as derived in Sect. 5 for global spillovers carry over to the case of local spillovers. Instead, when $\partial\lambda^*/\partial H_1 > 0$, the relative number of firms in region 1 is always increasing in H_1; whereas it also increases in region 2 only if the effect of knowledge spillovers on capital productivity in this region is sufficiently strong to overcome the reduction in its share.

The effect of public expenditure in region 1 on the total number of firms corresponding to the interior equilibrium is

$$\frac{\partial n^*}{\partial H_1} = \frac{\partial \lambda^*}{\partial H_1}\left(\frac{\alpha_2 - \alpha_1}{\alpha_1 \alpha_2}\right) K - \left(\frac{n_1^*}{\alpha_1}\frac{\partial \alpha_1}{\partial H_1} + \frac{n_2^*}{\alpha_2}\frac{\partial \alpha_2}{\partial H_1}\right) \tag{30}$$

This effect can also be non monotonic, depending on the relative strength of the demand and productivity effects acting on n^* (simulations, not presented here, show that for some parameter values it initially increases with H_1 until a maximum value is reached and decreases thereafter).

Finally, turning to the core-periphery equilibria when manufacturing is agglomerated in region i the number of firms in this region is K / α_i and zero in the other region. The total number of firms always increases with H_1, except when the manufacturing sector is fully agglomerated in region 2 and spillovers are perfectly local.[14]

6.2 Local Bifurcation Analysis

Due to its analytical complexity, we proceed with numerical simulations in order to study the local stability properties of the map $Z(\lambda_t)$ for the case of local spillovers.

Figure 1c, d presents bifurcation diagrams showing the impact of trade freeness ϕ on the long-term regional allocation of capital λ_t for (c) $\theta = 0.99$ and (d) $\theta = 0$. From the properties of the sustain values for the boundary fixed points it follows,

[14] The case of perfectly local spillovers is also explored in Commendatore et al. (2008b, 2009). In those papers, however, the number of firms does not change since the provision of the public good does not affect capital productivity.

since $a > \frac{1}{2\sqrt{s_E(1-s_E)}} > 1$, that no real sustain value for $\lambda^{CP(0)}$ exists; moreover, the sustain value $\phi_2^{S(1)} < 1$ for $\lambda^{CP(1)}$ is visible in both diagrams.

Comparing Fig. 1c with Fig. 1b, we can notice that for $0 < \phi \leq \phi^{bif}$ the behaviour of λ_t is qualitatively similar. The most notable changes, following the slight reduction of the parameter θ from 1 to 0.99, occur within the interval $\phi^{bif} < \phi \leq 1$ where even though the interior fixed point decreases over the range $\phi^{bif} < \phi < \tilde{\phi}_2$ as in Fig. 1b, unlike that diagram it increases over the range $\tilde{\phi}_2 < \phi < \phi_2^{S(1)}$. With increasing economic integration, the larger productivity rise in region 1 – induced by a larger provision of public goods and by the exploitation of localised spillovers – overcomes the disadvantage of a smaller market size and favours agglomeration in this region. Moreover, further increasing ϕ, as the sustain point $\phi_2^{S(1)}$ is crossed, the interior fixed point undergoes a transcritical bifurcation, involving an exchange of stability with the boundary fixed point $\lambda^{CP(1)}$, and exists the (0, 1) interval. Unlike the case of perfectly global spillovers presented in Fig. 1b, strong economic integration determines a "near catastrophic" agglomeration of capital in region 1.[15]

Figure 1d presents the case of perfectly local spillovers, $\theta = 0$. We notice that as all productivity improvements occur in region 1, the sustain value of ϕ for the core-periphery equilibrium $\lambda^{CP(1)}$ (corresponding to full agglomeration in region 1) is much smaller than in Fig. 1c. Moreover, the interior equilibrium is always increasing with ϕ within the interval $\phi^{bif} < \phi < \phi_2^{S(1)}$. Finally, the crossing of the flip bifurcation point, which is shifted to the left, is followed by narrower fluctuations in the interval $\phi^A < \phi < \phi^{bif}$ and below ϕ^A, due to the asymmetry of the map $Z(\lambda_t)$, by a more frequent emergence of the boundary equilibrium $\lambda^{CP(1)}$.

7 Global Dynamics

So far, we have analysed the local dynamics around the interior fixed point. In this section, we investigate the properties of the global dynamics for both cases of perfectly global and local spillovers. In particular, since our map involves two boundary conditions, local properties around the interior fixed point do not necessarily hold on a global level. It is possible to analytically derive conditions on the parameters for which no, one or both boundary conditions impact upon the first return map. Since those conditions are not explicitly solvable, Fig. 2 – which uses five different values of the spillovers parameter θ –[16] illustrates parameter regions that imply different properties for the first return map.

---

[15] This phenomenon has been detected for the case of perfectly local spillovers in Commendatore et al. (2009).

[16] Figure 2 is also drawn for $\gamma = 10$, $A = 4$, $L = 1$, $\mu = 0.6$, $\sigma = 3$, $H_2 = 0$ and by implication for $H_1^{\max} = 0.556$.

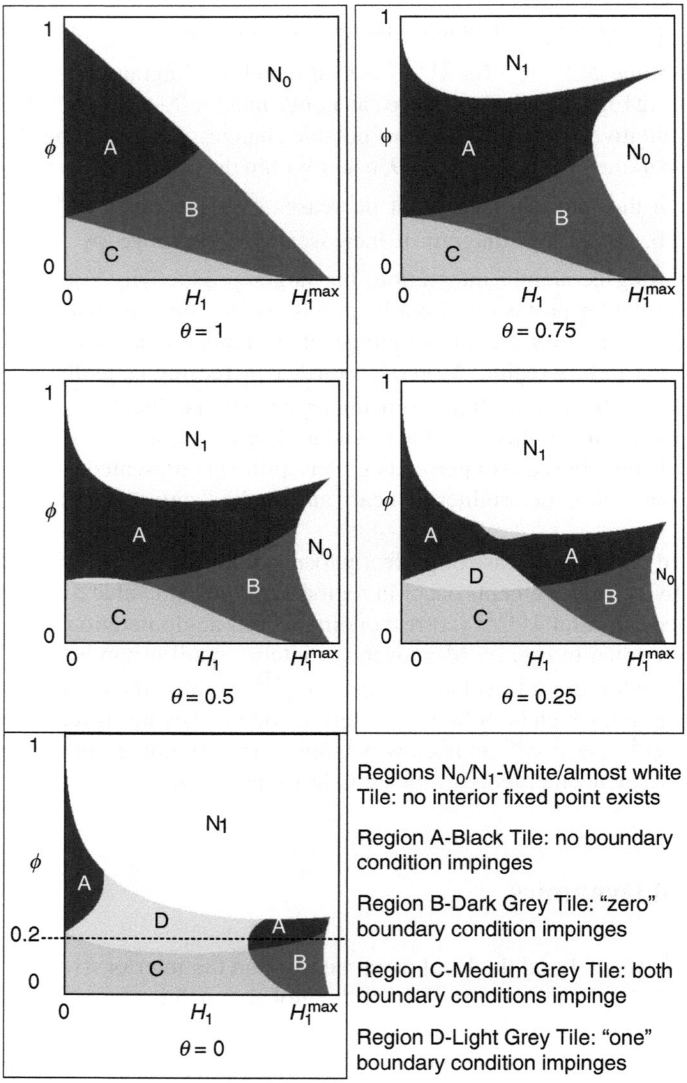

Fig. 2 Bifurcation on R&D public expenditure in Region 1 and trade freeness for different values of knowledge spillovers

In the regions N trade is sufficiently free, in particular the trade freeness parameter is greater than the respective sustain value and no interior fixed point exists. In region N_0 (N_1) the boundary fixed point $\lambda^{CP(0)}$ ($\lambda^{CP(1)}$) is locally and globally stable. The respective other boundary fixed point is locally and globally unstable. For all other parameter constellations, both boundary equilibria are locally unstable and an interior fixed point exists, which is locally stable as long as $\phi > \phi^{bif}$. It is particularly interesting to analyse parameter constellations for which the global dynamics is attracted to the (locally unstable) boundary fixed points. Decisive for that to occur is whether or not the boundary conditions impinge upon the dynamics, i.e. whether or not $0 \leq F(\lambda) \leq 1$.

In region A, no boundary condition affects the dynamics, i.e. $Min(F(\lambda)) \geq 0$ and $Max(F(\lambda)) \leq 1$ (see also Fig. 3, middle left panel for a typical first return map). Therefore, the two boundary fixed points, which are locally unstable, are also globally unstable and no time path starting at an arbitrary initial condition will be attracted to them. Either the interior fixed point or the periodic orbit (born after the flip bifurcation) is also globally stable – almost all initial conditions will be attracted to them.[17]

In region B only the lower boundary condition is binding, i.e. $Min(F(\lambda)) < 0$ and $Max(F(\lambda)) < 1$ (see Fig. 3, middle right panel for a representative first return map). In this region, the boundary fixed point $\lambda^{CP(0)}$ has a basin of attraction: initial conditions λ_0 for which $F(\lambda_0) \leq 0$ – i.e., initial conditions on the bold segment in the return map depicted in Fig. 3, middle right panel – and all pre-images of that range will be attracted to it. Some initial conditions will be attracted to a trapping set, which is constructed using $Max(F(\lambda))$ and its iterates (see Fig. 3, bottom right panel). Therefore, in region B both the boundary fixed point $\lambda^{CP(0)}$ and the interior fixed point have a basin of attraction. Figure 3 top left panel illustrates the respective basins of attraction.

Figure 3 is drawn for the same parameters as Fig. 2, in addition we assume $\phi = 0.2$ and $\theta = 0$. On the horizontal axes H_1 is varied between zero and its upper limit and the vertical axes represents possible values for the initial condition. A white (black) tile indicates that a time path emerging from the respective initial condition converges to the boundary fixed point $\lambda^{CP(0)}$ ($\lambda^{CP(1)}$); and a grey tile indicates that the time path is attracted to an interior fixed point. For region B this figure clearly shows the co-existing basins of attraction for $\lambda^{CP(0)}$ (white tiles) and for the interior fixed point (grey tiles).[18]

In Region C, both boundary conditions are binding, i.e. $Min(F(\lambda)) < 0$ and $Max(F(\lambda)) > 1$. Now almost all initial conditions are attracted to either of the boundary fixed points:[19] initial conditions λ_0 for which $F(\lambda_0) > 1$ ($F(\lambda_0) < 0$) and all pre-images of that range will be attracted to the boundary fixed point $\lambda^{CP(1)}$ ($\lambda^{CP(0)}$). Figure 3, top left panel and top right panel (which is an enlargement of the former) illustrate the highly fractal structure of the two basins of attraction: a white (black) tile indicates that for the respective λ_0–H_1 combination the time path converges to the boundary fixed point $\lambda^{CP(0)}$ ($\lambda^{CP(1)}$).

Region D is a mirror image of region B: now only the upper boundary condition is binding, i.e. $Min(F(\lambda)) > 0$ and $Max(F(\lambda)) > 1$. In this region, the boundary fixed

---

[17] There might exist unstable fixed points (with different periodicity); these and their pre-images are not attracted to the stable fixed points.

[18] As noted before, for $\theta = 1$ the model coincides from an analytical point of view with the one analysed in Commendatore and Kubin (2006), where this case is studied in greater detail.

[19] Exceptions are again the possibly existing unstable fixed points (with different periodicity) and their pre-images.

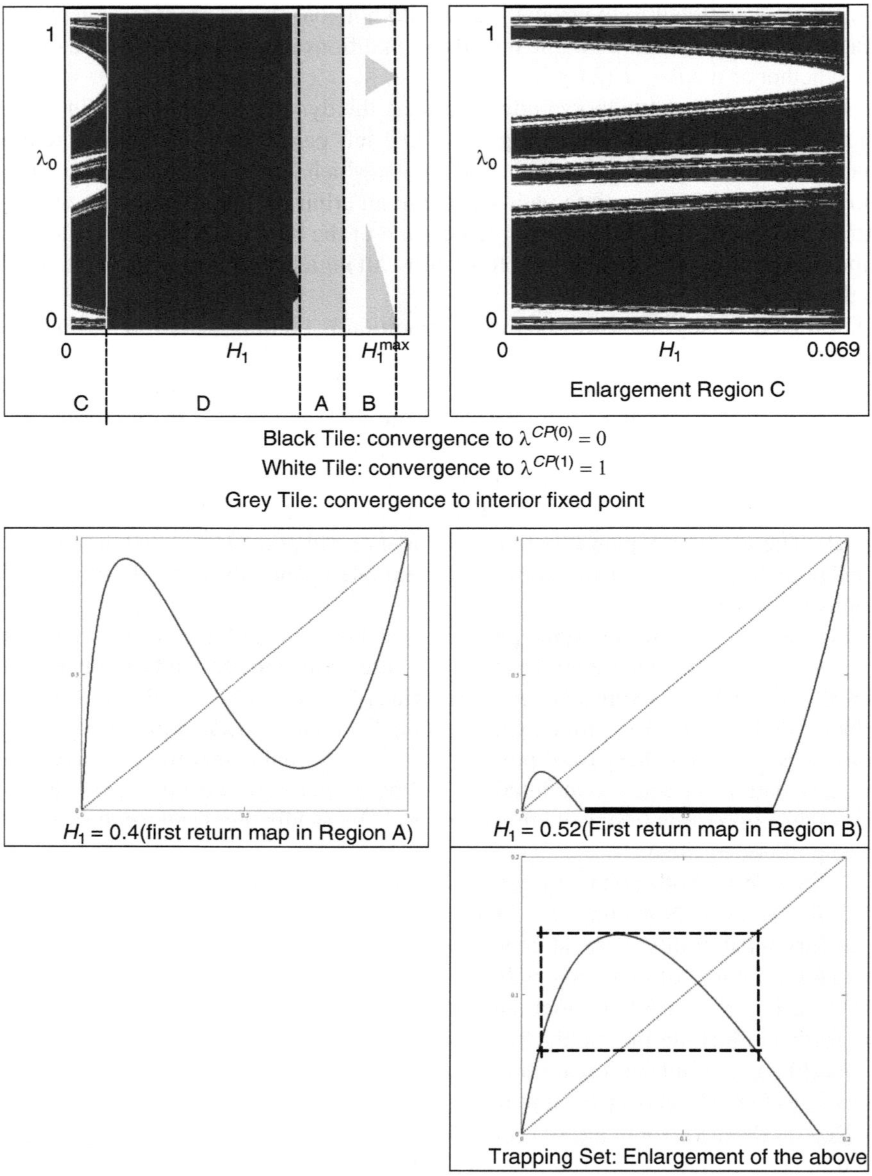

Fig. 3 Upper panels: Bifurcation on R&D public expenditure in Region 1 and the share of capital located in Region 1 at $t = 0$. Lower panels: Plot of the first return map for different values of R&D public expenditure in Region 1

point $\lambda^{CP(1)}$ has a basin of attraction: initial conditions λ_0 for which $F(\lambda_0) \geq 1$ and all pre-images of that range will be attracted to it. Some initial conditions will be attracted to a trapping set, which is constructed using $Min(F(\lambda))$ and its iterates. Therefore, in this region both the boundary fixed point $\lambda^{CP(1)}$ and the interior fixed point have a basin of attraction. Figure 3 top left panel illustrates the respective basins of attraction and shows the co-existing basins of attraction for $\lambda^{CP(1)}$ (black tiles) and for the interior fixed point (grey tiles).

To conclude, whether both regions co-exist or whether the agglomeration ends up in region 1 or 2 can be – depending upon the parameters – highly volatile. Of particular interest is the role of the spillover intensity θ. There are no analytic results available. However, based upon inspection of Fig. 2 it can be surmised that a lower θ, i.e. more localized spillovers, tends to have the following effects: region A shrinks – the parameter space for which both regions co-exist shrinks; regions N_1 and D increase – therefore the parameter space for which manufacturing will be agglomerated in region 1 increases; finally, Regions N_0 and B shrink – the parameter space for which manufacturing will be agglomerated in region 2 shrinks. Therefore, a lower θ, i.e. more localized spillovers, apparently tends to favour agglomeration; and it tends to favour agglomeration in region 1 (consistent with the fact that in the case under consideration $H_1 > H_2$).

8 Conclusions

Building upon our previous work, we have delivered further insights on the channels through which productive public expenditure might influence industrial location. We have presented a FC model with a public sector involved in R&D activities which have the effect of enhancing productivity in the manufacturing sector.

Results include (a) comparative statics on the impact of public policy on industrial location; (b) local dynamic analysis on the impact of trade freeness on the long-term regional allocation of capital; (c) global dynamic analysis.

The comparative static analysis of the impact of R&D public expenditure on industrial location is delivered under the alternative assumptions of global and (partially/perfectly) local knowledge spillovers. Results are in line with our previous work, as the demand and productivity effects occur. When knowledge spillovers are global, public policy can only affect the regional distribution of capital via the demand effect, whereas its impact on the spatial distribution of firms also depends upon the productivity effect. On the other hand, assuming either partially or perfectly local knowledge spillovers, the overall effect of an increase in R&D public expenditure on the spatial distribution of capital depends on the relative strength of the demand and productivity effects. The same holds for the spatial distribution of firms.

We have used bifurcation diagrams to depict the local stability properties of industrial location equilibria for both cases of global and local knowledge spillovers. Studying the impact of trade freeness on the long-term regional allocation of capital, we concluded that when knowledge spillovers are at least partially local, strong economic integration leads to a "near catastrophic" agglomeration of capital in the region with the relatively higher R&D effort.

Finally, global dynamic analysis has shown the crucial role played by the degree of knowledge diffusion across regions in determining the global stability properties of industrial location equilibria. Although not supported by analytical results, graphical analysis of the basins of attraction of boundary fixed points led us to conclude that equilibria are highly volatile depending on the spillover intensity parameter θ. In particular, more localized spillovers tend to favour agglomeration in the region with a higher R&D effort.

Acknowledgements We would like to thank Theresa Grafeneder-Weissteiner and Elisabetta Michetti for valuable comments. The usual caveat applies.

Appendix

Proof of Proposition 1

Consider that (a) the left-hand side defines a function of ϕ, $f(\phi) = \frac{\gamma-2}{\gamma} a s_E (1 - s_E)\frac{(1-\phi^2)^2}{\phi}$ which is positive for $\gamma > 2$ and $\phi \neq 1$, it has an asymptote at 0 and it is decreasing over the range $0 < \phi < 1$ (whereas it is increasing for $\phi > 1$ and tangent to the horizontal axis at $\phi = 1$); (b) the right-hand side, instead, defines a quadratic function of ϕ, $g(\phi) = s_E(a^2-1)(1-\phi^2)+(1-a\phi)^2$. It has a minimum at $\bar{\phi} \equiv \frac{a}{a^2(1-s_E)+s_E}$, that is, $g(\bar{\phi}) = \frac{s_E(1-s_E)(a^2-1)^2}{a^2(1-s_E)+s_E}$, which is always positive (zero) for $a \neq 1$ ($a = 1$) but finite (given the constraints over the parameters), $g(\bar{\phi}) \geq 0$. Given the behavior of $f(\phi)$ and $g(\phi)$, there exists a unique flip bifurcation value ϕ^{bif} in the range $0 < \phi < 1$. Finally, notice that for $a = 1$, (25) is also satisfied for $\phi = 1$.

Proof of Proposition 2

Statement A) corresponds to the case $a > 1 (H_1 > H_2)$. The derivative $\frac{\partial \lambda^*}{\partial \phi}$ is equal to zero at $\tilde{\phi}_{1,2} = \frac{a(1-2s_E)\pm(a^2-1)\sqrt{s_E(1-s_E)}}{a^2(1-s_E)-s_E}$. For $s_E < \frac{1}{1+a^2}$, we have that $0 < \tilde{\phi}_2 < \tilde{\phi}_1$. Since $a > 1$, it is also true that $\tilde{\phi}_2 < \phi_2^{S(1)}$. This can be verified considering that $\tilde{\phi}_2$ is a monotonically decreasing function of s_E within the range $0 \leq s_E < \frac{a^2}{1+a^2}$, with $\frac{a^2}{1+a^2} > \frac{1}{1+a}$; that also $\phi_2^{S(1)}$ is a monotonically decreasing function of s_E for $s_E > 0$; and that the former function lies always below the latter within the range $0 < s_E < \frac{a}{1+a^2}$. It follows that, $\lambda(\tilde{\phi}_2)$ is a minimum since the interior equilibrium is equal to s_E at $\phi = 0$ and it is equal to 1 at $\phi = \phi_2^{S(1)}$. Finally, λ^* does not cut the 0 line as long as $s_E < \frac{a-\sqrt{a-1}}{2a} < \frac{1}{1+a^2}$, condition which ensures that there is no real sustain point for $\lambda^{CP(0)} = 0$ (see Table 1). Statement B), instead, corresponds to the case $a < 1$ ($H_1 > H_2$). As before we have that $0 < \tilde{\phi}_2 < \tilde{\phi}_1$. Since $a < 1$, it is also true that $\tilde{\phi}_2 < \phi_2^{S(0)}$. This can be verified considering that $\tilde{\phi}_2$ is a monotonically increasing function of s_E within the range $\frac{a}{1+a^2} < s_E \leq 1$, with $\frac{a}{1+a^2} < \frac{1}{1+a}$; that also $\phi_2^{S(0)}$ is a monotonically increasing function of s_E for $s_E > 0$; and that the former function lies always below the latter within the range $\frac{a}{1+a^2} < s_E < 1$. It follows that, $\lambda(\tilde{\phi}_2)$ is a maximum since the interior equilibrium is equal to s_E at $\phi = 0$ and it is equal to 0 at $\phi = \phi_2^{S(0)}$. Finally, λ^* does not cut the 1 line as long as $s_E > \frac{1+\sqrt{1-a^2}}{2} > \frac{1}{1+a^2}$, condition which ensures that there is no real sustain point for $\lambda^{CP(1)} = 1$ (see Table 2).

References

Alligood, K. T., Sauer, T. D., & Yorke, J. A. (1997). *Chaos: An introduction to dynamical systems.* New York: Springer.

Baldwin, R. E., & Martin, P. (2004). Agglomeration and regional growth. In J. V. Henderson & J. F. Thisse (Eds.), *Handbook of regional and Urban economics: Volume 4 cities and geography* (pp. 2671–2711). Elsevier North-Holland: New York.

Baldwin, R. E., Forslid, R., Martin, P., Ottaviano, G., & Robert-Nicoud, F. (2003). *Economic geography and public policy*. Princeton, NJ: Princeton University Press.

Bischi, G. I., & Lamantia, F. (2002). Nonlinear duopoly games with positive cost externalities due to spillovers. *Chaos, Solitons and Fractals*, 13, 701–721.

Bischi, G. I., Dawid, H., & Kopel, M. (2003a). Gaining the competitive edge using internal and external spillovers: A dynamic analysis. *Journal of Economic Dynamics and Control*, 27, 2171–2193.

Bischi, G. I., Dawid, H., & Kopel, M. (2003b). Spillover effects and the evolution of firm clusters. *Journal of Economic Behavior and Organization*, 50, 47–75.

Brakman, S., Garretsen, H., & van Marrewijk, C. (2008). Agglomeration and government spending. In S. Brakman & H. Garretsen (Eds.), *Foreign direct investment and the multinational enterprise* (pp. 89–116). Cambridge, MA: MIT.

Commendatore, P., & Kubin I. (2006). *Taxation and agglomeration*. Working paper 06–08, Center for Nonlinear Dynamics in Economics and Finance (CeNDEF), University of Amsterdam.

Commendatore, P., Currie, M., & Kubin, I. (2007). Chaotic footloose capital. *Nonlinear Dynamics, Psychology, and Life Sciences*, 11(2), 267–289.

Commendatore, P., Currie, M., & Kubin, I. (2008a). Footloose entrepreneurs, taxes and subsidies. *Spatial Economic Analysis* 3, 115–141.

Commendatore, P., Kubin, I., & Petraglia, C. (2008b). Productive public expenditure in a new economic geography model. *Économie Internationale/International Economics*, 114(2), 133–160.

Commendatore, P., Kubin, I., & Petraglia, C. (2009). Footloose capital and productive public services. In N. Salvadori, P. Commendatore, & M. Tamberi (Eds.), *Geography, structural change and economic development: Theory and empirics* (pp. 3–28). Aldershot: Edward Elgar.

Duranton, G., & Puga, D. (2004). Micro-foundations of urban agglomeration economies. In J. Henderson & J. F. Thisse (Eds.), *Handbook of Urban and regional economics* (Vol. 4). North-Holland: Amsterdam.

Fujita, M., Krugman, P. R., & Venables, A. (1999). *The spatial economy: Cities, regions and international trade*. Cambridge, MA: MIT.

Jaffe, A., Traitenberg, M., & Henderson, R. (1993). Geographical localization of knowledge spillovers as evidenced by patent citations. *Quarterly Journal of Economics*, 108, 577–598.

Krugman, P. R. (1991). Increasing returns and economic geography. *Journal of Political Economy*, 99, 483–499.

Martin, P., & Rogers, C. A. (1995). Industrial location and public infrastructure. *Journal of International Economics*, 39, 335–351.

Mees, A. I. (1975). The revival of cities in medieval Europe. *Regional Science and Urban Economics*, 5, 403–425.

Neary, P. J. (2001). On hype and hyperbolas: Introducing the new economic geography. *Journal of Economic Literature*, 39, 536–561.

Nooteboom, B. (2000). *Learning and innovation in organizations and sconomies*. Oxford: Oxford University Press.

Puga, D. (1998). Urbanization patterns: European versus less developed countries. *Journal of Regional Science*, 38, 231–252.

Puu, T. (1981). Catastrophic structural change in a continuous regional model. *Regional Science and Urban Economics*, 11, 317–333.

Sachs, J. D., & McCord, G. C. (2008). Regional development, geography of. In S. N. Durlauf & E. Blume (Eds.), *The new palgrave dictionary of economics*, Second Edition. UK: Palgrave Macmillan. The New Palgrave Dictionary of Economics Online. Palgrave Macmillan. 08 April 2009 <http://www.dictionaryofeconomics.com/article?id=pde2008_R000077> doi: 10.1057/9780230226203.1413.

Samuelson, P. (1954). The transfer problem and transport costs, II: Analysis of effects of trade impediments. *Economic Journal*, 64, 264–289.

Venables, A. J. (2008). New economic geography. In S. N. Durlauf & E. Blume (Eds.), *The new palgrave dictionary of economics*, Second Edition. UK: Palgrave Macmillan. The New Palgrave Dictionary of Economics Online. Palgrave Macmillan. 08 April 2009 http://www.dictionaryofeconomics.com/article?id=pde2008_E000247 doi: 10.1057/9780230226203.1182.

Weibull, J. W. (1997). Evolutionary Game Theory. Cambridge, MA: MIT.

White, R. W. (1985). Transitions to chaos with increasing system complexity: The case of regional industrial systems. *Environment and Planning A*, 17, 387–396.

Dynamics in Non-Binding Procurement Auctions with Boundedly Rational Bidders

Domenico Colucci, Nicola Doni, and Vincenzo Valori

1 Introduction

Auction theory has always recognised that in many settings bidders' strategies can be influenced by the revelation of some information that is privately held by the auctioneer. Usually it is assumed that the auctioneer holds some information regarding the item put up for auction. As a consequence, its revelation can allow bidders to have a more accurate estimate of their valuation for the object and to make less uncertain their utility in case their bid is accepted.[1]

Some recent papers investigate the importance of a different kind of auctioneer's private information: in multidimensional auctions, bidders can be ignorant about the real awarding rule. Katok and Wambach (2008) define this competitive mechanism as "non-binding auctions". More specifically, it is often assumed that a buyer can rank different bids according not only to the prices, but also to the quality associated to each proposals. The qualitative assessment usually depends on buyer's preferences that can be her private information because they are related to her tastes or to her specific requirements.[2] In this case bidders can always calculate

[1] Milgrom and Weber (1982) represents the seminal paper on this issue. By analysing an affiliated values auction model they stated the celebrated linkage principle, according to which expected revenue increases if the auctioneer commits to reveal any information about the value of the object. More recently some authors have shown that this principle can be wrong in some different contexts. See Ganuza (2004), Board (2009).

[2] See Gal-Or et al. (2007), Rezende (2009), Katok and Wambach (2008) for an analysis of this issue in procurement settings. Cason et al. (2003) and Chan et al. (2003) emphasise how this issue affects the awarding of subsidies in natural resource management programs. Note that a secret award rule is often present also in the procedures for the privatization of previously State-owned enterprises: Governments usually compare different proposals and at the end they select one private firm on the basis not only of the economic offers, but also of different factors, like political, social and environmental considerations.

V. Valori (✉)
Facoltà di Economia, Università di Firenze Via Lombroso, 6/17 50134 Firenze,
e-mail: vincenzo.valori@dmd.unifi.it

G.I. Bischi et al. (eds.), *Nonlinear Dynamics in Economics, Finance and the Social Sciences*, DOI 10.1007/978-3-642-04023-8_10,

thoroughly the ex-post profit associated to each specific bid; however, the information policy adopted by the buyer influences their estimate of the probability to be the winner. When the buyer chooses to reveal privately (publicly) her information suppliers are involved in a standard auction setting, with independent private (public) values. Conversely, the case in which the buyer conceals her information represents a novelty in the auction literature, and that is why we want to explore in more depth the characteristics of this game and the properties of its Nash equilibrium.

In the next section we introduce the general model of an auction where the buyer conceals her private information, as proposed by Gal-Or et al. (2007) and we show that this specific setting is closely related to classic models of horizontal differentiation. In particular, we emphasise how, in the case with only two bidders, their equilibrium bidding strategies are equivalent to duopolists' pricing strategies in the Nash equilibrium of an Hotelling model with exogenous location. For this purpose we follow the recent generalization of the Hotelling model put forward by Kim (2007). In the general case of n bidders a multidimensional auction with concealment of buyer's private information is formally identical to the model of product differentiation studied by Perloff and Salop (1985): the only difference is the analysis of the strategic value of the buyer's private information in Gal-Or et al. (2007) with respect to the otherwise more general model of Perloff and Salop (1985).

In Sect. 3 we study a simple dynamic version of the above model. To this end we posit a sequence of auctions take place in time, to which a given set of suppliers participate without actually knowing the quality assessments held by the auctioneers. It is thus a situation in which a given set of suppliers compete repeatedly to procure a specific good. We simply assume that every buyer is characterized by a vector of quality assessments, one for each supplier.[3] From the suppliers standpoint, buyers' assessments correspond to independent random draws from a given probability distribution. Each supplier maintains some kind of expectation regarding their opponents' behaviour which we shall suppose to be wrapped up in an expectation about a mean of the opponents' prices. Clearly this hypothesis qualifies agents as having bounded rationality, in that the opponents are treated as if they were one, whereas a fully rational player would have to figure out the best response to the predicted bids of every other player. This depends on the fact that these auctions are non-binding, so that qualities as well as bids determine the winner. While considering a mean price is clearly suboptimal, it nonetheless has the property of depending on the entire set of choices by the competitors as implied by full rationality. On the contrary, concentrating only on the others' best price, as would be optimal in standard auctions, is not rational in this context because the contract is not necessarily awarded to the lowest bidder. In turn expectations about the opponents' mean price are updated adaptively. This model entails a moderate departure from rationality in the vicinity of the steady state, which turns out to be unique and

[3] We could also imagine the case in which there is a unique buyer who, perhaps due to frequent job rotation related to political evolution, has time-changing quality assessments.

implies coordination on the Nash equilibrium of the stage game, and our purpose is precisely to study local stability: these observations are the rationale for using the above limited rationality model of choice.

Alternatively the dynamic model can be seen in the Hotelling framework as a straightforward way to model the repeated situation of market competition in a horizontally differentiated oligopoly with boundedly rational sellers. The vast literature on oligopoly dynamics focuses mainly on the Cournot model and is neatly surveyed by Barkley-Rosser (2002). Relatively fewer papers examine the dynamics in the Hotelling setting (for an example, see Puu and Gardini (2002)). However they focus on the spatial competition, in which firms choose both a price and a location. Conversely, in the present paper we analyse the dynamics of (price-only) competition in the framework formalized by Perloff and Salop (1985). The stability of the dynamical system is not easy to assess in general because the reaction functions may be non-differentiable at the steady state. In such a case the analysis can resort to the derivative of the reaction function in a neighborhood of the steady state. We will show how the stability of the Nash equilibrium of this game is affected by the distribution from which bidders' qualities are drawn. More specifically if $n = 2$ then the Nash equilibrium is always stable whatever the quality distribution. Conversely when $n > 2$ stability can be violated. In fact, if bidders' qualities are drawn from a specific class of densities then in equilibrium reaction functions are negatively sloped and the system may fail to converge.

2 Auctions with Horizontally Differentiated Suppliers

2.1 Equilibrium in Non-Binding Auctions

Assume there is a unique buyer wishing to procure a single unit of a specific product by means of an auction procedure. There are n firms, competing to supply the item. Both the buyer and the suppliers are assumed to be risk-neutral. We allow the buyer to value the specific product provided by each seller differently. Let q_i denote the buyer's evaluation of the quality associated to the bid of supplier i. We assume that the quality parameters are independent and identically distributed (i.i.d.) random variables with a continuous density f (with cumulative distribution function F) over the support $\left[\underline{q}, \bar{q}\right]$ and that their realisations are privately known by the buyer only.

The utility the buyer can obtain contracting with a specific supplier depends on the quality of his product and the price asked to provide it:

$$U(q_i, p_i) = q_i - p_i \qquad i = 1, \dots, n.$$

A multidimensional auction is held in order to select a supplier, and we assume that the score function used to rank alternative bids is the same as the buyer's utility function.[4] Suppliers are characterized by identical production costs, normalized to be 0. Every competing bidder submits an economic bid p_i in order to maximise his expected profit, equal to his ex-post profit times the probability of being the selected contractor:

$$\max_{p_i} p_i \Pr\{q_i - p_i \geq \max_{j \neq i} q_j - p_j\}$$

By taking into account how the quality of each competitor is distributed we can rewrite the maximization problem as follows:

$$\max_{p_i} p_i \int_{\underline{q}}^{\bar{q}} \left(\prod_{j \neq i} F\left(q_i + p_j - p_i\right) \right) f\left(q_i\right) dq_i \tag{1}$$

Further, restricting the attention only to symmetric equilibria and assuming the common bid submitted by competitors other than i equals $\bar{p}$ the above rewrites as

$$\max_{p_i} H\left(p_i, \bar{p}\right) \tag{2}$$

$$H\left(p_i, \bar{p}\right) \equiv p_i \int_{\underline{q}}^{\bar{q}} \left(F\left(q_i + \bar{p} - p_i\right)\right)^{n-1} f\left(q_i\right) dq_i$$

Notice that, defining $V(x) = \Pr\{\max_{j \neq i}\{q_j\} - q_i \leq x\}$ and $v(x) = V'(x)$ (i.e. the density of the difference between the highest quality of i's competitors and i's own quality, which can be written explicitly using convolutions) the maximization problem can also be written as

$$\max_{p_i} p_i V\left(\bar{p} - p_i\right)$$

Optimising with respect to p_i the first order condition $\frac{\partial H(p_i, \bar{p})}{\partial p_i} = 0$ can be expressed as

$$V\left(\bar{p} - p_i\right) - p_i v\left(\bar{p} - p_i\right) = 0 \tag{3}$$

Imposing $p_i = \bar{p} = p^*$ we obtain the (candidate) Nash equilibrium of this game:

$$p^* = \frac{V(0)}{v(0)} = \frac{1/n}{(n-1)\int_{\underline{q}}^{\bar{q}} \left[F\left(q_i\right)\right]^{n-2} f^2\left(q_i\right) dq_i} \tag{4}$$

Obviously, the second order condition $\frac{\partial^2 H(p^*, p^*)}{\partial p_i^2} < 0$ or, $2v(0) - \frac{V(0)}{v(0)} v'(0) > 0$ needs to hold for the above equilibrium price to be the solution of problem (2): the density $v(\cdot)$ has to be differentiable in zero for this to make sense, otherwise

[4] Notice that the buyer's utility might be negative even with her best buy: we are implicitly ruling out the outside option of not purchasing the good at all.

we need the function appearing in (3) to be decreasing around p^*. A condition which bypasses the possible non-smoothness of $v(\cdot)$ and which ensures that p^* is a Nash equilibrium, is that the distribution $V(\cdot)$ be log-concave.[5] In turn, a condition which ensures this is that $f(x)$ be log-concave[6] (see An (1998) and Bagnoli and Bergstrom (2005)). Interestingly, it also turns out to be sufficient for stability, as we shall explain in Sect. 3.

Formula (4) emphasises the way in which the optimal bidding strategy is affected by the (common) beliefs of suppliers over the buyer's preferences, represented by the distribution function F. This solution is the same as the equilibrium price of the model of monopolistic competition proposed by Perloff and Salop (1985) (see 12 and 13, p. 110).They analyse the outcome of competition in a differentiated market with n firms and L consumers. Each consumer is identified by specific tastes, represented by an n-dimensional vector of values, one for each firm, and corresponding to independent draws from the same probability function F. Consumers maximize their net utility, given by the difference between each firm's value and the correspondent price. So the oligopolist's maximization problem in this product differentiated market is formally equivalent to the bidder's maximization problem in a non-binding auction. The only exception is that oligopolists face L consumers, while bidders compete to serve a unique buyer. However this difference does not affect the price equilibrium that is identical for these two games.

Perloff and Salop (1985) prove that there can be at most one symmetric price equilibrium. However, in the case of $n > 2$ the possibility of equilibria in asymmetric prices is not ruled out, even though costs are identical for each firm. Conversely, in the case in which $n = 2$ the existence of a multiplicity of equilibria is not admissible, so the symmetric equilibrium is surely unique. The case with only two

---

[5] The log-concavity of V means that the ratio $\frac{v}{V}$ is decreasing. This implies that, if the first order condition holds, i.e.

$$V(\bar{p} - p_i)\left(1 - p_i \frac{v(\bar{p} - p_i)}{V(\bar{p} - p_i)}\right) = 0$$

then, for a positive quantity δ

$$1 - (p_i + \delta)\frac{v(\bar{p} - p_i - \delta)}{V(\bar{p} - p_i - \delta)} < 0$$

and

$$1 - (p_i - \delta)\frac{v(\bar{p} - p_i + \delta)}{V(\bar{p} - p_i + \delta)} > 0$$

which guarantees that the first order condition selects indeed a maximum. Notice that the above means that the objective function in (2) is pseudo-concave (so its critical point is a global maximum).

[6] Indeed log-concavity of the density f_X implies the log-concavity of the density of the mirror image f_{-X} and of the distribution function F (see Bagnoli and Bergstrom (2005), Theorem 8 and Theorem 1 respectively). In turn it is easy to see that the distribution function F^n and its associated density $nF^{n-1}f$ are both log-concave. The convolution of $nF^{n-1}f$ and f_{-X} gives the density v which is again log-concave (see An (1998), Corollary 1). Finally, Theorem 1 of Bagnoli and Bergstrom (2005) implies that V is also log-concave.

firms is particularly interesting because it has a formal correspondence with the model of price competition in the classic duopoly à la Hotelling. In the next section we will show how the distribution representing buyers' preferences is very close to the distribution of consumers along the Hotelling line.

2.2 *Optimal Price Strategy in an Hotelling Game with Non-Uniform consumers*

Imagine there are two suppliers, A and B, having production costs equal to 0 and located at either end of a Hotelling line of unit length. Consumers are characterized on the basis of their location parameter $\theta \in [0, 1]$, and they are distributed on this Hotelling line according to a cumulative distribution $G(\theta)$, having a strictly positive density $g(\theta)$ over the interior of the support. Their utility function when they buy the product from supplier i is equal to:

$$U = v - p_i - bd_i \qquad i = A, B$$

where v is their reservation price for each good, p_i is the price charged by supplier i, d_i is the distance from supplier i, where $d_A = \theta$, $d_B = 1 - \theta$, b is the linear cost of transport. In such a setting, given a price pair (p_A, p_B), we can define $\widetilde{\theta}$ as the consumer indifferent between supplier A and B, where:

$$\widetilde{\theta} = \frac{b + p_B - p_A}{2b}$$

As a consequence all the consumers on the left of $\widetilde{\theta}$ prefer supplier A, while those on the right prefer supplier B. Therefore, the maximization problems of both suppliers are:

$$\max_{p_A \leq v-b} p_A(G(\widetilde{\theta})) \qquad \text{and} \qquad \max_{p_B \leq v-b} p_B(1 - G(\widetilde{\theta}))$$

If we assume that consumers' reservation price, v, is sufficiently high, so that in equilibrium all of them buy some product, we have that in the Nash equilibrium of this game suppliers' optimal price strategies are:

$$p_A = 2b\frac{G(\theta^*)}{g(\theta^*)} \qquad \text{and} \qquad p_B = 2b\frac{1 - G(\theta^*)}{g(\theta^*)} \tag{5}$$

where θ^* must satisfy the following implicit equation

$$\theta^* = \frac{1}{2} + \frac{1 - G(\theta^*)}{g(\theta^*)} \tag{6}$$

Remark that this solution is also valid if we assume that suppliers A and B compete in a market with a single buyer, whose position is unknown to them. If $G(\theta)$ represents their common beliefs over his possible position the maximization problems are unchanged, except for the fact that in this case suppliers maximize their expected profit. On the basis of this new interpretation of the Hotelling game we can state the following proposition:

Proposition 1. *The Nash-equilibrium of the auction game with concealment of buyer's private information when there are only two suppliers is equivalent to the Nash-equilibrium of the Hotelling game if*

(1) buyer's position θ is a function of suppliers' quality in the auction game, $\theta = \frac{b+q_B-q_A}{2b}$ and
(2) suppliers' beliefs over θ are consistent with their beliefs over the initial qualities.

Proof. In order to prove the result we need to derive $g(\theta)$. First we define $z = q_B - q_A$. z is then distributed as the difference between suppliers' qualities and its density has positive values on the support $[-b, +b]$ according to the convolution

$$h(z) = \int_{-\infty}^{+\infty} f(z+q)\, f(q)\, dq$$

where we are assuming $f(\cdot)$ to be identically zero outside the support $[\underline{q}, \bar{q}]$. Now we can note that θ is a monotone transformation of the random variable z. In fact:

$$\theta = \frac{1}{2} + \frac{z}{2b}$$

As a consequence, θ is distributed over the support $[0, 1]$ according to the following density function:

$$g(\theta) = 2bh(2b\theta - b)$$

It is easy to note that h, and consequently g, are symmetric functions. This fact implies that $G(1/2) = 1/2$ and consequently condition (6) is satisfied for $\theta = 1/2$ Substituting this value in (5) we obtain:

$$p_A = p_B = \frac{2bG\left(\frac{1}{2}\right)}{g\left(\frac{1}{2}\right)} = \frac{1}{2h(0)} = \frac{1}{2\int_{\underline{q}}^{\bar{q}} f^2(q)\, dq}$$

But this solution is coincident with formula (4) when $n = 2$.

Therefore the conclusion of Perloff and Salop (1985) and Gal-Or et al. (2007), according to which firms' optimal price strategy depends on the distribution of buyers' tastes, corresponds to the result achieved by Kim (2007), that the optimal price strategy of duopolists in an Hotelling setup depends on the distribution of consumers' location along Main Street.

3 Economic Dynamics Under Bounded Rationality

In this section we shall embed the above analysis into a simple dynamic framework, meant to be a rough indicator of whether the Nash equilibrium derived above is bound to be actually reached if agents are either boundedly rational and/or are unsure about the other players' rationality. Each seller is supposed to be participating in a sequence of auctions in which, at each time, the seller's quality parameter is drawn from the same distribution. In the stage game each seller has a best strategy which depends on the bids of his competitors. The reaction function of a generic seller can be derived by solving for p_i in the first order condition applied to the objective function in (1):

$$\int_{\underline{q}}^{\bar{q}} \left(\prod_{j \neq i} F\left(q_i + p_j - p_i\right) \right) f\left(q_i\right) dq_i - p_i \int_{\underline{q}}^{\bar{q}} \sum_{j \neq i} f\left(q_i + p_j - p_i\right) \times \prod_{h \neq j,i} F\left(q_i + p_h - p_i\right) f\left(q_i\right) dq_i = 0 \tag{7}$$

Notice that the best response of seller i, p_i^*, does not depend on the competitors' bids which have no chance of winning. More precisely $p_h - \min_j \{p_J\} > \bar{q} - \underline{q}$ implies that $\frac{\partial p_i^*}{\partial p_h} = 0$. While the converse is not always true we can observe that, generically, a change in the strategy of a possibly winning competitor affects the optimal bid p_i^*. So when price dispersion is sufficiently low, a bidder's best response is influenced by the full vector of competitors' bids.

Therefore in this context each bidder's optimal strategy is not focused on beating the best price, as in standard first price auction. However, even with a restricted number of competitors it is quite difficult to calculate explicitly the best response function (see (7)). For this reason we hypothesise that agents simplify their problem by treating their opponents as if they were all bidding a price equal to a mean of the full vector of competitors' prices and we let sellers update their expectations about its value adaptively. Results are robust to having subject-specific weighted averages; due to the heavier required notation we shall stick to simple means in the following (see Footnote 7).

Summing up, sellers at each time t, solve the same optimization problem using the first order conditions in (3) given a different value for the opponents' mean price, and will therefore be using the same reaction function evaluated at these different values. The symmetry imposed over the sellers implies that each shall have the same reaction function, $R()$. So seller i at time t will choose $p_{t,i} = R\left(\bar{p}^e_{t,-i}\right)$ and compute the expected mean price of the opponents according to

$$\bar{p}^e_{t,-i} = \bar{p}^e_{t-1,-i} + \alpha_i \left(\bar{p}_{t-1,-i} - \bar{p}^e_{t-1,-i}\right) \tag{8}$$

$$\bar{p}_{t,-i} = \sum_{j \neq i} \frac{p_{t,j}}{n-1}$$

$$p_{t,i} = R\left(\bar{p}^e_{t,-i}\right) \tag{9}$$

Notice that we allow some behavioral heterogeneity in that α_i may vary across different sellers. These equations define the following n-dimensional discrete dynamical system in $\bar{p}^e_{t,-i}$

$$\begin{cases} \bar{p}^e_{t+1,-1} = \bar{p}^e_{t,-1} + \alpha_1 \left(\frac{\sum_{j\neq 1} R\left(\bar{p}^e_{t,-j}\right)}{n-1} - \bar{p}^e_{t,-1} \right) \\ \qquad \cdots \\ \bar{p}^e_{t+1,-n} = \bar{p}^e_{t,-n} + \alpha_n \left(\frac{\sum_{j\neq n} R\left(\bar{p}^e_{t,-j}\right)}{n-1} - \bar{p}^e_{t,-n} \right) \end{cases} \tag{10}$$

which possesses a single steady state whereby

$$\bar{p}^e_{t,-i} = p^* \quad i = 1, \ldots, n$$

where p^* is the Nash equilibrium derived above in (4). Stability of the steady state can be characterised as usual studying the Jacobian matrix of the system evaluated at the steady state, provided the reaction function is differentiable at p^*: in this case we have

$$J_n = \begin{pmatrix} 1-\alpha_1 & \alpha_1 \frac{R'(p^*)}{n-1} & \cdots & \alpha_1 \frac{R'(p^*)}{n-1} \\ \cdots & \cdots & \ddots & \cdots \\ \alpha_n \frac{R'(p^*)}{n-1} & \alpha_n \frac{R'(p^*)}{n-1} & \cdots & 1-\alpha_n \end{pmatrix}$$

We aim at giving conditions on the underlying parameters of the model that ensure that the spectral radius of J_n is less than one for any choice of the vector of gain parameters $\alpha_1, \ldots, \alpha_n$ in $(0,1)^n$ or, symmetrically, provide conditions under which a suitable choice of such parameters implies instability of the steady state and therefore that the Nash equilibrium will not be reached as t grows. In particular it is well known that if $\|\cdot\|$ is a matrix norm on J_n and $\rho(J_n)$ is its spectral radius then

$$\rho(J_n) \leq \|J_n\|$$

Consider for example $\left\| \underset{n\times n}{A} \right\| = \max_i \sum_{j=1}^n |a_{ij}|$. In the case of the above matrix J_n it is

$$\|J_n\| = \max_i 1 - \alpha_i + \alpha_i \left|R'\left(p^*\right)\right| \tag{11}$$

Therefore $|R'(p^*)| \leq 1$ implies $\rho(J_n) \leq 1$ for all possible choices of α_i and therefore stability of the steady state[7]. So it is interesting to establish conditions ensuring a bound on the (absolute value of the) derivative of the reaction function in

[7] Note that this would not change if in (10) heterogeneous weighted averages replaced the arithmetic means. Indeed such generalisation would not alter the matrix norm (11) and therefore the stability conditions.

p^*. This is what we do in Proposition 2. But first, we need to set conditions granting differentiability of R on the steady state.

Lemma 1. *For the problem (2) differentiability of the reaction function at p^* holds if and only if*

$$\begin{aligned} f(\bar{q}) = f\left(\underline{q}\right) &= 0 \text{ when } n = 2 \\ f(\bar{q}) = 0 & \quad \text{ when } n > 2 \end{aligned} \tag{12}$$

Proof. We now want to ascertain the differentiability of v i.e.

$$v(z) = \begin{cases} 0 & if \quad z < \underline{q} - \bar{q} \\ \int_{\underline{q}-z}^{\bar{q}} (n-1)[F(q+z)]^{n-2} f(q+z) f(q)\, dq & if \ \underline{q} - \bar{q} \le z < 0 \\ \int_{\underline{q}}^{\bar{q}-z} (n-1)[F(q+z)]^{n-2} f(q+z) f(q)\, dq & if \ 0 \le z \le \bar{q} - \underline{q} \\ 0 & if \quad z > \bar{q} - \underline{q} \end{cases}$$

Let $n = 2$. Using Leibnitz's rule we get

$$v'(0^-) = f^2\left(\underline{q}\right) + \int_{\underline{q}}^{\bar{q}} f'(q) f(q)\, dq = \frac{f^2(\bar{q})}{2} + \frac{f^2\left(\underline{q}\right)}{2}$$

$$v'\left(0^+\right) = -f^2(\bar{q}) + \int_{\underline{q}}^{\bar{q}} f'(q) f(q)\, dq = -\left(\frac{f^2(\bar{q})}{2} + \frac{f^2\left(\underline{q}\right)}{2}\right)$$

therefore $v'(0)$ exist if and only if $f^2\left(\underline{q}\right) = f^2(\bar{q}) = 0$, in which case $v'(0) = 0$.

Instead, when $n > 2$

$$v'(0^-) = \int_{\underline{q}}^{\bar{q}} (n-1)(n-2)[F(q)]^{n-3} f^3(q) + (n-1)[F(q)]^{n-2} f'(q) f(q)\, dq$$

$$v'\left(0^+\right) = -(n-1) f^2(\bar{q}) + v'(0^-)$$

so $v'(0)$ exist if and only if $f^2(\bar{q}) = 0$ and $v'(0^-) \neq \pm\infty$.

Proposition 2. *Consider the problem (2) under condition (12). When $n = 2$ we have $R'(p^*) = \frac{1}{2}$; in this case the dynamical system (10) is always locally stable. When $n > 2$ condition $R'(p^*) < 1$ always holds, while $R'(p^*) > -1$ (and therefore local stability) holds if and only if*

$$\frac{v'(0)}{v(0)^2} < \frac{3}{2} n \tag{13}$$

Proof. Under differentiability of the density v we can apply implicit differentiation to (3) to get:

$$R'\left(p^*\right) = \frac{v\left(0\right) - p^* v'\left(0\right)}{2v\left(0\right) - p^* v'\left(0\right)} \tag{14}$$

Therefore, when $n = 2$, Lemma 1 shows that $v'(0) = 0$, so (14) implies $R'(p^*) = \frac{1}{2}$ as stated.

When $n > 2$ we have

$$R'\left(p^*\right) < 1 \Leftrightarrow \frac{v\left(0\right) - p^* v'\left(0\right)}{2v\left(0\right) - p^* v'\left(0\right)} < 1$$

which always holds, given the second order condition $2v(0) - p^* v'(0) > 0$ and the fact that $v(0) = \int_{\underline{q}}^{\bar{q}} (n-1)(F(q))^{n-2} f^2(q)\, dq > 0$.

Finally:

$$R'\left(p^*\right) > -1 \Leftrightarrow \frac{v\left(0\right) - p^* v'\left(0\right)}{2v\left(0\right) - p^* v'\left(0\right)} > -1$$
$$\Leftrightarrow \frac{v'\left(0\right)}{v^2\left(0\right)} < \frac{3}{2}n$$

The above result, implying that the $n = 2$ case is a threshold above which stability is not necessarily granted, is reminiscent of classic results from the literature on dynamics in the Cournot model such as Theocharis (1960). Remark that the conditions for stability when $n > 2$ can be violated only if the reaction function at p^* is decreasing. Notice that $R'(p^*) = -\frac{\partial^2 H(p^*,p^*)}{\partial \bar{p} \partial p_i} / \frac{\partial^2 H(p^*,p^*)}{\partial p_i^2}$ can be negative only if $\frac{\partial^2 H(p^*,p^*)}{\partial \bar{p} \partial p_i}$ is negative, given the second order condition $\frac{\partial^2 H(p^*,p^*)}{\partial p_i^2} < 0$. This means that only under strategic substitutability at equilibrium can the system fail to converge to the Nash equilibrium. In other words it has to be the case that a more aggressive strategy by suppliers $j \neq i$ (i.e. a lower bid) raises i's marginal profit. Vice versa strategic complementarity at equilibrium is always associated with a (dynamically) stable Nash equilibrium.

A specific example in which (13) fails is as follows: let $n = 3$ and consider the beta density function with parameters $a = 3/4, b = 3$

$$f\left(x\right) = \frac{x^{-\frac{1}{4}}\left(1-x\right)^2}{\int_0^1 x^{-\frac{1}{4}}\left(1-x\right)^2 dx}$$

We get $\frac{v'(0)}{v^2(0)} \simeq 5.4604$, thus violating condition (13), and $p^* = \frac{V(0)}{v(0)} \simeq \frac{1/3}{1.5595} = 0.21374$ implying

$$R'\left(p^*\right) = \frac{v\left(0\right) - p^* v'\left(0\right)}{2v\left(0\right) - p^* v'\left(0\right)} \simeq -4.5591$$

In this case[8] the system's Jacobian at the steady state is

$$J_3 = \begin{pmatrix} 1-\alpha_1 & \alpha_1 \frac{R'(p^*)}{2} & \alpha_1 \frac{R'(p^*)}{2} \\ \alpha_2 \frac{R'(p^*)}{2} & 1-\alpha_2 & \alpha_2 \frac{R'(p^*)}{2} \\ \alpha_3 \frac{R'(p^*)}{2} & \alpha_3 \frac{R'(p^*)}{2} & 1-\alpha_3 \end{pmatrix}$$

which for the above specific value of $R'(p^*)$ has eigenvalues outside the unit circle for suitable values[9] of $\alpha_1, \alpha_2, \alpha_3$.

Our last point regards dynamics in the $n = 2$ case when the density $v(\cdot)$ is not differentiable in zero. In this case, which happens if either $f(\bar{q})$ or $f\left(\underline{q}\right)$ are non-zero, $v'(0^-) > 0$ and $v'\left(0^+\right) < 0$ so in turn $R'(p^{*-}) < 0 < R'\left(p^{*+}\right) < 1$. This implies that, locally, a perturbation (either positive or negative) from p^* will eventually lead the dynamics to a decreasing path towards p^* which therefore turns out to be locally stable.

4 Conclusions

We have shown that the non-binding auction model analysed by Gal-Or et al. (2007) is formally equivalent to the differentiated market studied by Perloff and Salop (1985) and, for the $n = 2$ case, to the generalisation of the Hotelling duopoly recently proposed by Kim (2007). We have examined more in depth the symmetric equilibrium of this class of games. In particular, we have emphasised how a sufficient condition for both the existence and the stability of such equilibrium requires the log-concavity of the probability density of bidders' (oligopolists') qualities. However, there exist distribution functions for which stability can be violated. This is the result of an extension of these models to a dynamic framework in which bidders behave according to some expectation over the prices of their competitors (summarized by their mean), and update these expectations adaptively on the basis of the data from previous auctions. The steady state of such a system is unstable only if, in equilibrium, bidders' reaction functions are negatively sloped, i.e. under strategic substitutability. The dynamic analysis of oligopoly thus far has been focused mainly on the Cournot model. This work then represent an attempt to adopt

---

[8] Notice that a case like this would not be possible if f were log-concave. Indeed that would make V log-concave as well and therefore, given $p^* = \frac{V(0)}{v(0)}$,

$$v^2(0) - V(0)\, v'(0) > 0 \Rightarrow v(0) - p^* v'(0) > 0$$

and $R'(p^*) > 0$ as a consequence. This argument also shows that, under log-concavity the reaction function has positive slope and, because such slope cannot exceed 1, local stability is ensured.

[9] For example $\alpha_1 = 0.3, \alpha_2 = 0.4, \alpha_3 = 0.5$ imply the following eigenvalues: 1.5777, 1.4285, −1.2062.

the same approach in analysing other forms of market imperfections. Moreover, we have shown how this methodology can be applied also in an auction framework, at least when bidders have no private information and the Nash equilibrium is in pure strategies. Future research can be devoted to investigate the potentiality of such analysis in these kind of settings.

References

AN, M.Y. (1998) Logconcavity versus logconvexity: a complete characterization. *Journal of Economic Theory, 50*, 350–369.

Bagnoli, M. & Bergstrom, T. (2005) Log-concave probability and its applications. *Economic Theory, 26*, 445–469.

Barkley Rosser, J., Jr. (2002) The development of complex oligopoly dynamics theory. In T. Puu & I. Sushko (Eds.), *Oligopoly dynamics: models and tools*. Berlin: Springer.

Board, S. (2009) Revealing information in auctions: the allocation effect. *Economic Theory, 38*, 125–135.

Cason, T., Gangadharan, L., & Duke, C. (2003) A laboratory study of auctions for reducing non-point source pollution. *Journal of Environmental Economics and Management, 46*, 446–471.

Chan, C., Laplagne, P., & Appels, D. (2003) The role of auctions in allocating public resources. *Productivity Commission Staff Research Paper*. Productivity Commission, Melbourne.

Gal-Or, E., Gal-Or, M., & Dukes, A. (2007) Optimal information revelation in procurement schemes. *Rand Journal of Economics, 38*, 400–418.

Ganuza, J.J. (2004) Ignorance promotes competition: an auction model with endogenous private valuations. *Rand Journal of Economics, 35*, 583–598.

Katok, E. & Wambach, A. (2008) Collusion in dynamic buyer-determined reverse auctions. CSCR Working Papers n. 08–13.

Kim, J. (2007) The intensity of competition in the hotelling model: a new generalisation and applications. MPRA Paper n. 6876.

Milgrom, P. & Weber, R. (1982) A theory of auctions and competitive bidding. *Econometrica, 50*, 1089–1122.

Perloff, J. & Salop, S. (1985) Equilibrium with product differentiation. *Review of Economic Studies, 52*, 107–120.

Puu, T. (1998) The chaotic duopolists revisited. *Journal of Economic Behavior and Organization, 33*, 385–394.

Puu, T. & Gardini, L. (2002) Hotelling type duopoly and oligopoly. In T. Puu & I. Sushko (Eds.) *Oligopoly dynamics: models and tools*. Berlin: Springer.

Rezende, L. (2009) Biased procurement auctions. *Economic Theory, 38*, 169–185.

Theocharis, R. (1960) On the stability of the cournot solution on the oligopoly problem. *Review of Economic Studies, 27*, 133–134.

Delay Differential Nonlinear Economic Models

Akio Matsumoto and Ferenc Szidarovszky

1 Introduction

The asymptotical behavior of dynamic economic systems has been the focus of a large number of studies with both discrete and continuous time scales. They are based on the qualitative theory of difference or ordinary differential equations (Bellman, 1969; Goldberg, 1958). It has been shown by many authors that the introduction of information delay into the dynamic models significantly changes their asymptotical properties. For example, Chiarella and Szidarovszky (2004) consider dynamic oligopolies with partial information on the price function and Huang (2008) examines the role of information lag in economic dynamics, to name a few. There is a significant difference between models with fixed time lags and models with continuously distributed delays. In the first case there is an infinite spectrum, and in the second case with gamma-function type kernel functions, the spectrum is finite. An important special case of continuously distributed time lags is given by exponentially decreasing kernel functions.

In this paper we compare dynamics generated by fixed time lags and continuously distributed delay with exponential kernel function. We will first show that these two types of models generate the same local dynamics if the delay is sufficiently small. This is, however, not true if the delay becomes large.

The theoretical findings are illustrated by three well known economic models: the Goodwin model, the Kaldor–Kalecki model and the Cournot oligopoly model.

This paper is organized as follows. Section 2 introduces the main mathematical results, and the particular models are discussed in Sect. 3. Conclusions are drawn in Sect. 4.

A. Matsumoto (✉)
Department of Economics, Chuo University, 742-1, Higashi-Nakano,
Hachioji, Tokyo 192-0393, Japan,
e-mail: akiom@tamacc.chuo-u.ac.jp

G.I. Bischi et al. (eds.), *Nonlinear Dynamics in Economics, Finance and the Social Sciences*, DOI 10.1007/978-3-642-04023-8_11,

2 Mathematical Results

Consider first the general linear differential-difference equation

$$\sum_{k=0}^{n} \alpha_k y^{(k)}(t) + \sum_{k=0}^{n} \beta_k y^{(k)}(t+\theta) = 0 \tag{1}$$

with a single delay θ, where

$$y^{(k)}(t) = \frac{d^k}{dt^k} y(t) \quad \text{and} \quad y^{(k)}(t+\theta) = \frac{d^k}{dt^k} y(t+\theta).$$

Assuming small θ, linearization with respect to θ gives the approximation

$$\left(\sum_{k=0}^{n} \alpha_k y^{(k)}(t) + \sum_{k=0}^{n} \beta_k y^{(k)}(t) \right) + \left(\sum_{k=0}^{n} \beta_k y^{(k+1)}(t) \right) \theta = 0.$$

This is a linear homogeneous equation. As usual, looking for the solution in an exponential form $y(t) = ve^{\lambda t}$ gives

$$\sum_{k=0}^{n} (\alpha_k + \beta_k) \lambda^k e^{\lambda t} v + \sum_{k=0}^{n} \beta_k \lambda^{k+1} e^{\lambda t} v\theta = 0,$$

and after simplification the characteristic polynomial of the system becomes

$$\sum_{k=0}^{n} \alpha_k \lambda^k + \left(\sum_{k=0}^{n} \beta_k \lambda^k \right) (1 + \lambda\theta) = 0, \tag{2}$$

which is a polynomial of degree $n + 1$ in λ.

Consider next the equivalent delayed equation,

$$\sum_{k=0}^{n} \alpha_k y^{(k)}(t-s) + \sum_{k=0}^{n} \beta_k y^{(k)}(t) = 0. \tag{3}$$

Assuming continuously distributed lag with exponential kernel function,

$$w(t-s) = \frac{1}{\theta} e^{-\frac{t-s}{\theta}}$$

and taking delay expectation, a Volterra-type integro-differential equation is obtained

$$\int_0^t w(t-s) \sum_{k=0}^{n} \alpha_k y^{(k)}(s) ds + \sum_{k=0}^{n} \beta_k y^{(k)}(t) = 0. \tag{4}$$

In the first factor we can introduce the new variable $z = t - s$ to have

$$\int_0^t w(z) \sum_{k=0}^{n} \alpha_k y^{(k)}(t-z)dz + \sum_{k=0}^{n} \beta_k y^{(k)}(t) = 0.$$

If we seek the solution in the usual exponential form $y(t) = ve^{\lambda t}$ and substitute it into the above equation, we get

$$\int_0^t \frac{1}{\theta} e^{-\frac{z}{\theta}} \sum_{k=0}^{n} \alpha_k \lambda^k e^{\lambda(t-z)} v dz + \sum_{k=0}^{n} \beta_k \lambda^k e^{\lambda t} v = 0.$$

By dividing both sides by $e^{\lambda t}v$ and letting $t \to \infty$ we have a simplified expression for the first term:

$$\int_0^\infty \frac{1}{\theta} e^{-z(\lambda+\frac{1}{\theta})} dz \sum_{k=0}^{n} \alpha_k \lambda^k = \frac{1}{\theta} \left. \frac{e^{-z(\lambda+\frac{1}{\theta})}}{-(\lambda+\frac{1}{\theta})} \right]_{z=0}^{\infty} \sum_{k=0}^{n} \alpha_k \lambda^k = \frac{1}{\lambda\theta + 1} \sum_{k=0}^{n} \alpha_k \lambda^k,$$

so the equation further simplifies as

$$\frac{1}{\lambda\theta + 1} \sum_{k=0}^{n} \alpha_k \lambda^k + \sum_{k=0}^{n} \beta_k \lambda^k = 0, \tag{5}$$

which is equivalent to (2). Therefore the local asymptotic behavior of the two dynamics is identical. We summarize this result:

Theorem 1. *Local dynamics generated by the general delay differential equation with a single and small delay is the same as the dynamics by the general differential equation with continuously distributed time lag with exponential kernel function.*

In the case of the general kernel function

$$w(t-s) = \frac{1}{n!} \left(\frac{n}{\theta}\right)^{n+1} (t-s)^n e^{-\frac{n(t-s)}{\theta}},$$

we know that as $\theta \to \infty$ or $n \to \infty$, the function converges to the Dirac-delta function centered at $t - s = 0$ and $t - s = \theta$, respectively. Therefore, in this limiting case the integro-differential equation (4) converges to the deterministic case with fixed delay. It is very interesting that in the exponential kernel function ($n = 0$) case, the two processes are even equivalent concerning the local behavior of the equilibrium. This is not true however for larger values of n, as it is demonstrated in Matsumoto and Szidarovszky (2009).

3 Economic Examples

We confirm Theorem 1 by examining various delay economic models when the time delay is small and investigate the global dynamics of the delay models with continuously distributed time delay when the time delay is large.

3.1 Goodwin Model with Investment Lag

Goodwin (1951) constructed a business cycle model with nonlinear acceleration principle of investment and showed that the model gives rise to cyclic oscillations when its stationary state is locally unstable. Goodwin's basic model is summarized as a 1D nonlinear differential equation,

$$\varepsilon\dot{y}(t) - \varphi(\dot{y}(t)) + (1-\alpha)y(t) = 0,$$

where a time variable y is national income, α the marginal propensity to consume, which is a positive constant and less than unity, ε a positive adjustment coefficient of y and $\varphi(\dot{y}(t))$ denotes the induced investment that is dependent on the rate of change in national income. The dot stands for differentiation with respect to time t. Goodwin's model adopts the nonlinear acceleration principle, according to which investment is proportional to the change in national income in a neighborhood of the equilibrium income but becomes inflexible for the extremely larger or smaller values of income.

"In order to come close to reality" (Goodwin, 1951, p. 11), the production lag θ between decisions to invest and the corresponding outlays is introduced into the above model and then the modified model becomes

$$\varepsilon\dot{y}(t) - \varphi(\dot{y}(t-\theta)) + (1-\alpha)y(t) = 0. \tag{6}$$

This is a *neutral delay nonlinear differential equation* in which θ is the fixed time lag. Since it is difficult to analytically solve this delay nonlinear model, it is a natural way to use a tractable approximation of (6). In particular, to investigate dynamics, we rewrite the equation as

$$\varepsilon\dot{y}(t+\theta) - \varphi(\dot{y}(t)) + (1-\alpha)y(t+\theta) = 0,$$

and expands it with respect to θ around $\theta = 0$ to obtain the following second-order nonlinear differential equation:

$$\varepsilon\theta\ddot{y}(t) + [\varepsilon + (1-\alpha)\theta]\dot{y}(t) - \varphi(\dot{y}(t)) + (1-\alpha)y(t) = 0.$$

Clearly, $y(t) = 0$ for all t is a stationary state of this equation. Its asymptotic behavior is determined by the eigenvalues, which are the solutions of the characteristic equation,

$$\varepsilon\theta\lambda^2 + [\varepsilon + (1-\alpha)\theta - \upsilon]\lambda + (1-\alpha) = 0, \tag{7}$$

where $\upsilon = \varphi'(0)$. The characteristic roots are

$$\lambda_{1,2} = \frac{-k \pm \sqrt{k^2 - 4\varepsilon\theta(1-\alpha)}}{2\varepsilon\theta},$$

where $k = \varepsilon + (1-\alpha)\theta - \upsilon$. It follows that the product of the characteristic roots is positive since $0 < \alpha < 1$ and both ε and θ are positive:

$$\lambda_1\lambda_2 = \frac{1-\alpha}{\varepsilon\theta} > 0,$$

which excludes the possibility of saddle stationary point. It also follows that the sum of the characteristic roots can be of either sign,

$$\lambda_1 + \lambda_2 = -\frac{\varepsilon + (1-\alpha)\theta - \upsilon}{\varepsilon\theta} \gtreqless 0.$$

Given the values of α and ε, the indeterminacy of the sign of the last expression means that the (υ, θ)-space is divided into two parts by the partition line

$$\upsilon = \varepsilon + (1-\alpha)\theta.$$

For all υ above this line, the sum of the characteristic roots is positive, hence the stationary state is locally unstable. In the same way, the stationary state is locally asymptotically stable for all υ below this line.

Continuously distributed time delay is an alternative approach to deal with time delay in investment. If we adopt it and denote the expected change of national income at time t by $\dot{y}^e(t)$, then Goodwin's delayed equation (6) can be written as the system of Volterra-type integro-differential equations:

$$\begin{cases} \varepsilon\dot{y}(t) - \varphi(\dot{y}^e(t)) + (1-\alpha)y(t) = 0, \\ \dot{y}^e(t) = \int_0^t \frac{1}{\theta} e^{-\frac{t-s}{\theta}} \dot{y}(s)ds, \end{cases} \tag{8}$$

where θ is a positive real parameter which is associated with the length of the delay. The second equation of (8) shows that the weighting function of the past changes in national income gives the most weight to the most recent income change and the weight is exponentially declining afterwards. Before turning to a closer examination of this model, we rewrite it as a system of ordinary differential equations. The time-differentiation of the second equation of (8) gives a simple equation for the new variable $z = \dot{y}^e$:

$$\dot{z}(t) = \frac{1}{\theta}\left(\dot{y}(t) - z(t)\right). \tag{9}$$

Solving the first equation for $\dot{y}$, replacing $\dot{y}^e$ with z, replacing $\dot{y}$ in (9) with the new expression of $\dot{y}$ and then adding the new dynamic equation of z will transform the system of the integro-differential equations to the following 2D system of ordinary differential equations:

$$\begin{cases} \dot{y}(t) = -\dfrac{1-\alpha}{\varepsilon} y(t) + \dfrac{1}{\varepsilon}\varphi(z(t)), \\[2ex] \dot{z}(t) = \dfrac{1}{\theta}\left(-\dfrac{1-\alpha}{\varepsilon} y(t) + \dfrac{1}{\varepsilon}\varphi(z(t)) - z(t)\right). \end{cases} \tag{10}$$

The Jacobian matrix of this system at $y = z = 0$ has the form

$$\mathbf{J_G} = \begin{pmatrix} -\dfrac{1-\alpha}{\varepsilon} & \dfrac{\nu}{\varepsilon} \\[2ex] -\dfrac{1-\alpha}{\varepsilon\theta} & \dfrac{1}{\theta}\left(\dfrac{\nu}{\varepsilon} - 1\right) \end{pmatrix}. \tag{11}$$

The corresponding characteristic equation is quadratic in λ:

$$\lambda^2 + \frac{\varepsilon + (1-\alpha)\theta - \nu}{\varepsilon\theta}\lambda + \frac{1-\alpha}{\varepsilon\theta} = 0.$$

Notice that this characteristic equation is equivalent to the characteristic equation (7). It follows that the local stability conditions are also identical. This means that the two delay dynamic systems generate the same dynamics in the neighborhood of $\theta = 0$.

We now turn our attention to the dynamics of (8) when θ is large. It is well-known that the Goodwin model generates a limit cycle when its stationary point is locally unstable. Goodwin (1951) assumed a piecewise linear investment function in his simulations. We numerically confirm his result but for the sake of analytical convenience, we assume a hyperbolic tangent type investment function:

$$\varphi(\dot{y}) = \delta\left(\tanh(\dot{y} - a) - \tanh(-a)\right), \quad \delta > 0 \text{ and } a = 1. \tag{12}$$

We perform numerical simulations with the parameter values $\varepsilon = 0.5$ and $\alpha = 0.6$ as in Goodwin (1951). To make the stationary point locally unstable, we take $\theta = 0.8$ and $\delta = (1 + a^2)(\varepsilon + (1-\alpha)\theta) + 0.01$. The numerical result is illustrated in Fig. 1 in which two trajectories, one continuous line starting at point a and the other dotted line at point b, are seen to converge to the limit cycle.

Recently, Matsumoto (2009) reexamined Goodwin's model and showed the coexistence of multiple limit cycles, a stable cycle surrounding a unstable cycle when the stationary state is locally stable. This is illustrated in Fig. 2 in which there are two limit cycles depicted as bold curves and the two trajectories starting at

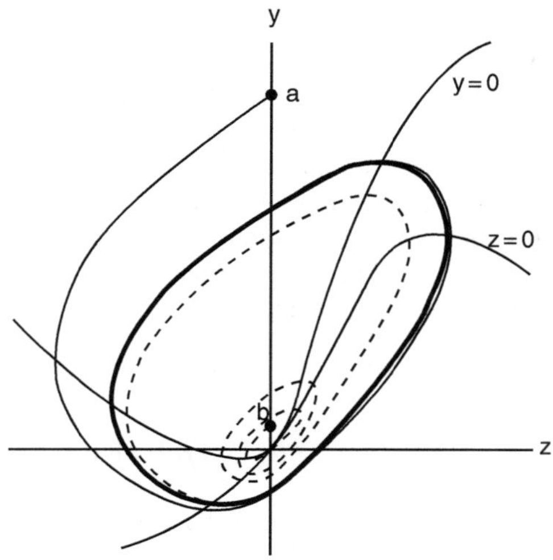

Fig. 1 Existence of a stable limit cycle

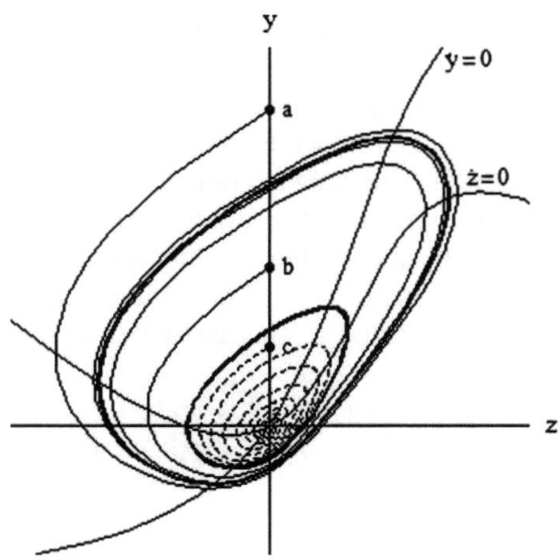

Fig. 2 Co-existence of a stable and an unstable limit cycles

points a and b converge to the outer limit cycle whereas a trajectory starting at point c approaches the stable stationary point. A parametric difference between the first simulation and the second simulation is that only the value of δ is changed to $(1 + a^2)(\varepsilon + (1 - \alpha)\theta) - 0.01$ from $(1 + a^2)(\varepsilon + (1 - \alpha)\theta) + 0.01$.

3.2 Kaldor–Kalecki Model with Investment Lag

Kaldor (1940) presented a business cycle model in which investment was positively related to the levels of income via a nonlinear relationship. Kalecki (1935) added a lag between the investment decision and the installation of investment goods. His model used a linear difference-differential equation to generate cyclic dynamics. The Kaldor–Kalecki model is a combination of nonlinear investment and a time lag in the capital accumulation. Let Y be the national income and K the capital stock. Then the Kaldor–Kalecki model can be written as

$$\begin{cases} \dot{Y}(t) = \alpha\left[I(Y(t), K(t)) - S(Y(t))\right], \\ \dot{K}(t) = I(Y(t-\theta), K(t)) - \delta K(t), \end{cases} \tag{13}$$

where $I(Y, K)$ is an investment function and $S(Y)$ is the saving function. Investment depends positively on income and negatively on capital, so $dI/dY = I_Y > 0$ and $dI/dK = I_K < 0$. Furthermore, it takes a S-shaped profile with respect to Y indicating that investment becomes inflexible for low as well as high levels of income. Savings depends on income in the usual way, i.e., $0 < dS/dY = S_Y < 1$. We assume also that $I_Y - S_Y > 0$ at the fixed point of (13), that is, investment increases faster than savings as national income increases in a neighborhood of the fixed point, following Kaldor. In addition, $\alpha > 0$ is the adjustment coefficient and $\delta > 0$ is the depreciation rate of the capital. The first equation of (13) states that income changes proportionally to the excess demand in the goods market. The second equation is a standard capital accumulation equation but includes a time lag θ.

Consider first the local stability of (13) without time delay (i.e., $\theta = 0$), which is equivalent to the original Kaldor model. The Jacobian matrix has the form

$$\mathbf{J_K} = \begin{pmatrix} \alpha(I_Y - S_Y) & \alpha I_K \\ I_Y & I_K - \delta \end{pmatrix}$$

with the determinant

$$det\mathbf{J_K} = \alpha(I_Y - S_Y)(I_K - \delta) - \alpha I_K I_Y$$

and the trace

$$tr\mathbf{J_K} = \alpha(I_Y - S_Y) + (I_K - \delta).$$

Kaldor (1940) made two basic assumptions: $det\mathbf{J_K} > 0$ in order to exclude the possibility that a stationary point is saddle and $tr\mathbf{J_K} < 0$ to make the stationary point unstable. As seen in Chang and Smyth (1971), the gist of Kaldor's argument can be translated to show an existence of an endogenously persistent fluctuation by applying the Poincaré–Bendixson theorem. For this end, the local instability of the stationary point is the first requirement. Figure 3 illustrates the birth of a Kaldorian limit cycle with the following configuration of the model: The investment function

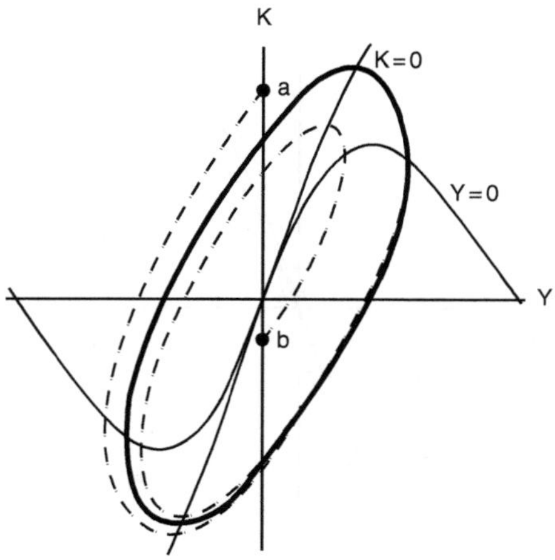

Fig. 3 Existence of a Kaldorian limit cycle

is separable with respect to Y and K,

$$I(Y, K) = \phi(Y) + \beta K, \ \beta < 0,$$

where $\phi(Y)$ is assumed to be a symmetric S-shaped function,

$$\phi(Y) = \frac{A}{1 + e^{-BY}} - \frac{A}{2}, \quad A > 0 \text{ and } B > 0,$$

and the parameters are specified as $A = 4$, $B = 1$, $c = 0.6$, $\alpha = 0.8$, $\beta = -0.2$ and $\delta = 0.05$. It can be seen that the limit cycle attracts two different trajectories, one starting at point a and the other starting at point b in the neighborhood of the stationary point.

Although we numerically confirm the existence of the Kaldorian limit cycle when the stationary point is locally unstable, we are interested in the destabilizing effect caused by a delay in investment so that the stationary point becomes asymptotically stable when $tr\mathbf{J_K} < 0$. In Fig. 4, two trajectories belonging to the two different initial points a and b spiral toward the stationary point when $\beta = -0.4$ and $\delta = 0.2$ with the other parameters being unchanged.

Now we are back to the delay Kaldor–Kalecki model (13). We first rewrite the capital accumulation equation as

$$\dot{K}(t + \theta) = I(Y(t), K(t + \theta)) - \delta K(t + \theta).$$

If the time lag is small enough, then linearizing it with respect to θ around $\theta = 0$ gives

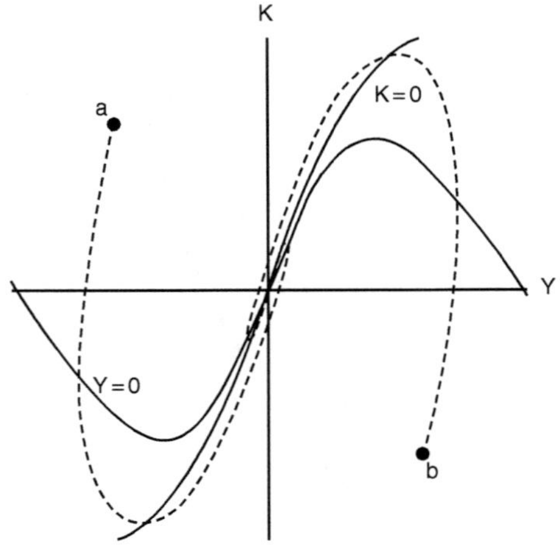

Fig. 4 A stable Kaldorian stationary point

$$\dot{K}(t) - \{I(Y(t), K(t)) - \delta K(t)\} + \left\{\ddot{K}(t) - I_K \dot{K}(t) + \delta \dot{K}(t)\right\} \theta = 0.$$

Introducing the new variable, $Z(t) = \dot{K}(t)$, the delayed Kaldor–Kalecki model is reduced to a 3D system of ordinary differential equations:

$$\begin{cases} \dot{Y}(t) = \alpha(\, I(Y(t), K(t)) - S(Y(t))\,), \\ \dot{K}(t) = Z(t), \\ \dot{Z}(t) = \dfrac{1}{\theta}\{I(Y(t), K(t)) - \delta K(t)\} + \{(I_K - \delta) - \dfrac{1}{\theta}\}Z(t). \end{cases} \tag{14}$$

The Jacobian matrix is

$$\mathbf{J_D} = \begin{pmatrix} \alpha(I_Y - S_Y) & \alpha I_K & 0 \\ 0 & 0 & 1 \\ \dfrac{1}{\theta} I_Y & \dfrac{1}{\theta}(I_K - \delta) & (I_K - \delta) - \dfrac{1}{\theta} \end{pmatrix}$$

with the determinant,

$$det\mathbf{J_D} = -\frac{det\mathbf{J_K}}{\theta} < 0$$

and the trace

$$tr\mathbf{J_D} = tr\mathbf{J_K} - \frac{1}{\theta} < 0,$$

where the inequalities are due to the assumptions $det\mathbf{J_K} > 0$ and $tr\mathbf{J_K} < 0$ in the Kaldor model. The characteristic equation of $\mathbf{J_D}$ is

$$\lambda^3 + a_1\lambda^2 + a_2\lambda + a_3 = 0, \tag{15}$$

where the coefficients are

$$a_1 = -tr\mathbf{J_D} > 0,$$

$$a_2 = \alpha(I_Y - S_Y)(I_K - \delta) - \frac{1}{\theta}\left(\alpha(I_Y - S_Y) + (I_K - \delta)\right),$$

$$a_3 = -det\mathbf{J_D} > 0.$$

If we assume continuously distributed time lag in the capital accumulation process, then $Y(t-\theta)$ is replaced by the expected income $Y^e(t)$, which is defined as the weighted average of the past realized incomes from zero to time t,

$$Y^e(t) = \int_0^t \frac{1}{\theta} e^{-\frac{t-s}{\theta}} Y(s)ds.$$

The delay 2D Kaldor–Kalecki model (13) can be reduced to a 3D system of ordinary differential equations:

$$\begin{cases} \dot{Y}(t) = \alpha\left[I(Y(t), K(t)) - S(Y(t))\right], \\ \dot{K}(t) = I(Y^e(t), K(t)) - \delta K(t), \\ \dot{Y}^e(t) = \frac{1}{\theta}\left(Y(t) - Y^e(t)\right), \end{cases} \tag{16}$$

where the last equation is obtained by time differentiation of $Y^e(t)$. The Jacobian matrix at the stationary point is

$$\mathbf{J_C} = \begin{pmatrix} \alpha(I_Y - S_Y) & \alpha I_K & 0 \\ 0 & I_K - \delta & I_Y \\ \frac{1}{\theta} & 0 & -\frac{1}{\theta} \end{pmatrix}.$$

It can be easily checked that the Jacobian matrix $\mathbf{J_C}$ has the same characteristic equation as (15). Hence the two different dynamic systems (14) and (16) generate the same dynamics in a neighborhood of the stationary point if θ is sufficiently small. According to the Routh–Hurwitz stability criterion, a necessary and sufficient condition that all roots of the cubic characteristic equation (15) have negative real parts is that $a_1 > 0$, $a_2 > 0$, $a_3 > 0$ and $a_1a_2 - a_3 > 0$. Notice that $a_1 > 0$ and $a_3 > 0$ are already shown to be positive due to Kaldor's assumptions. For sufficiently small θ, a_2 could be positive because its second term $-tr\mathbf{J_K}/\theta > 0$ is positive and can dominate the first term. By the same token $a_1a_2 - a_3$ can be positive for a small θ. Hence it is safe to presume that the delay Kaldor–Kalecki system is stable when the investment delay is sufficiently small. Since a small θ means a small lag effect, this result is reasonable under the assumption that the original Kaldor model is stable as shown in Fig. 4. The next question which we raise is whether or not the stability of the stationary state changes as the lengths of delays increase. We consider this question in model (16) only, since (14) is inappropriate for a large θ.

Now we turn our attention to the dynamic behavior of the delay Kaldor–Kalecki model with a large θ. As seen above, the coefficients a_1 and a_3 of the characteristic equation are positive. However, the sign of a_2 is not determined. Solving $a_2 = 0$ for θ yields the critical value of θ,

$$\theta_2 = \frac{\alpha(I_Y - S_Y) + (I_K - \delta)}{\alpha(I_Y - S_Y)(I_K - \delta)} > 0$$

implying that a_2 is positive for $\theta < \theta_2$. By the definitions of the coefficients of (15), we have

$$a_1 a_2 - a_3 = \frac{-A\theta^2 + B\theta - C}{\theta^2},$$

where

$$A = \alpha(I_Y - S_Y)(I_K - \delta)\left[\alpha\left(I_Y - S_Y\right) + (I_K - \delta)\right] > 0,$$

$$B = [\alpha(I_Y - S_Y) + (I_K - \delta)]^2 + \alpha I_K I_Y \gtreqless 0,$$

$$C = \alpha(I_Y - S_Y) + (I_K - \delta) < 0.$$

Let us denote the numerator of the last equation by $f(\theta)$. Since $f(\theta)$ is a concave quadratic polynomial, $f(0) = -C > 0$ implies that $f(\theta) = 0$ has one positive root, θ^*,

$$\theta^* = \frac{B + \sqrt{B^2 - 4AC}}{2A}.$$

Since $f(\theta^*) = 0$, $f(\theta) < 0$ for $\theta > \theta^*$. Furthermore $f(\theta_2) = (\theta_2)^2(-a_3) < 0$ and $\theta^* < \theta_2$ imply that $a_2 > 0$ at $\theta = \theta^*$. To emphasize the dependency of the coefficients on θ, we denote $a_i(\theta)$ for $i = 1, 2, 3$. For $\theta = \theta^*$, $a_1(\theta^*)a_2(\theta^*) - a_3(\theta^*) = 0$. By replacing $a_3(\theta^*)$ of the characteristic equation with $a_1(\theta^*)a_2(\theta^*)$, we are able to factor the characteristic equation,

$$(\lambda + a_1(\theta^*))(\lambda^2 + a_2(\theta^*)) = 0$$

that can be explicitly solved for λ. One of the three roots is real and negative whereas the other two are pure imaginary,

$$\lambda_1 = -a_1(\theta^*) < 0 \quad \text{and} \quad \lambda_{2,3} = \pm i\sqrt{a_2(\theta^*)} = \pm i\xi.$$

In order to apply the Hopf bifurcation theorem, we have to show that the real parts of the complex roots are sensitive to a change in the bifurcation parameter,

θ. Suppose that λ is a function of θ. By implicitly differentiating the characteristic equation with respect to θ we have

$$\frac{d\lambda(\theta^*)}{d\theta} = \frac{1}{\theta^{*2}} \frac{\lambda(\theta^*)^2 - tr\mathbf{J_K}\lambda(\theta^*) + det\mathbf{J_K}}{3\lambda(\theta^*)^2 + 2a_1(\theta^*)\lambda(\theta^*) + a_2(\theta^*)}.$$

Substituting $\lambda = \pm i\xi$ and arranging terms yield

$$\text{Re}\left(\frac{d\lambda(\theta^*)}{d\theta}\right) = \frac{1}{\theta^{*2}} \frac{\xi^2 - det\mathbf{J_K} - a_1(\theta^*)tr\mathbf{J_K}}{2(\xi^2 + a_1(\theta^*)^2)},$$

where the denominator is positive. We can show that the numerator is never zero. Substituting

$$a_1(\theta^*) = -tr\mathbf{J_K} + \frac{1}{\theta^*}$$

and

$$\xi^2 = a_2(\theta^*) = det\mathbf{J_K} + \alpha I_K I_Y - \frac{1}{\theta} tr\mathbf{J_K}$$

into the numerator and assuming that the resultant expression is zero yield

$$(tr\mathbf{J_K})^2 + \alpha I_K I_Y - \frac{2}{\theta} tr\mathbf{J_K} = 0.$$

However $a_2(\theta^*)a_1(\theta^*) = a_3(\theta^*)$ means that

$$\left(\frac{1}{\theta^*}\right)\left(det\mathbf{J_K} + \alpha I_K I_Y - \frac{1}{\theta^*} tr\mathbf{J_K}\right) = \frac{1}{\theta^*} det\mathbf{J_K}$$

which can be rewritten as

$$tr\mathbf{J_K}\left[\frac{1}{\theta^{*2}} - \alpha\left(I_Y - S_Y\right)\left(I_k - \delta\right)\right] = 0,$$

where the equality is impossible, since $tr\mathbf{J_K} < 0$, $I_Y - S_Y > 0$, $I_k - \delta < 0$ and $\theta^* > 0$. Therefore we have

$$\text{Re}\left(\frac{d\lambda(\theta^*)}{d\theta}\right) \neq 0.$$

This implies that the real parts of the complex roots change signs as $\theta - \theta^*$ changes from negative to positive values. That is, it guarantees the existence of Hopf bifurcation.

Theorem 2. *The Kaldor–Kalecki model with continuously distributed lags having an exponential kernel function is locally asymptotic stable for* $0 \leq \theta < \theta^*$ *while it loses the stability at* $\theta = \theta^*$ *via a Hopf bifurcation.*

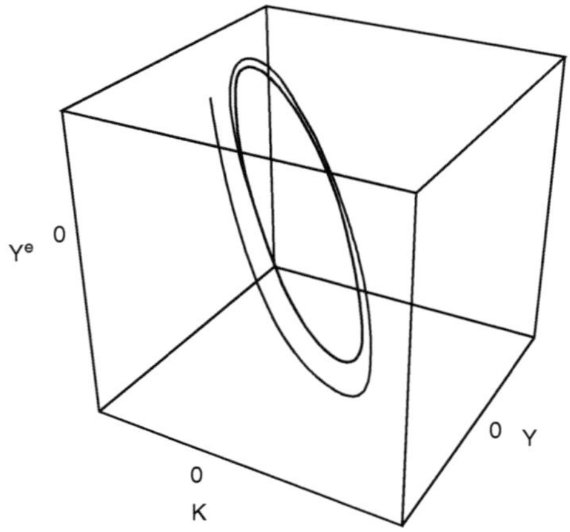

Fig. 5 Existence of a limit cycle in the delay Kaldor–Kalecki model

It is uncertain whether the limit cycle is subcritical or supercritical. In Fig. 5, simulation results are shown with $\theta = 0.7$ and parameter values $c = 0.6$, $\alpha = 0.8$, $\beta = -0.4$ and $\delta = 0.2$, which are the same as in the simulation study presented in Fig. 4. The critical value is $\theta^* \simeq 0.37$. The delay Kaldor–Kalecki model generates a supercritical limit cycle due to the destabilizing effect of the investment lag.

3.3 Delay Nonlinear Cournot Model

We will examine a dynamic Cournot duopoly game when a firm has an information lag in the receipt of information about its competitor's output. We assume that each firm adaptively adjusts its output to the desired level of output:

$$\begin{cases} \dot{x}_1(t) = k_1 \left\{ R_1(x_2(t-\theta_1)) - x_1(t) \right\}, \\ \\ \dot{x}_2(t) = k_2 \left\{ R_2(x_1(t-\theta_2)) - x_2(t) \right\}, \end{cases} \tag{17}$$

where x_i, k_i, θ_i and $R_i(x_j)$ are output, a positive adjustment coefficient, a time lag and the best reply function of firm i for $i, j = 1, 2$ and $i \neq j$. Special duopoly models such as the classical Cournot model with a linear price function and a nonlinear Cournot model with a unit-elastic price function will be considered later to specify the best reply functions.

To consider a linearization of the system, we suppose that the information lags are sufficiently small and an advance θ_1 time in the first equation of (17) and an advance θ_2 time in the second one:

$$\dot{x}_1(t+\theta_1) = k_1\{R_1(x_2(t)) - x_1(t+\theta_1)\},$$

$$\dot{x}_2(t+\theta_2) = k_2\{R_2(x_1(t)) - x_2(t+\theta_2)\}.$$

Define the difference between the left-hand side and the right-hand side by

$$F_1(\theta_1) = \dot{x}_1(t+\theta_1) - k_1\{R_1(x_2(t)) - x_1(t+\theta_1)\}$$

and

$$F_2(\theta_2) = \dot{x}_2(t+\theta_2) - k_2\{R_2(x_1(t)) - x_2(t+\theta_2)\}.$$

Differentiating each function with its lag at $\theta_i = 0$ and arranging terms yield

$$\theta_1\ddot{x}_1(t) = -k_1\theta_1\dot{x}_1(t) - \dot{x}_1(t) + k_1\{R_1(x_2(t)) - x_1(t)\}$$

and

$$\theta_2\ddot{x}_2(t) = -k_2\theta_2\dot{x}_2(t) - \dot{x}_2(t) + k_2\{R_2(x_1(t)) - x_2(t)\}.$$

Introducing the new variables $y_1(t) = \dot{x}_1(t)$ and $y_2(t) = \dot{x}_2(t)$, we can transform the 2D delay differential equation system (17) into the following 4D system of ordinary differential equations:

$$\begin{cases} \dot{x}_1(t) = y_1(t), \\ \dot{x}_2(t) = y_2(t), \\ \dot{y}_1(t) = \dfrac{k_1}{\theta_1}\{R_1(x_2(t)) - x_1(t)\} - \left(k_1 + \dfrac{1}{\theta_1}\right)y_1(t), \\ \dot{y}_2(t) = \dfrac{k_2}{\theta_2}\{R_2(x_1(t)) - x_2(t)\} - \left(k_2 + \dfrac{1}{\theta_2}\right)y_2(t). \end{cases} \tag{18}$$

The Jacobian matrix is

$$\mathbf{J_L} = \begin{pmatrix} 0 & 0 & 1 & 0 \\ 0 & 0 & 0 & 1 \\ -\dfrac{k_1}{\theta_1} & \dfrac{k_1}{\theta_1}\gamma_1 & -\left(k_1 + \dfrac{1}{\theta_1}\right) & 0 \\ \dfrac{k_2}{\theta_2}\gamma_2 & -\dfrac{k_2}{\theta_2} & 0 & -\left(k_2 + \dfrac{1}{\theta_2}\right) \end{pmatrix},$$

where γ_i is the derivative of $R_i(x_j)$ evaluated at the stationary point. The characteristic equation of $\mathbf{J_L}$ can be written as

$$a_0\lambda^4 + a_1\lambda^3 + a_2\lambda^2 + a_3\lambda + a_4 = 0, \tag{19}$$

where

$$\begin{aligned} a_0 &= \theta_1\theta_2, \\ a_1 &= \theta_1 + \theta_2 + (k_1 + k_2)\theta_1\theta_2, \\ a_2 &= 1 + k_1k_2\theta_1\theta_2 + (k_1 + k_2)(\theta_1 + \theta_2), \\ a_3 &= k_1 + k_2 + k_1k_2(\theta_1 + \theta_2), \\ a_4 &= k_1k_2(1 - \gamma_1\gamma_2). \end{aligned} \tag{20}$$

The above procedure is suitable for a situation in which the information lag is fixed and sufficiently small. If the lags are uncertain, we can model time lags in a continuously distributed manner. If firm 1's expectation of the competitor's output is denoted by $x_2^e(t)$ and firm 2's expectation of the competitor's output is denoted by $x_1^e(t)$ and both expectations are based on the entire history of the outputs from zero up to t with exponentially decreasing weights, then the delay differential equation system (17) can be written as the 2D system of integro-differential equations:

$$\begin{cases} \dot{x}_1(t) = k_1 \left\{ R_1(x_2^e(t)) - x_1(t) \right\}, \\ \dot{x}_2(t) = k_2 \left\{ R_2(x_1^e(t)) - x_2(t) \right\}, \end{cases} \tag{21}$$

with

$$x_1^e(t) = \int_0^t \frac{1}{\theta_1} e^{-\frac{t-s}{\theta_1}} x_1(s)ds$$

$$x_2^e(t) = \int_0^t \frac{1}{\theta_2} e^{-\frac{t-s}{\theta_2}} x_2(s)ds.$$

This system is equivalent to the following 4D system of ordinary differential equations:

$$\begin{cases} \dot{x}_1(t) = k_1 \left\{ R_1(x_2^e(t)) - x_1(t) \right\}, \\ \dot{x}_2(t) = k_2 \left\{ R_2(x_1^e(t)) - x_2(t) \right\}, \\ \dot{x}_1^e(t) = \dfrac{1}{\theta_1}(x_1(t) - x_1^e(t)), \\ \dot{x}_2^e(t) = \dfrac{1}{\theta_2}(x_2(t) - x_2^e(t)). \end{cases}$$

The Jacobian of this system can be written as

$$\begin{pmatrix} -k_1 & 0 & 0 & k_1\gamma_1 \\ 0 & -k_2 & k_2\gamma_2 & 0 \\ \frac{1}{\theta_1} & 0 & -\frac{1}{\theta_1} & 0 \\ 0 & \frac{1}{\theta_2} & 0 & -\frac{1}{\theta_2} \end{pmatrix}.$$

Simple calculation shows that the characteristic equation of this matrix can be written as a quartic equation in λ:

$$a_0\lambda^4 + a_1\lambda^3 + a_2\lambda^2 + a_3\lambda + a_4 = 0$$

with the same coefficients as defined in (20). The identical characteristic equation means that (17) and (21) exhibit the same dynamics in a neighborhood of the stationary point as Theorem 1 claims.

If $\gamma_1\gamma_2 < 1$, then all coefficients of the characteristic equation are positive, and the Routh–Hurwitz theorem implies that the roots have negative real parts if and only if

$$\begin{vmatrix} a_1 & a_0 \\ a_3 & a_2 \end{vmatrix} > 0 \quad \text{and} \quad \begin{vmatrix} a_1 & a_0 & 0 \\ a_3 & a_2 & a_1 \\ 0 & a_4 & a_3 \end{vmatrix} > 0.$$

The first condition is satisfied because the second-order determinant is always positive,

$$(k_1+k_2)(1+k_1\theta_2)(1+k_2\theta_2)\theta_1^2+\theta_2(1+(k_1+k_2)\theta_1)+\theta_1(1+(k_1+k_2)\theta_2) > 0.$$

The second condition depends on the value of $\gamma_1\gamma_2$. Expanding the third-order determinant, and solving the inequality gives a lower bound for $\gamma_1\gamma_2$, and by combining it with the upper bound $\gamma_1\gamma_2 < 1$, we get the following condition for the local asymptotic stability of the stationary state:

$$1 > \gamma_1\gamma_2 > -\frac{(k_1+k_2)(1+k_1\theta_1)(1+k_2\theta_1)(\theta_1+\theta_2)(1+k_1\theta_2)(1+k_2\theta_2)}{k_1k_2(\theta_1+\theta_2+\theta_1\theta_2(k_1+k_2))^2}. \tag{22}$$

In the case of the linear Cournot model, the price function is given by

$$p = a - b(x_1 + x_2)$$

and so the profit function of firm i is defined as

$$\pi_i = (a - b(x_1 + x_2))\, x_i - c_i x_i,$$

where c_i is the constant marginal cost. The best reply function and its derivative are

$$R_i(x_j) = \frac{a - c_i - bx_j}{2b} \quad \text{and} \quad \gamma_i = -\frac{1}{2}.$$

Since $1 > \gamma_1\gamma_2 = 1/4 > 0$, (22) is satisfied. Hence the delay linear Cournot model is always stable for any values of information lags, θ_i.

In the case of the unit-elastic demand, the price function is given by

$$p = \frac{1}{x_1 + x_2}$$

and the profit function of firm i is defined as

$$\pi_i = \frac{x_i}{x_1 + x_2} - c_i x_i.$$

Assuming an interior solution, the profit maximization yields a bell-shaped best reply function,

$$R_i(x_j) = \sqrt{\frac{x_j}{c_i}} - x_j.$$

Cournot outputs are determined by an intersection of the best reply curves,

$$x_1^C = \frac{c_2}{(c_1 + c_2)^2} \quad \text{and} \quad x_2^C = \frac{c_1}{(c_1 + c_2)^2}.$$

The derivatives of the best response functions evaluated at the Cournot point are derived as

$$\gamma_1 = -\frac{c_1 - c_2}{2c_1} \quad \text{and} \quad \gamma_2 = \frac{c_1 - c_2}{2c_2}.$$

If there are no time lags, the dynamic system is represented by (17) with $\theta_i = 0$. The asymptotic properties of the trajectories $x_1(t)$ and $x_2(t)$ depend on the location of the eigenvalues of the Jacobian matrix of the system. The eigenvalues are obtained by solving the associated characteristic equation,

$$\lambda^2 + (k_1 + k_2)\lambda + k_1 k_2(1 - \gamma_1\gamma_2) = 0.$$

Here $k_1 + k_2 > 0$ by the definition of the adjustment coefficient and $\gamma_1\gamma_2 < 1$, since

$$\gamma_1\gamma_2 = -\frac{(1-c)^2}{4c} \quad \text{with} \quad c = \frac{c_2}{c_1}.$$

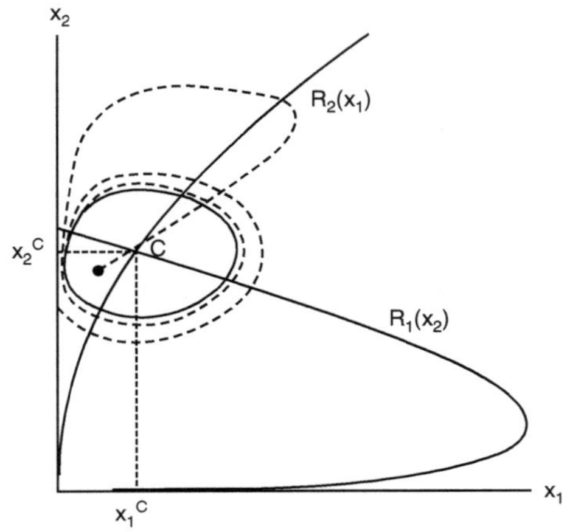

Fig. 6 The birth of a Cournot cycle

The roots of the characteristic equation have negative real parts. Hence the nonlinear Cournot model with no information lags is always asymptotically stable.[1]

Now we examine the asymptotic behavior of the delay nonlinear Cournot model. The value of $\gamma_1\gamma_2$ can be any negative number between $-\infty$ and zero by the appropriate choice of the cost ratio c. Notice that the stability condition (22) is violated if $\gamma_1\gamma_2$ is negative with large absolute value. In particular, Fig. 6 illustrates the dynamic behavior of the trajectories when the stability condition is violated, in which the parameters are specified as $k_1 = k_2 = 0.8$, $\theta_1 = \theta_2 = 2$, $c_1 = 1$ and $c_2 = 0.045$. It can be seen that a trajectory starting at the dot point converges to a limit cycle surrounding a locally unstable Cournot point.[2]

4 Concluding Remarks

Delay models with fixed lags and models with continuously distributed delays were compared in this paper. By selecting exponential kernel function, we first proved that with small delays the two types of dynamics generate identical local asymptotic properties. However with large delays this interesting equivalence was not true anymore.

[1] The discrete-time version of the nonlinear Cournot model has been extensively studied, and it is demonstrated that simple nonlinear best reply functions can generate a very rich dynamics involving chaos and multistability (Bischi et al., 2009; Puu, 2003; Puu and Sushko, 2002). The delay differential Cournot model with product differentiation is considered in Matsumoto and Szidarovszky (2007).

[2] A trajectory seems to cross itself as dynamics generated in a 4D space is projected to a 2D space.

Three particular economic models (Goodwin's business cycle model, Kaldorian business cycle model with Kaleckian investment lag and the Cournot oligopoly model) illustrated the theoretical results, and computer simulations showed that more complex dynamics emerged if a large value of time delay was selected.

Acknowledgements This paper was done when the first author visited the Department of Systems and Industrial Engineering of the University of Arizona. He appreciated its hospitality over his stay. The authors wish to express their gratitude to an anonymous referee for useful comments and also want to acknowledge the encouragement and support by Kei Matsumoto for the research leading to this paper. They appreciate financial support from Chuo University (Joint Research Project 0981) and the Japan Ministry of Education, Culture, Sports, Science and Technology (Grant-in-Aid for Scientific Research (C) 21530172). The usual disclaimer applies.

References

Bellman, R. (1969). *Stability theory of differential equations*. New York: Dover.

Bischi, G., Chiarella, C., Kopel, M., & Szidarovszky, F. (2009). *Nonlinear oligopolies: Stability and bifurcations*. Berlin: Springer.

Chang, W., & Smyth, D. (1971). The existence and persistence of cycles in a non-linear model: Kaldor's 1940 model reexamined. *Review of Economic Studies, 38*, 37–44.

Chiarella, C., & Szidarovszky, F. (2004). Dynamic oligopolies without full information and with continuously distributed time lags. *Journal of Economic Behavior and Organization, 54*, 495–511.

Goldberg, S. (1958). *Introduction to difference equations*. New York: Wiley.

Goodwin, R. (1951). The nonlinear accelerator and the persistence of business cycle. *Econometrica, 19*, 1–17.

Huang, W. (2008). Information lag and dynamic stability. *Journal of Mathematical Economics, 44*, 513–529.

Kaldor, N. (1940). A model of the trade cycle. *Economic Journal, 50*, 78–92.

Kalecki, M. (1935). A macroeconomic theory of the business cycle. *Econometrica, 3*, 327–344.

Matsumoto, A. (2009). Note on Goodwin's 1951 nonlinear accelerator model with an investment lag. *Journal of Economic Dynamics and Control, 33*, 832–842.

Matsumoto, A., & Szidarovszky, F. (2007). Delayed nonlinear cournot and bertrand dynamics with product differentiation. *Nonlinear Dynamics, Psychology and Life Sciences, 11*, 367–395.

Matsumoto, A., & Szidarovszky, F. (2009). *Dynamic Goodwin's business cycle model with continuously distributed time lags* (DP 115). Institute of Economic Research, Chuo University.

Puu, T. (2003). *Attractors, bifurcations and chaos* (2nd edn.). Berlin: Springer.

Puu, T., & Sushko, I. (2002). *Oligopoly dynamics*. Berlin: Springer.

Imperfect Competition, Learning and Fluctuations

Piero Ferri and Anna Maria Variato

1 Introduction

The strategic roles of imperfect competition and monopolistic competition have undergone considerable changes since their inception in the thirties, when (Robinson 1933) and (Chamberlin 1933) first and independently formalized the two similar concepts. For instance, their field of application has transcended microeconomics to occupy an important place in macroeconomics, where they are supposed to be superior with respect to perfect competition. This happened some time ago (see Kalecki 1971; Minsky 1954, 2004), and it has been recently reformulated in the so called "new" Keynesian literature (see Blanchard and Kiyotaki 1987).

This potential superiority, however, has been strongly limited by the presence of two hypotheses, namely the symmetry hypothesis and the rational expectation assumption, which are instead untenable in a world of imperfect knowledge economics (IKE, according to the definition of Frydman and Goldberg 2007). Specifically, in such an environment, heterogeneity among agents rules out the symmetry hypothesis. Furthermore, the presence of imperfect knowledge suggests that the interdependence processes characterizing heterogeneous agents must be dealt with in a simplified macro approach. Finally, agents have to learn the dynamic process from the data.

The aim of the present paper is to analyze the dynamic effects that arise from the simultaneous presence of imperfect competition and uncertainty in a medium-run perspective. One of the most important contributions of our approach lies in the choice of this span of time. On the surface, the emphasis on medium-run prevents our analysis taking into account cycles showing either too high or too low frequencies. In other words, the fluctuations we deal with are not business cycles as traditionally meant. More deeply, at a methodological level, the focus on medium-run requires consideration of a model in which aggregate demand and aggregate

P.E. Ferri (✉)
Università di Bergamo, Dipartimento di Economia via dei Caniana, 2 24127 Bergamo, Italy,
e-mail: pietro-enrico.ferri@unibg.it

G.I. Bischi et al. (eds.), *Nonlinear Dynamics in Economics, Finance and the Social Sciences*, DOI 10.1007/978-3-642-04023-8_12,

supply are intertwined: neither aggregate demand nor aggregate supply can take the lead, as conventionally assumed when studying short-run as opposed to long-run.

Further contributions appear in the mechanisms generating the dynamics of the model. In particular, the simultaneous adoption of non-linear equations and a regime switching mechanism, implies the use of simulations and gives rise to one of the engines of fluctuations.

Given this theoretical framework, we focus on three specific issues. First of all we try to assess whether fluctuations generated in the model are persistent. Secondly, we try to evaluate the effects on dynamics of the presence of uncertainty and the consequent need of agents to learn about the environment they face. Finally, we check the robustness of the model to significant changes in the parameters involving the strategic variables related to competition.

The structure of the paper is the following. Section 2 illustrates the methodological aspects. Section 3 defines the role of imperfect competition in a world with uncertainty. Sections 4 and 5 deal respectively with aggregate supply and aggregate demand problems. Section 6 introduces the remaining equations of the model. Section 7 explains the learning process. Section 8 shows the persistent fluctuations generated by the model, while Sect. 9 illustrates its robustness. Section 10 concludes.

2 Medium-run Fluctuations and Regime Switching

When considering stylized facts in advanced industrialized countries there is a tendency to concentrate on polar cases: either the short-run vibrations of the economy are considered (as it happens in the short run forecast of the economy) or very long-run spans of time are taken into account, as usually happens when dealing with technical change. (Solow 2000) has insisted on the importance of a medium-run approach (see also Blanchard 1997). In this perspective, the stylized facts featuring fluctuations become particularly revealing (see Comin and Gertler 2006).

The oscillations between growth and stagnation can be studied by adopting the hypothesis of multiple equilibria. In this perspective, the economy is characterized either by a "bad" state (state 1), where debt is high and growth is low, or by a "good" state (state 2), marked by the situation of sustained growth accompanied by low debt. In order to obtain multiple equilibria, nonlinear relationships are usually introduced. Alternatively, one might refer to piecewise linear techniques, which assume that certain functions change discontinuously when they reach a threshold.[1] In order to implement this approach, one needs to move along three steps. First of all, a threshold must be identified. In the present case, it is represented by a particular value of the rate of income growth (gth), supplemented by a stochastic term to

[1] This is the strategy followed in this paper (see also Ferri et al. 2001).

soften the determinism of the model:[2]

$$g^{th} = \frac{g_{01} + g_{02}}{2} + \varepsilon_t$$

In other words, the threshold is given by the average of the two steady state values of the rate of growth g_{0j} (where the suffix 0 stands for a steady state value and $j = 1, 2$ represents respectively regime 1 and 2); ε t is a stochastic variable, normally distributed with a zero mean and a constant variance.

Secondly, one has to identify the equations undergoing changes when the threshold is reached.[3] In the present paper, the supply equations undergo changes, i.e. the price equation, the productivity function and income distribution. It is important to stress that these changes may refer either to the values of the parameters or to the steady state values.

The final step of a regime switching model consists in considering the dynamics. It is important to stress that growth cycles depends on what happens: (a) within each state, (b) between the two states and (c) the time spent in each regime. In other words, history matters (see also Day and Walter 1989).

3 Imperfect Competition and Uncertainty

Any macro analysis that focuses attention on the role of imperfect competition is bound to face a fundamental challenge: its presumed superiority to perfect competition (see Solow 1998). The main difficulty associated with the choice of this market structure lies in the openness of its definition, as opposed to the uniqueness of features characterizing perfect competition. There is no way to define univocally a macro model grounded on imperfect competition, as different markets may be imperfect in different ways (or degrees). Such a limit does not apply to perfect competition where, in addition, the properties of microeconomic equilibrium extend to the macroeconomic one without variations. In contrast, the macro achievements of imperfect competition depend on the overall hypothesis that have been put forward. Symmetric equilibrium is one of these examples. The hypothesis of symmetric equilibrium seems too strong for a macro analysis that intends to analyze short to medium-run phenomena characterized by some degree of uncertainty. If one drops this hypothesis by allowing agents heterogeneity, the analysis becomes extremely challenging (see also Delli Gatti and Gallegati 2002; Aoki and Yoshikawa 2007).

The basic departure from perfect competition is represented by the violation of the hypothesis related to the amount and distribution of information among

[2] According to (Frydman and Goldberg 2007), the IKE environment is not sufficiently illustrated by the above hypothesis.

[3] Changes can be also smooth as happens in the so called STAR models. See (Tong 1990; Ferri 2008). Furthermore, this is a macro threshold. For a micro interpretation, see, for instance, (Yokoo and Ishida 2008).

individuals. Broadly speaking, the presence of uncertainty tends, on one hand, to favor heterogeneity, (except probably in those extreme situations such as the Keynesian liquidity trap that sometime happen in the economy) and to impose, on the other, limitations on the study of the process of the interdependence between the various agents.[4] More specifically, in an imperfect knowledge economy, two tenets are worth stressing. To begin with, differently from what is assumed by general equilibrium, information is asymmetric as a result of agents (or sectors) heterogeneity and uncertainty. It follows that most interactions can only be dealt with by rough approximations, which are typical in macro approaches. In the second place, in the presence of uncertainty, which is a stronger version of the concept of risk as shown by (Knight 1921), the hypothesis of rational expectations that underlies the so called new neoclassical synthesis models faces increasing difficulties. The hypothesis of bounded rationality, where agents acts as econometricians and learn, are becoming more popular (see Evans and Honkapohja 2001).

Even though uncertainty implies individual heterogeneity, in the present paper we shall abstract from this dimension in order to be able to emphasize the specific impact of other aspects of imperfect competition. In particular, we focus on the effects of introducing a unique learning mechanism, which represents a significant departure from the rational expectation hypothesis.

One implication of these considerations is that the supply equation in the macro model becomes less straightforward than is usually supposed (see Howitt 2006). In particular, in the model at hand, an inflation equation of the following type is considered:

$$\pi_t = \varphi[\varphi_1 \overline{E_t} \pi_t + (1-\varphi_1)\pi_{t-1}] + (1-\varphi)\pi_{mt} - \sigma_1 \left(u_t - u_{0j}\right) \tag{1}$$

where π_t and π_{mt} represent inflation respectively referred to final products and the raw materials, u_t is the unemployment rate, while u_{0j} is the NAIRU in the prevailing regime. It is worth stressing that all the variables should be indexed by $j = 1, 2$. However, in order to simplify the notations, the index j is only used for steady states and for parameters that change from one regime to the other.

The presence of π_{mt} is based on the assumption of separability in the production function, stressed by (Rotemberg and Woodford 1996). This equation is compatible with a so called new-Keynesian Phillips curve (see Woodford 2003). However, in this case, expectations (E) are not rational. Finally, the dynamics of π_{mt} are supposed to be the following:

$$\pi_{mt} = \pi_{mt-1}[1 + \lambda_j(g_t - g_{0j})] \tag{2}$$

This equation epitomizes the difficulties of agents to have precise information on the market for raw materials and hence the necessity of referring to some rule of thumb in order to make forecasts.[5] According to (De Grauwe 2008) this is just an example

[4] There are at last four ways in which these processes can be analyzed: (a) the game approach; (b) the statistical physics approach suggested by (Aoki and Yoshikawa 2007); (c) so called agent- based computational approach (ACE) based upon the simulation of interacting agents (see Tesfatsion 2006; Leijonhufvud 2006); (d) the macro approach followed in this paper.

[5] According to IMF (2008) the future on the oil price ranges from 64$ to 145$ per barrel for 2009.

of simple rule or heuristic that characterizes a world of imperfect information and cognitive limitation of individuals. In this equation, $\lambda_2 > \lambda_1$, which implies that the price of raw materials accelerates in regime 2, when the threshold rate of growth gth is crossed.

4 The Supply Equations

In order to consider further aspects of the supply side of the model, a fundamental equation is represented by the dynamics of labour productivity, which can be expressed in the following way:

$$\tau_t = \tau_{1j} + \tau_2 g_{k,t} \quad (j = 1, 2) \tag{3}$$

where τ_{1j} represents the exogenous component and where g_k stands for the rate of growth of capital. The hypothesis

$$\tau_{12} > \tau_{11}$$

implies that the exogenous component of technical change increases when a threshold rate of growth is reached.[6] In order to introduce the link between g_t and $g_{k,t}$ one must introduce the capital-output ratio (v), which is given by the following equation:

$$v_t = v_{t-1} \frac{1 + g_{k,t}}{1 + g_t} \tag{4}$$

The growth rate of capital is set by the following relation:

$$g_{k,t} = \frac{i_t}{v_{t-1}} - \delta \tag{5}$$

where δ represents the exogenous depreciation rate, while $i_t = I_t/Y_{t-1}$ is the investment ratio to be determined by the aggregate demand side of the model.

In this context, labour demand is given by:

$$l_t = l_{t-1} \frac{(1 + g_t)}{(1 + \tau_t)} \tag{6}$$

where l_t represents the employment ratio, referred to a normalized labor supply. It follows that unemployment (u_t) is given by:

$$u_t = 1 - l_t \tag{7}$$

[6] On the relationship with Kaldorian hypothesis, see (Ferri 2007).

5 The Structure of Aggregate Demand

According to a review article by (Blanchard 2008, p.5): "One major fact is that shifts in the aggregate demand for goods affect output substantially more than we would expect in a perfectly competitive economy." In the new neoclassical synthesis models, the persistent role of aggregate demand in shaping the dynamics of fluctuations have been mainly explained in terms of rigidities.[7] Instead of assuming the existence of such rigidities one can see them as the result of deeper causes. Uncertainty and imperfect competition may be among them. From a methodological point of view, this choice overcomes the common "ad-hoc assumption" critique addressed to new neoclassical synthesis models. While imperfect competition determines directly rigidities operating on the supply side, uncertainty causes rigidities on the demand side too. Furthermore uncertainty helps to enlighten how nominal and real rigidities, if one insists on this terminology, are interdependent. Specifically, this interdependence is shown by the relationship between consumption, investment and nominal debt. Instead of emphasizing the impact of debt on cash flow and investment (see Fazzari et al. 2008), the present paper stresses the relationship with consumer spending because of its relevance to the present economic conditions.

In particular, consumption is a positive function of some target income and a negative function of the interest rate on the accumulated debt (deflated by expected inflation):

$$C_t = cY_t^* - c_3 \frac{R_t D_t}{\overline{E}_t P_t}$$

where Y_t^* is the target income, while the remaining part represents the debt service deflated by expected inflation. R_t, i.e. the nominal rate of interest, appears in nominal terms because is predetermined with respect to prices. In the present paper: (a) it is assumed that the target income is a weighted average of expected and last year income; (b) the equation is normalized by Y_{t-1} on both sides; (c) furthermore, since the debt ratio is defined as:

$$d_t = \frac{R_t D_t}{P_{t-1} Y_{t-1}}$$

one obtains:

$$c_t = c_1(1 + \overline{E}_t g_t) + c_2 - c_3 \frac{R_t d_t}{(1 + \overline{E}_t \pi_t)} \tag{8}$$

[7] (Fazzari et al. 1998) have shown, in a static model, that the role of aggregate demand in a model with imperfect competition can be more direct. On the essential role of aggregate demand in a monetary economy of production, see also (Pasinetti 2007).

Inserting the consumption function into the equilibrium condition that defines aggregate demand equals supply in an open economy, one gets the following expression:

$$g_t = i_t + c_1\left(1 + \overline{E}_t g_t\right) + c_2 - c_3 \frac{R_t d_t}{1 + \overline{E}_t \pi_t} - 1 - \frac{(1 + \pi_{mt})}{(1 + \pi_t)} m_{0j} \tag{9}$$

where c_1 and c_2 represent the propensity to consume past and forecast income, while c_3 measures the impact of debt service. Two aspects are worth considering. First of all, in dynamic terms, this equality implies that 1 plus the rate of growth of output (g_t) must be equal to the sum of the investment ratio ($i_t = I_t / Y_{t-1}$) and the consumption ratio, diminished by the share of imports. The last term illustrates the impact of the import ratio (m), corrected by changes in the terms of trade. In the second place, the consumption function (i.e. the second, third and fourth components on the r.h.s. of (9)), stresses the relationship between income distribution, financial aspects and institutional factors. More precisely, in this formulation, debt increases from interest and consumption and diminishes because of wages received. The debt ratio[8] evolves according to the following formula:

$$d_t = \frac{d_{t-1}\left(1 + R_{t-1}\right)}{\left(1 + g_{t-1}\right)\left(1 + \pi_{t-1}\right)} + \frac{c_{t-1}}{\left(1 + g_{t-1}\right)} - \omega_{0j} \tag{10}$$

where ω_{0j} stands for the labor income share indexed by the regime status. Since debts contracts are predetermined in nominal terms, inflation can affect them. This is why π appears in the denominator.[9]

The interdependence between real and financial aspects is mainly concentrated in the consumption function. Consequently, the investment function has been rather simplified; it depends on both the accelerator and the (simplified) cost of capital (r) (see Fazzari et al. 2008)[10]:

$$i_t = \eta_1 + \eta_2 \overline{E} g_t - \eta_3\left(r_t - r_{0j}\right) \tag{11}$$

6 Completing the Model

The model tries to integrate aggregate demand and supply aspects (see also Asada et al. 2006) in a medium-run perspective, where labor supply has been normalized. Other equations must be presented in order to close the model.

---

[8] The equation for nominal debt at the beginning of the period is $D_t = D_{t-1}\left(1 + R_{t-1}\right) + P_{t-1}C_{t-1} - W_{t-1}N_{t-1}$. Dividing by $P_{t-1}Y_{t-1}$, one obtains the formula in the text.

[9] This formula is different when referred to debt of the firms.

[10] On the attempt of microfounding this equation, see (Minsky 2004).

In detail, monetary authorities fix the nominal rate of interest (R) according to a Taylor rule of the type:

$$R_t = R_{0j}^* + \psi_1 \left(\overline{E}_t \pi_t - \pi_{0j}\right) + \psi_2 \left(\overline{E}_t g_t - g_{0j}\right) \tag{12}$$

The monetary authorities are supposed to try to reach the nominal rate target ($R_{0j}{}^*$), along with the inflation and growth objectives according to the prevailing regime j.

The real rate of interest is related to the nominal rate by the Fisher formula:

$$r_t = \frac{(1 + R_t)}{\left(1 + \overline{E}_t \pi_t\right)} - 1 \tag{13}$$

Finally, income distribution equation is given by:

$$\omega_t = \omega_{0j} \tag{14}$$

Income distribution is exogenous within each regime, but varies between regimes. In other words, one can assume that the mark-up of imperfect competitive firms varies according to the prevailing regime. This implies that the labour share (ω) has different steady state values in the two regimes:

$$\omega_{01} \neq \omega_{02}$$

For given expectations, each regime contains the following 14 unknowns in 14 equations:

$$c_t, d_t, i_t, g_t, l_t, u_t, \pi_t, \pi_{mt}, \omega_t, R_t, r_t, g_k, v_t, \tau_t$$

In economic terms, the steady state of the model is defined by the fulfilment of expectations, and the constancy of the unknowns (rates of growth and ratios).[11]

7 Learning

Macro relationships simplify a complex world that is subject to change: it follows that the values of the parameters, along with the equations themselves, are bound to change. In this context, a sophisticated approach to expectations is needed. The assumption is made that in forming expectations firms assume a simple recursive least square device (see Branch and Evans 2006).[12] In more detail, assume that the

[11] The steady state of growth is given by $g_{0j} = \frac{\tau_{11} + \tau_2 \delta}{1 - \tau_2}$, while the steady state value of inflation is given by $\pi_{0j} = \frac{B_j + (1 + R_{0j})}{1 + g_{0j}} - 1$, where B_j is obtained by referring to (1), (10) and (13).

[12] Others have assumed a Markov process as (Ferri 2007).

economic law of motion takes the form of these two equations:

$$\begin{cases} y_{j,t} = b_{j,t}^T x_t + \varepsilon_{j,t} & j = 1, \dots, n \\ b_{j,t} = b_{j,t-1} + \theta_{j,t} \end{cases}$$

where $b_{j,k}$ is the (Kx1) parameter vector and x_i is the (Kx1) vector of explanatory variables; $\varepsilon_{j,t}$ and $\theta_{j,t}$ are assumed to zero mutually independent random sequences with variances equal respectively to $R_{j2,t}$ and $R_{j1,t}$. In a vector autoregression (VAR) specification, conditional forecasts of $y_{j,t}$ are given by:

$$y_{j,t|t-1} = \hat{b}_{j,t-1}^T x_t$$

The forecasting problem consists in selecting an appropriate procedure for constructing the sequence of $\hat{b}_t$. One such procedure is the Kalman filter recursion. As shown by (Sargent 1999), the Kalman filter is equivalent to recursive least squares (RLS) for particular restrictions of the variances R_1 and R_2. RLS is simply a recursive formulation of ordinary least squares (see Evans and Honkapohja 2001).

In what follows, agents are supposed to form expectations according to a RLS device. (Hommes and Sorger 1998) argue that expectations must be consistent with data in the sense that agents do not make systematic errors. This criterion implies that, at least, the forecasts and the data should have similar means and autocorrelations.[13]

8 Persistent Fluctuations

The dynamics of the model are generated by a nonlinear system of equations supplemented by the regime switching mechanisms just described. Technically, one should presents two systems of equations, one for each state. However, in order to economize space, only the meta system will be presented, indexed by $j = 1, 2$. Furthermore, only parameters that switch, along with the steady states, will be indexed, while the endogenous variable is only indexed by time.

The system of structural equations, along with the forecasting rules, is nonlinear and can be solved only by means of simulations.

The results of the simulations ($N = 200$)[14] are illustrated in Fig. 1, while the values of the parameters are shown in Table 1.

[13] The assumption that all the agents have the same learning process simplify both the aggregation (underlined by De Grauwe 2008) and the coordination problem (stressed by Howitt 2006).

[14] These simulations have been reiterated 20 times due to the presence of a stochastic components in the threshold value. The variance has been put equal to 2.5.

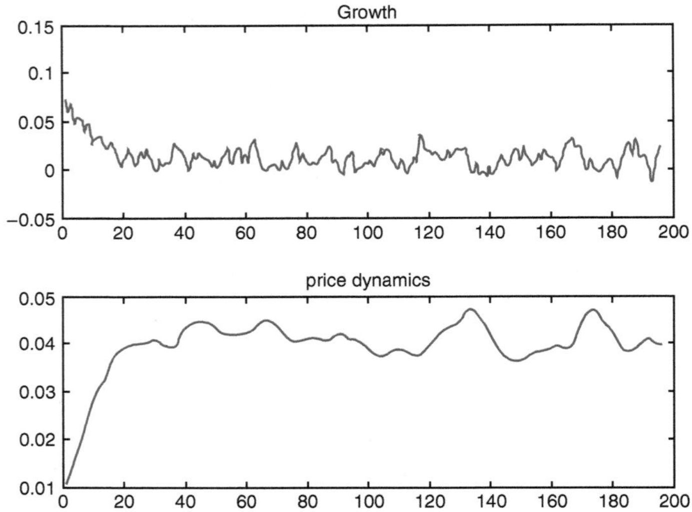

Fig. 1 The dynamics of the model

Table 1 Parameters of the model

$u_{0j} = .075$	$\varphi_1 = 0.8$	$\sigma_1 = 0.01$
	$\varphi = 0.7$	
$\tau_{11} = 0.002$	$\tau_2 = 0.03$	$\eta_1 = 0.20$
$\tau_{12} = 0.01$		
$\eta_2 = 0.40$	$\eta_3 = 0.60$	$c_1 = 0.40$
$c_2 = 0.405$	$c_3 = 0.10$	$\psi_1 = 1.80$
		$\psi_2 = 0.45$
$\lambda_1 = 0.95$	$\omega_{01} = 0.75$	$R_{01}{}^* = 0.005$
$\lambda_2 = 1.30$	$\omega_{02} = 0.78$	$R_{02}{}^* = 0.08$
		$m_0 = 0.001$

The dynamics are fuelled by two kinds of forces. The first is represented by aggregate demand and supply interaction that takes place when the threshold is randomly crossed. The second kind of forces refers to learning.

These forces generate fluctuations that do not explode but remain bounded for a long period of time. In other words, the switching of the economy is a persistent phenomenon. This result[15] depends on many factors, some of which are worth considering. At the outset, it depends on the presence and the nature of the two regimes. In the present case, the values of the parameters guarantee the existence of two steady states with the desired characteristics.

In the second place, the dynamics are a function of the value of the threshold. The benchmark value of the threshold growth rate has been set equal to the average of the two steady state values of growth rates. However, in order to soften the determinism

[15] On the different asymptotic results in the case of nonlinear system, see (Kuznestov 2004)

of the model, a random term has been added. Under this hypothesis, g spends almost 55% of the time in regime 2. If the threshold is changed, also the relative time spent in the two regimes is different and this affect the average rate of growth. (And this happens also when the standard deviation of the error term is modified).

In the third place, the dynamics are also a function of expectations.[16] Since expected values are very close to the actual, the learning mechanism is working in a satisfactory way.[17]

9 The Robustness of the Model

At this stage of the analysis, one must test the robustness of the model. A first aspect is the role of income distribution.[18] In fact, in our model imperfect competition rules income distribution that affects debt and hence consumption. The hypothesis of the benchmark case is that

$$\omega_{01} < \omega_{02}$$

This implies that markups are counter-cyclical. However, one can refer to the opposite hypothesis (on both hypotheses, see (Blanchard 2008), and check whether the model is robust enough in order to maintain fluctuations. The answer is positive. The graphs are similar to Fig. 1, while the parameters used are shown in Table 2.

Two observations are worth making at this stage of the analysis. The first is that the model is robust with respect to the different hypotheses concerning the behaviour of the mark-ups and hence of income distribution in the two regimes. In the second place, the range of variations of the different values is more limited with respect to other parameters and this stimulates the search for an improved specification of the model.

A particular attention must also be paid the parameter c_3 that links debt to consumption and that is playing an important role in the events that characterize the evolution of the world economy in recent times. Three observations are worth making. First of all, the system maintains its qualitative properties for changes in the value of this parameter of 50%. In the second place, c_3 affects the dynamics of inflation in particular. This also happens when one modifies m_{0j}, the weight of raw material in GDP. In the third place, particularly high values can destabilize the system. These results witness the interdependence that links aggregate demand to aggregate supply in the present model.

[16] Cellarier (2008) insists on the role of the information set, the memory length and the technique used on the stability of the model. In the present model, the contemporaneous information is not available for expectations, while the system fluctuates for different memory length.

[17] The mean values of g and Eg are practically the same. The same holds true for π and $E\pi$.

[18] One might ask whether the presence of an exogenous income distribution is compatible with the existence of an endogenous real rate of interest. The answer is affirmative because the latter is just a cost of capital that, in a medium run perspective, can diverge from the profit rate. Furthermore, in the long run the two still differs because of the risk premium.

Table 2 Different hypotheses about income distribution

Benchmark	Opposite
$\omega_{01} = 0.75$	$\omega_{01} = 0.795$
$\omega_{02} = 0.78$	$\omega_{02} = 0.78$

10 Concluding Remarks

The introduction of imperfect competition and uncertainty affects the structure of the macro models both at methodological and analytical level. Uncertainty stresses the importance of dealing with information in the double dimension of agents heterogeneity and expectation formation. Imperfect competition underlines the role of strategic interaction at micro level. Furthermore, it also leads to the issues of price formation ad income distribution at macro level.

In the present paper, the complexities emerging by the simultaneous adoption of uncertainty and imperfect competition have been simplified in a number of ways. As a starting point, we set aside both agents heterogeneity and strategic interaction. Analytically, the choice of non-linear specification leads to the need for further simplifications in order to reach meaningful results. We adopted a stochastic regime switching mechanism that allows the system to move between two states. We also introduced a learning mechanism based upon recursive least squares.

Even provisional, the results of the analysis are interesting at least from two points of view. On one hand, they show how the existence and persistence of fluctuations are strictly related to the analytical specification of the dynamic structure of the model, which includes both the choice of the regime switching mechanisms and the expectation formation. On the other, they show how the macroeconomic dimension is intrinsically integrated (a concept that opposes both to sequential adjustment and to the independence of parts). Integration comes forward in different ways. First of all, there is a strict interdependence between aggregate demand and aggregate supply. In the second place, uncertainty not only alters the structure of the macro model on both supply and demand sides mainly through expectations, but it also creates the foundation for a monetary production economy, which is a system where the financial dimension is not a mere veil for transactions. Such a condition is essential for aggregate demand to have a lasting role in aggregate dynamics. Finally, there is the relationship between growth and distribution where imperfect competition affects directly distribution through the dynamics of mark-ups and therefore indirectly the dynamics of nominal debts. In this way, it contributes to integrate nominal and real aspects.

The analysis can be extended in various ways. One way consists in modifying the model so that the steady state values can also be affected by aggregate demand and not only the transient dynamics as in the present model. Another way is to consider in an explicit way the dynamic relationship between imperfect competition and uncertainty. Furthermore, one can increase the complexity of the expectational process by allowing the presence of heterogeneity (see De Grauwe 2008). One can also examine the impact of learning on the formulations of the equations themselves (see, for instance, Cellarier 2008). Last but not least, it is also possible to improve the empirical results of the model. The aim of our simulations has been more in the spirit of stress testing than to mimic real data. The reasons for this attitude depend on the fact that the econometric parameters are mainly biased against turbulent periods. For these periods, there are not enough data yet. And this is the reason why simulations and scenarios are chosen.

References

Akerlof, G. A. (2007). The missing motivation in macroeconomics. *American Economic Review,* 97, 5–36

Aoki, M., & Yoshikawa, H. (2007). *A Perspective from statistical physics and combinatorial stochastic processes.* Cambridge (UK): Cambridge University Press

Asada, T. P., Chen, C., Chiarella, C., & Flaschel, P. (2006). Keynesian dynamics and the wage-price spiral: A baseline disequilibrium model. *Journal of Macroeconomics,* 28, 90–130

Blanchard, O. J. (1997). The medium-run. *Brookings Papers on Economic Activity*, 2, 89–158

Blanchard O. J. (2008). The state of macro. *NBER Working Papers*, n.14259, Cambridge (MA)

Blanchard, O. J., & Kiyotaki, N. (1987). Monopolistic competition and the effects of aggregate demand. *American Economic Review*, 77, 647–666

Branch, W. A., & Evans, G. W. (2006). A simple recursive forecasting model. *Economic Letters*, 91, 158–166

Cellarier, L. L. (2008). Least squares learning and business cycles. *Journal of Economic Behavior and Organization*, 68, 553–564

Chamberlin, E. H. (1933). *The theory of monopolistic competition.* Cambridge (Mass): Harvard University Press

Comin, D., & Gertler, M. (2006). Medium- term business cycles. *American Economic Review,* 96, 523–551

Day, R. H., & Walter, J. L. (1989). Economic growth in the very long-run: On the multiple-phase interaction of population, technology and social infrastructure. In W. A. Barnett, J. Geweke, & K. Shell (Eds.), *Economic complexity.* Cambridge (UK): Cambridge University Press

De Grauwe, P. (2008). Animal spirits and monetary policy. *CeSifo Working Paper*, 2418, Munich, Germany

D. Delli Gatti & M. Gallegati (Ed.) (2002). *Eterogeneità degli agenti economici e interazione sociale: Teorie e verifiche empiriche.* Bologna: Il Mulino

Evans, G. W., & Honkapohja, S. (2001). *Learning and expectations.* Princeton (NJ): Princeton University Press

Fazzari, S., Ferri, P., & Greenberg, E. (1998). Aggregate demand and firm behavior: A new perspective on Keynesian microfoundations. *Journal of Post Keynesian Economics*, 20 (Summer), 527–558

Fazzari, S., Ferri, P., & Greenberg, E. (2008). Cash flow, investment and Keynes-Minsky cycles. *Journal of Economic Behavior and Organization*, 65, 555–572

Ferri, P. (2007). The labor market and technical change in endogenous cycles. *Metroeconomica*, 58 (Winter), 609–633

Ferri, P. (2008). The economics of nonlinearity. In R. Scazzieri, A. K. Sen, & S. Zamagni (Eds.), *Markets, money and capital. Hicksian economics for the 21st century.* Cambridge(UK): Cambridge University Press, forthcoming

Ferri, P., Greenberg, E., & Day, R. H. (2001). The Phillips curve, regime switching, and the Nairu. *Journal of Economic Behavior and Organization,* 46, 23–37

Frydman, R., & Goldberg, M. D. (2007). *Imperfect knowledge economics.* Princeton (NJ): Princeton University Press

Hommes, C., & Sorger, G. (1998). Consistent expectations equilibria. *Macroeconomic Dynamics,* 2, 287–321

Howitt, P. (2006). Coordination issues in the long-run growth. In L. Tesfatsion & I. Judd (Eds.), *Handbook of computational economics*, (Vol. 2, pp. 1606–1624)

IMF (2008). Annual Report, Washington

Kalecki, M. (1971). *Selected essays on the dynamics of a capitalist economy*. Cambridge (UK): Cambridge University Press

Knight, F. H. (1921). *Risk, uncertainty and profit*. Boston (Mass): Houghton Mifflin

Kuznestov, Y. A. (2004). *Elements of applied bifurcation theory*. New York: Springer

Leijonhufvud, A. (2006). Agent-based macro. In L. Tesfatsion & K. I. Budd (Eds.), *Handbook of computational economics*, (Vol. 2, pp. 1626–1637). Amsterdam: Elsevier

Minsky, H. P. (1954, 2004). *Induced Investment and Business Cycles*. Cheltenham (UK): Edward Elgar

Pasinetti, L. (2007). *Keynes and the cambridge keynesians*. Cambridge (UK): Cambridge University Press

Rotemberg, J., & Woodford, M. (1996). Imperfect competition and the effects of energy price increases on economic activity. *Journal of Money, Credit and Banking*, 28, 549–577

Robinson, J. (1933). *The economics of imperfect competition*. London: Macmillan

Sargent, T. J. (1999). *The conquest of American inflation*. Princeton (NJ): Princeton University Press

Solow, R. (1998). *Monopolistic competition and macroeconomic theory*. Cambridge (UK): Cambridge University Press

Solow, R. (2000). Towards a macroeconomics of the medium run. *Journal of Economic Perspectives,* 14, 151–58

Tesfatsion, L. (2006). Agent-based computational economics: a constructive approach to economic theory. In L. Tesfatsion & K. I. Budd (Eds.), *Handbook of computational economics*, (Vol. 2, pp. 831–880). Amsterdam: Elsevier

Tong, H. (1990). *Non-linear time series*. Oxford: Oxford University Press

Woodford, M. (2003). *Interest and prices*. Princeton (NJ): Princeton University Press

Yokoo, M., & Ishida, J. (2008). Misperception-driven chaos: Theory and policy implications. *Journal of Economic Dynamics and Control*, 32, 1732–1753

Persistent Disequilibrium Dynamics and Economic Policy

Luca Colombo and Gerd Weinrich

1 Introduction

Disequilibrium phenomena seem to be common occurrences of many advanced economies. Starting with the Great Depression, the high unemployment rates in Europe spanning over a decade since the mid-1970s, or the high inflation over the 1980s are just prominent examples. The Japanese recession started in the mid-1990s and not yet fully resolved has put the combination of unemployment and deflation on the spotlight, making clear that a liquidity trap cannot be easily discarded as a purely theoretical possibility. Most recently, the current global recession prompted by the financial sector crisis seems to share similar features, and many observers and policy makers are invoking Keynesian type remedies.

Nonetheless, many economists still consider the representative agent flexible price model as the workhorse of macroeconomics, despite its conclusions are irremediably at odds with all the evidence on the persistence of disequilibrium phenomena (see, e.g., Blanchard, 2000).

For instance, a key result of the flexible price approach is that of money neutrality, which goes against the observed long lasting effects on output and employment of monetary shocks. The New Keynesian literature that developed over the 1990s (see, e.g., Ball and Romer, 1990 and Blanchard, 1990) has emphasized the role of nominal and real rigidities in the wage and price adjustment processes in determining large aggregate effects of monetary shocks. However, most economists maintain that the price level eventually adjusts so that money neutrality is restored. We claim instead that the result of money neutrality in the long run should not be taken for granted, and that the economy may remain stuck in a quasi-stationary state of permanent unemployment, in the absence of appropriate policy interventions.[1]

[1] A state is *stationary* if all variables are constant; it is *quasi-stationary* if all real variables are constant but the nominal variables may change.

G. Weinrich (✉)
Facoltà di Economia, Università Cattolica di Milano Largo Gemelli 1 20123 Milano,
e-mail: gerd.weinrich@unicatt.it

G.I. Bischi et al. (eds.), *Nonlinear Dynamics in Economics, Finance and the Social Sciences*, DOI 10.1007/978-3-642-04023-8_13,

Building on previous work (e.g., Bignami et al., 2004; Colombo and Weinrich, 2003a), we develop a conceptual framework that endogenously allows for the emergence of disequilibrium situations (such as unemployment or deflation), and provides therefore for an ideal setup to investigate the effects and persistency of shocks, and the effectiveness of different economic policies. In particular, we consider an economy consisting of overlapping generations consumers, firms producing by means of an atemporal production function, and a government financing public expenditure through a tax on firms' profits. Within each period prices are fixed, and consistent allocations are obtained by means of temporary equilibrium with stochastic rationing. Prices are then adjusted between successive periods according to the strength of rationing on each market in the previous period.[2]

The gradual adjustment of prices and wages (and hence their inability to function as instantaneous and perfect allocation devices and the need for quantity adjustments to complement them in making trades feasible) is the primary mechanism to explain the propagation and the persistent real effects of shocks. It is worth stressing that we do not account endogenously for the reasons behind different degrees of wage and price stickiness, but we rather rely on exogenously given rules. Although in this perspective our approach is obviously ad hoc, it is on the other hand consistent with several underlying conceptual models of price rigidities (be they in the New Keynesian or in the Neoclassical tradition), and it allows us to effectively parametrize the different degrees of wage and price stickiness that are observed in reality.[3]

The role of price and wage rigidities is complemented by that of other factors amplifying the importance of the spillovers among markets and affecting the reaction of consumers and firms to shocks and policy interventions. Following Colombo and Weinrich (2006, 2008), we focus in particular on the role of consumers' expectations and firms' inventories. Expectations are especially important since they influence consumers' choices and hence the response of the economy to a shock. For instance, a restrictive shock may determine an aggregate demand deficit and lead the economy into a state of (Keynesian) unemployment. Finding ways to convince consumers to hold inflationary expectations – in this way increasing their current consumption and hence aggregate demand – may prove a powerful

[2] A natural idea is to relate the adjustment of prices to the size of the dissatisfaction of agents with their (foregone) trades. A reliable measure of such a dissatisfaction requires stochastic rationing, since – as opposed to deterministic rationing – it is compatible with manipulability of the rationing mechanism and therefore provides an incentive for rationed agents to express demands that exceed their expected trades, as argued by Green (1980), Svensson (1980), Gale (1979, 1981) and Weinrich (1982, 1984, 1988). For a definition of manipulability see for example Böhm (1989) or Weinrich (1988).

[3] The New Keynesian literature, in particular, has investigated many possible causes for the real rigidity of prices and wages ranging from efficiency wages (see, for example, Shapiro and Stiglitz, 1984) and countercyclical mark-ups (e.g., Rotemberg and Woodford, 1991), to coordination failure (e.g., Ball and Romer, 1991) and credit markets imperfections (e.g., Bernake and Gertler, 1995 and Kiyotaki and Moore, 1997). Attention has been devoted as well to the sources of nominal stickiness focusing, for instance, on menu cost (e.g., Mankiw, 1985), near rationality (e.g., Akerlof and Yellen, 1985) and staggered contracts (Calvo, 1983).

tool for recovering from the recession. At the same time, the explicit consideration of firms' inventories is important for fully assessing the impact of a shock on the economy. Focusing again on a restrictive shock, inventories have in fact an obvious reinforcement effect: by increasing the reduction in labor demand following the shock, inventories contribute to further depress the aggregate demand and to favor the convergence of the economy to a quasi-stationary state with permanent unemployment.

In the second part of the paper, we will discuss in details the direct effects and the interplay of expectations and inventories in the propagation of shocks and in the explanation of their persistence, as well as their influence on the outcomes of different economic policies aimed at resolving or mitigating the effects of deflationary recessions.

The paper is organized as follows. In Sect. 2 we outline our base model, define a temporary equilibrium, show its existence and uniqueness, and study the dynamics of the economy. In Sect. 3 we investigate by means of numerical simulations the dynamic behavior of the economy, focusing on the effects of the adjustment of prices and wages, and showing the possibility of chaotic dynamic behavior as well as the emergence of a Phillips curve as an attractor of our dynamic system. In Sect. 4 we extend our base model to encompass the role of consumers' expectations and firms' inventories, and in Sect. 5 we study the dynamic behavior of the extended economy by focusing especially on the possible policy remedies to deflationary recessions. Section 6 summarizes and suggests avenues for future research.

2 The Base Model

Following Colombo and Weinrich (2003a) and Bignami et al. (2004), we focus on an economy composed of n OLG-consumers offering labor inelastically when young and consuming a composite consumption good in both periods of their life. The consumption good is produced by n' firms, using an atemporal production function whose only input is labor. The public sector of the economy is represented by a government that levies a proportional tax on firms' profits to finance its expenditure for goods. Budget deficits and surpluses may arise through money creation or destruction.

The timing of the model is such that the aggregate profit Π_{t-1} realized by firms in period $t-1$ is distributed at the beginning of period t in part as tax to the government ($tax\Pi_{t-1}$) and in part to young consumers ($(1-tax)\,\Pi_{t-1}$), where $0 \leq tax \leq 1$. Also at the beginning of period t old consumers hold a total quantity of money M_t – consisting of savings generated in period $t-1$ – that allows households to transfer purchasing power between periods.

We denote with X_t the aggregate quantity of the good purchased by young consumers in period t, p_t its price, w_t the nominal wage and L_t the aggregate quantity of labor. Then we get $M_{t+1} = (1-tax)\,\Pi_{t-1} + w_t L_t - p_t X_t$.

Letting G be the quantity of goods purchased by the government and taking into account that old households want to consume all their money holdings in period t, the aggregate consumption is $Y_t = X_t + \frac{M_t}{p_t} + G$. Since $\Pi_t = p_t Y_t - w_t L_t$, denoting with $\Pi_t - \Pi_{t-1} = \Delta M_t^P$ and $\Delta M_t^C = M_{t+1} - M_t$ the variation in the money stock held by producers before they distribute profits and by consumers, respectively, we obtain $\Delta M_t^C + \Delta M_t^P = p_t G - tax \Pi_{t-1} =$ budget deficit.

Young households first visit the labor market where they either can sell their inelastic labor supply ℓ^s, or they are rationed to zero. Then on the goods market they may be rationed according to the stochastic rule

$$x_t = \begin{cases} x_t^d & \text{with prob. } \rho\gamma_t^d, \\ c_t x_t^d & \text{with prob. } 1 - \rho\gamma_t^d, \end{cases}$$

where x_t^d is the quantity demanded, $\rho \in [0, 1]$ a fixed structural parameter of the rationing mechanism, $\gamma_t^d \in [0, 1]$ a rationing coefficient which the household perceives as given but which will be determined in equilibrium and $c_t = \frac{\gamma_t^d - \rho\gamma_t^d}{1-\rho\gamma_t^d}$. These settings are chosen such that the expected value of x_t is $\gamma_t^d x_t^d$, that is, $E x_t = \gamma_t^d x_t^d$.

The effective demand x_t^{di}, $i = 0, 1$, is obtained from solving

$$\max_{x_t} \rho\gamma_t^d u\left(x_t, \frac{\omega_t^i - x_t}{\theta_t^e}\right) + \left(1 - \rho\gamma_t^d\right) u\left(c_t x_t, \frac{\omega_t^i - c_t x_t}{\theta_t^e}\right) \tag{1}$$

subject to the constraint $0 \leq x_t \leq \omega_t^i$, where $\omega_t^0 = \frac{1-tax}{p_t}\frac{\Pi_{t-1}}{n}$ and $\omega_t^1 = \omega_t^0 + \frac{w_t}{p_t}\ell^s$ are real income in case of rationing and no rationing on the labor market, respectively, and $\theta_t^e = p_{t+1}^e / p_t$ is the expected relative price for period t.

The aggregate supply of labor is $L^s = n\ell^s$. Denoting with L_t^d the aggregate demand of labor and with $\lambda_t^s = \min\left\{\frac{L_t^d}{L^s}, 1\right\}$ the fraction of young consumers that will be employed, the aggregate demand of goods of young consumers is

$$X_t^d = \lambda_t^s n x_t^{d1} + \left(1 - \lambda_t^s\right) n x_t^{d0} \equiv X^d\left(\lambda_t^s; \frac{w_t}{p_t}, \frac{(1 - tax)\,\Pi_{t-1}}{p_t}\right),$$

where x_t^{d0} and x_t^{d1} are the effective quantities demanded in case of rationing and no rationing, respectively, on the labor market.

Problem (1) can not be solved analytically for a generic utility function. In order to derive analytic results, we assume that $u(x_t, x_{t+1}) = x_t^h x_{t+1}^{1-h}$ and $\rho = 1$ (i.e., zero/one rationing). Therefore, problem (1) requires to maximize the expected utility $\gamma_t^d x_t^h \left(\left(\omega_t^i - x_t\right)/\theta_t^e\right)^{1-h}$, from which we obtain that the effective demand x_t^{di} is equal to $h\omega_t^i$, which is independent of γ_t^d and θ_t^e, although it depends on the real income ω_t^i. The total aggregate demand of the consumption sector is then obtained by adding old consumers' aggregate demand M_t/p_t and government demand G:

$$Y_t^d = X^d\left(\lambda_t^s;\alpha_t,(1-tax)\,\pi_t\right)+m_t+G_t,$$

where $\alpha_t = w_t/p_t, \pi_t = \Pi_{t-1}/p_t$ and $m_t = M_t/p_t$.

Each of the n' identical firms uses an atemporal production function $y_t = f(\ell_t)$. As with consumers, firms too may be rationed, by means of a rationing mechanism analogue to that assumed for the consumption sector. Denoting the single firm's effective demand of labor by ℓ_t^d, the quantity of labor effectively transacted is

$$\ell_t = \begin{cases} \ell_t^d & \text{with prob. } \lambda_t^d, \\ 0 & \text{with prob. } 1-\lambda_t^d, \end{cases}$$

where $\lambda_t^d \in [0,1]$. On the goods market the rationing rule is

$$y_t = \begin{cases} y_t^s & \text{with prob. } \sigma\gamma_t^s, \\ d_t y_t^s & \text{with prob. } 1-\sigma\gamma_t^s\ , \end{cases} \tag{2}$$

where $\sigma \in (0,1)\,, \gamma_t^s \in [0,1]$ and $d_t = \frac{(\gamma_t^s-\sigma\gamma_t^s)}{(1-\sigma\gamma_t^s)}$. σ is a fixed parameter of the mechanism whereas λ_t^d and γ_t^s are perceived rationing coefficients taken as given by the firm the effective value of which will be determined in equilibrium. The definition of d_t ensures that $Ey_t = \gamma_t^s y_t^s$. It is obvious that $E\ell_t = \lambda_t^d\ell_t^d$.

The firm's effective demand $\ell_t^d = \ell^d\left(\gamma_t^s;\alpha_t\right)$ is obtained from the expected profit maximization problem

$$\max_{\ell_t^d} \gamma_t^s f\left(\ell_t^d\right) - \alpha_t\ell_t^d$$

subject to

$$0 \le \ell_t^d \le \frac{d_t}{\alpha_t} f\left(\ell_t^d\right).$$

The aggregate labor demand is $L_t^d = n'\ell_t^d\left(\gamma_t^s;\alpha_t\right) \equiv L^d\left(\gamma_t^s;\alpha_t\right)$ and, because only a fraction λ_t^d of firms can hire workers, the aggregate supply of goods is

$$Y_t^s = \lambda_t^d n' f\left(\ell^d\left(\gamma_t^s;\alpha_t\right)\right) \equiv Y^s\left(\lambda_t^d,\gamma_t^s;\alpha_t\right).$$

2.1 Temporary Equilibrium Allocations

For any given period t a feasible allocation can be described as a temporary equilibrium with rationing as follows.

Definition 1. *Given a real wage α_t, a real profit level π_t, real money balances m_t, a level of public expenditure G and a tax rate tax, a list of rationing coefficients*

$\left(\gamma_t^d, \gamma_t^s, \lambda_t^d, \lambda_t^s, \delta_t, \varepsilon_t\right) \in [0,1]^6$ *and an aggregate allocation* $\left(\overline{L}_t, \overline{Y}_t\right)$ *constitute a* temporary equilibrium with rationing *if the following conditions are fulfilled:*

(C1) $\overline{L}_t = \lambda_t^s L^s = \lambda_t^d L^d\left(\gamma_t^s; \alpha_t\right);$

(C2) $\overline{Y}_t = \gamma_t^s Y^s\left(\lambda_t^d, \gamma_t^s; \alpha_t\right) = \gamma_t^d X^d\left(\lambda_t^s; \alpha_t, (1 - tax)\,\pi_t\right) + \delta_t m_t + \varepsilon_t G;$

(C3) $\left(1 - \lambda_t^s\right)\left(1 - \lambda_t^d\right) = 0; \left(1 - \gamma_t^s\right)\left(1 - \gamma_t^d\right) = 0;$

(C4) $\gamma_t^d\left(1 - \delta_t\right) = 0; \delta_t\left(1 - \varepsilon_t\right) = 0.$

Conditions (C1) and (C2) require that expected aggregate transactions balance. This means that all agents have correct perceptions of the rationing coefficients. Equations (C3) formalize the short-side rule according to which at most one side on each market is rationed. The meaning of the coefficients δ_t and ε_t is that also old households and/or the government can be rationed. However, according to condition (C4) this may occur only after young households have been rationed (to zero).

Depending on which market sides are rationed, we can characterize different types of equilibrium. More precisely, we indicate with *Keynesian Unemployment* $[K]$ an equilibrium in which there is excess supply on both markets $\left(\lambda_t^s < 1,\ \gamma_t^s < 1\right)$; with *Repressed Inflation* $[I]$ one in which there is excess demand on both markets $(\lambda_t^d < 1, \gamma_t^d < 1)$; with *Classical Unemployment* $[C]$ one where there is excess supply on the labor market and excess demand on the goods market $(\lambda_t^s < 1, \gamma_t^d < 1)$; and finally with *Underconsumption* $[U]$ an equilibrium with excess demand on the labor market and excess supply on the goods market $(\lambda_t^d < 1, \gamma_t^s < 1)$.

The existence and uniqueness of temporary equilibrium is shown in Colombo and Weinrich (2003a, pp. 9–12). In particular it is shown that the notion of temporary equilibrium with rationing defines a unique temporary equilibrium allocation given by $\left(\overline{L}_t, \overline{Y}_t\right) = \left(\mathcal{L}\left(\alpha_t, \pi_t, m_t, G, tax\right), \mathcal{Y}\left(\alpha_t, \pi_t, m_t, G, tax\right)\right)$.

2.2 Dynamics

In order to investigate the dynamic behavior of the economy, we need to link successive periods, which is done by the adjustment of prices and by the changes in the stock of money and in profits. As for the latter, by definition of these variables one immediately obtains

$$\Pi_t = p_t \mathcal{Y}\left(\alpha_t, \pi_t, m_t, G, tax\right) - w_t \mathcal{L}\left(\alpha_t, \pi_t, m_t, G, tax\right)$$

and

$$\begin{aligned} M_{t+1} &= (1 - tax)\,\Pi_{t-1} + w_t \overline{L}_t - p_t \overline{X}_t \\ &= (1 - tax)\,\Pi_{t-1} + w_t \overline{L}_t - p_t \overline{Y}_t + \delta_t M_t + \varepsilon_t p_t G \\ &= (1 - tax)\,\Pi_{t-1} - \Pi_t + \delta_t M_t + \varepsilon_t p_t G. \end{aligned}$$

Regarding the adjustment of prices we make the standard assumption that whenever an excess of demand (supply) is observed, the price rises (falls). In terms of the rationing coefficients observed in period t, this amounts to

$$p_{t+1} < p_t \Leftrightarrow \gamma_t^s < 1; \qquad p_{t+1} > p_t \Leftrightarrow \gamma_t^d < 1;$$

$$w_{t+1} < w_t \Leftrightarrow \lambda_t^s < 1; \qquad w_{t+1} > w_t \Leftrightarrow \lambda_t^d < 1.$$

More precisely, in our numerical analysis, these adjustments have been specified by means of the non-linear rules:[4]

$$p_{t+1} = \left(\gamma_t^s\right)^{\mu_1} p_t, \text{ if } \gamma_t^s < 1; \qquad p_{t+1} = \left(\frac{\gamma_t^d + \delta_t + \varepsilon_t}{3}\right)^{-\mu_2} p_t, \text{ if } \gamma_t^d < 1;$$

$$w_{t+1} = \left(\lambda_t^s\right)^{\nu_1} w_t, \text{ if } \lambda_t^s < 1; \qquad w_{t+1} = \left(\lambda_t^d\right)^{-\nu_2} w_t, \text{ if } \lambda_t^d < 1,$$

where μ_1, μ_2, ν_1 and ν_2 are nonnegative parameters for the speeds of adjustment.

Then the adjustment equations for the real wage are

$$\alpha_{t+1} = \frac{\left(\lambda_t^s\right)^{\nu_1}}{(\gamma_t^s)^{\mu_1}}\alpha_t \quad \text{if } \left(\overline{L}_t, \overline{Y}_t\right) \in K \cup U,$$

$$\alpha_{t+1} = \frac{\left(\lambda_t^d\right)^{-\nu_2}}{\left(\frac{\gamma_t^d + \delta_t + \varepsilon_t}{3}\right)^{-\mu_2}}\alpha_t \quad \text{if } \left(\overline{L}_t, \overline{Y}_t\right) \in I,$$

$$\alpha_{t+1} = \frac{\left(\lambda_t^s\right)^{\nu_1}}{\left(\frac{\gamma_t^d + \delta_t + \varepsilon_t}{3}\right)^{-\mu_2}}\alpha_t \quad \text{if } \left(\overline{L}_t, \overline{Y}_t\right) \in C,$$

whereas $\theta_t = p_{t+1}/p_t$ is given by

$$\theta_t = \left(\gamma_t^s\right)^{\mu_1} \quad \text{if } \left(\overline{L}_t, \overline{Y}_t\right) \in K \cup U,$$

$$\theta_t = \left(\frac{\gamma_t^d + \delta_t + \varepsilon_t}{3}\right)^{-\mu_2} \quad \text{if } \left(\overline{L}_t, \overline{Y}_t\right) \in I \cup C.$$

The dynamics of the model in real terms is given by the sequence $\{(\alpha_t, m_t, \pi_t)\}_{t=1}^{\infty}$, where α_{t+1} is as above and

[4] The rules we consider here are the same used in Bignami et al. (2004). In Colombo and Weinrich (2003a) we consider instead linear adjustment rules. The different formulations of the adjustment mechanisms are without implications in terms of the qualitative results emerging from the numerical analysis of the economy dynamics.

$$\pi_{t+1} = \frac{[\mathcal{Y}(.) - \alpha_t \mathcal{L}(.)]}{\theta_t},$$

$$m_{t+1} = \frac{1}{\theta_t}\left[\delta_t m_t + \varepsilon_t G + (1 - tax)\,\pi_t\right] - \pi_{t+1}.$$

3 Complex Dynamics and the Phillips Curve as an Attractor

The dynamic behavior of the economy outlined above is described by a non-linear three-dimensional dynamical system with state variables α_t, m_t and π_t that entails three subsystems (corresponding to the three nondegenerate equilibrium regimes) each of which may become effective through endogenous regime switching.[5] In order to investigate the model dynamics one needs therefore to use numerical simulations.[6] We use the utility function $u(x_t, x_{t+1}) = x_t^h x_{t+1}^{1-h}$ and the production function $f(\ell) = a\ell^b$, and we specify the following parameter set, corresponding to a stationary Walrasian equilibrium as a benchmark case:

$$\begin{gathered} a = 1, \quad b = 0.85, \quad h = 0.5, \quad L^s = 100, \quad n' = 100, \quad \alpha_0 = 0.85, \\ m_0 = 46.25, \quad \pi_0 = 15, \quad G = 7.5, \quad tax = 0.5. \end{gathered} \tag{3}$$

As shown by Bignami et al. (2004), the dynamic behavior of the system is very sensitive to the choice of relevant parameters, such as the economic policy instruments. The complexity of the economy dynamics is clearly illustrated by the bifurcation diagram in Fig. 1, showing the periodic (cycles of different orders) and non-periodic (chaotic) long-run characteristics of the system dynamics for different values of the government public expenditure G.[7] It is immediate to note that for G smaller than 7.5 (the Walrasian value) the economy converges to quasi-stationary states with (Keynesian) unemployment, which illustrates the possibility of permanent unemployment although prices and wages are flexible. It is also worth stressing that, as expected from textbook theory, an expansionary fiscal policy can help driving the economy towards full employment. However, a too expansionary fiscal policy ends up being destabilizing, as it induces an highly cyclical and irregular dynamic behavior, a feature that has to be added to the risk of inflation that is traditionally associated with expansionary policy measures.

---

[5] The underconsumption regime is degenerate in the sense that it can be seen as a limiting case of both the Keynesian and the inflationary regime.

[6] Our numerical simulations are based on the software MACRODYN that has been developed by Volker Böhm at the University of Bielefeld.

[7] The figure has been obtained by using the benchmark parameter set, except for G that is allowed to change, and by letting the nominal wages to be rigid downwards, i.e., $\nu_1 = 0$. All other adjustment speeds are set equal to $\mu_1 = \mu_2 = \nu_2 = 0.4$.

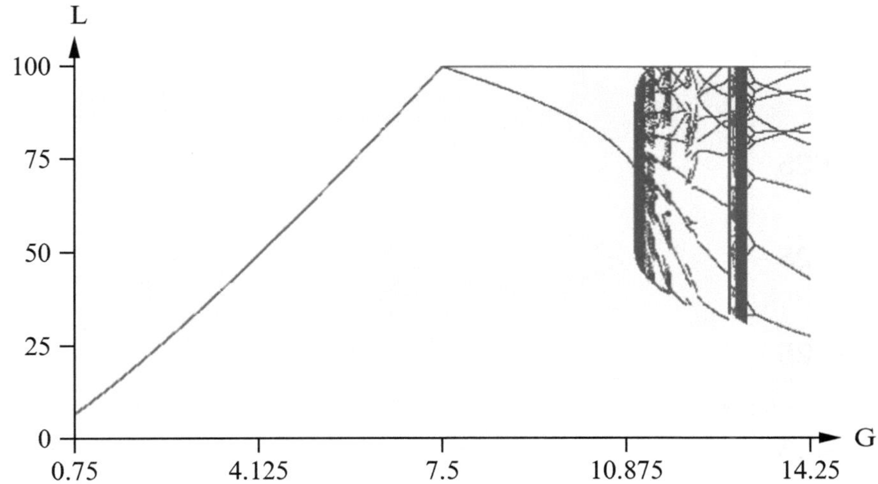

Fig. 1 Bifurcation diagram for employment over government demand

The dynamic response of our economy to a shock also crucially depends on the values of prices and wage adjustment speeds. For instance, Bignami et al. (2004) have shown that following a restrictive monetary shock (e.g., a reduction to $m_0 = 40$) to allow for some wage flexibility downwards helps the system to return to the Walrasian equilibrium, as expected from textbook theory. However, further increasing the downward flexibility of wage over a certain threshold gives rise to irregular (chaotic) behavior with frequently high unemployment rates.[8] As analyzed in detail by Colombo and Weinrich (2003a), a too high downwards adjustment speed of the wage has striking implications in terms of the dynamic behavior of employment and prices. This is evident from Fig. 2, showing an attractor in the unemployment rate ($u_t = \left(L^s - \overline{L}_t\right)/L^s$) – inflation rate ($v_t = (w_{t+1} - w_t)/w_t$) plane. In an economic perspective, it is apparent that this attractor describes a Phillips curve.

However, it is impossible to interpret this Phillips curve as a policy instrument in terms of a trade-off between unemployment and inflation as is commonly done. Any point on the curve is in fact but one element of a trajectory of pairs of rates of unemployment and wage inflation, and successive points of this trajectory may lie far away one from the other. Thus, even if the government tried to select a specific point on the curve in one period, in the next period already the system may go to a very different point on the curve.

[8] The threshold level of ν_1 at which irregular behavior appears is about 0.14 (see Bignami et al., 2004 for a more detailed analysis).

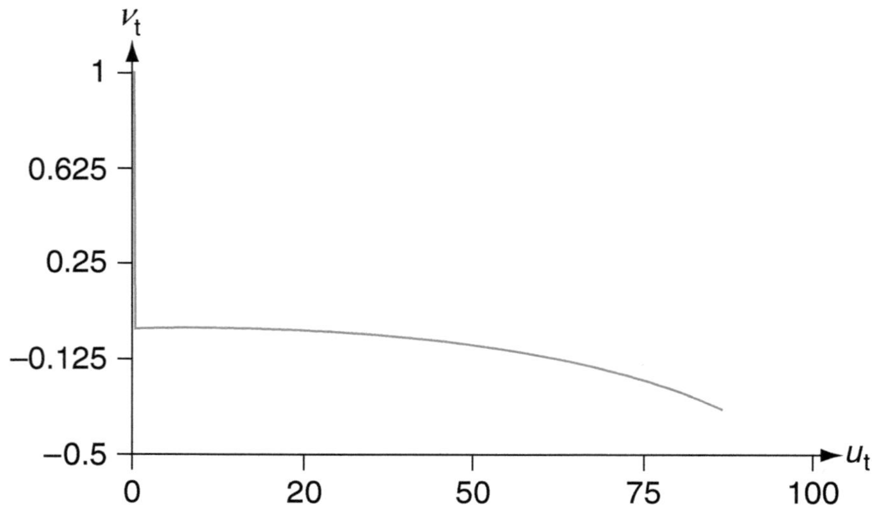

Fig. 2 A long run Phillips-curve

4 A Model with Consumers' Expectations and Firms' Inventories

The main contribution of the setup presented in Sect. 2 consists in developing (and characterizing the dynamic behavior of) a framework in the Keynesian tradition able to account for the emergence of endogenous business cycles, and for the possible convergence of the economy to quasi-stationary states characterized by disequilibrium situations (e.g., long-run unemployment or capacity under-utilization), even when prices and wages are allowed to adjust over time. Although providing useful results both in a methodological and in an economic perspective, this basic framework neglects several features that would be useful in using the above setup to investigate real world situations, and evaluate the effects of different economic policies. Improving over the modelling shortcomings of the basic setup becomes especially important if one has the ambition to use the above approach as a lens to interpret and investigate the effects of crises like the one we are experiencing these days, for which Keynesian type remedies are invoked and being adopted.

In a series of recent papers, that in a policy perspective are motivated mainly by the study of deflationary recessions (as the one that hit Japan over the 1990s), we extend our framework to embed features allowing us to investigate the effects of variables that provide important channels for the propagation and the persistency of shocks, and play a prominent role in the current policy debate. In particular, in Colombo and Weinrich (2006) we add to the basic model presented in Sect. 2 the possibility of inventories holding by firms, and in Colombo and Weinrich (2008) we focus on the role played by consumers expectations. In the following, we outline a model combining both these features.

We consider the same economy as in the basic setup introduced in Sect. 2. However, as we now explicitly consider the role of consumers' expectations, whether old consumers hold a total quantity of money M_t (consisting of the savings generated in period $t-1$) at the beginning of period t depends on their price expectations for their second period of life. Since consumers may store the consumption good bought in the first period, they will voluntarily hold money only if they expect the good's price to decrease. They may be forced, however, to do this in case they are rationed in their consumption goods purchases in the first period. Firms too may now transfer unsold units of the consumption good into the future, as we allow for inventories holding by firms. Denoting with S_t the aggregate amount of inventories carried over by firms to period t, with Y_t^p the aggregate amount of goods produced and with Y_t the quantity sold in period t, there results $S_{t+1} = Y_t^p + S_t - Y_t$.

Taking expectations into account, young consumers have to decide whether to buy the quantities x_t and x_{t+1} in periods t and $t+1$, respectively, or buy the total quantity $x_t + x_{t+1}$ in period t and transfer x_{t+1} to period $t+1$. This in turn depends on the price expectation $\theta_t^e \equiv p_{t+1}^e / p_t$. If $\theta_t^e < 1$, then the consumer expects a decrease in the goods price and hence prefers to buy x_{t+1} in his second period of life, while if $\theta_t^e > 1$ he buys everything in his first period.

The case $\theta_t^e < 1$ is identical to the consumer's problem discussed in the base model of Sect. 2. As for the case $\theta_t^e > 1$, the consumer wants to buy the total quantity $x_t + x_{t+1} \equiv \widehat{x}_t$ in his first period of life, and thus has to meet the budget constraint

$$x_t + x_{t+1} \leq \omega_t^i\,, \qquad i = 0, 1\,.$$

Monotonicity of the utility function implies that his effective demand is $\widehat{x}_t^{di} = \omega_t^i$. Hence, the aggregate demand of goods by young consumers in case of deflationary expectations $\theta_t^e < 1$ is

$$\begin{aligned} X_t^d &= \lambda_t^s n x_t^{d1} + \left(1 - \lambda_t^s\right) n x_t^{d0} \\ &= h\left[(1 - tax)\,\frac{\Pi_{t-1}}{p_t} + \frac{w_t}{p_t}\lambda_t^s L^s\right] \equiv X^d\left(\lambda_t^s; \frac{w_t}{p_t}, \frac{(1 - tax)\,\Pi_{t-1}}{p_t}, h\right) \end{aligned} \qquad (4)$$

whereas in case of inflationary expectations $\theta_t^e > 1$ it is

$$\begin{aligned} \widehat{X}_t^d &= \lambda_t^s n \widehat{x}_t^{d1} + \left(1 - \lambda_t^s\right) n \widehat{x}_t^{d0} \\ &= (1 - tax)\,\frac{\Pi_{t-1}}{p_t} + \frac{w_t}{p_t}\lambda_t^s L^s = X^d\left(\lambda_t^s; \frac{w_t}{p_t}, \frac{(1 - tax)\,\Pi_{t-1}}{p_t}, 1\right). \end{aligned} \qquad (5)$$

From (4) and (5) it is evident that the only difference in the aggregate effective demand by young consumers implied by different expectations $\theta_t^e <$ or > 1 lies in the multiplicative factor $\tau \in \{h, 1\}$. Therefore, we identify the value of τ with the corresponding expectation type.

The total effective aggregate demand of the consumption sector is now obtained, as in our base setup, by adding old consumers' aggregate demand $m_t = M_t/p_t$ and government demand G:

$$Y_t^d = X^d\left(\lambda_t^s; \alpha_t, (1 - tax)\,\pi_t, \tau\right) + m_t + G,$$

where $\alpha_t \equiv w_t/p_t$ and $\pi_t \equiv \Pi_{t-1}/p_t$.

Turning now to the production sector, we continue to assume that all firms are identical and produce according to the production function $y_t^p = f(\ell_t) = a\ell_t^b, a, b > 0$. Denoting with s_t the inventories held at the beginning of period t, the total amount supplied by a firm is $y_t^s = y_t^p + s_t$. As for firms' rationing, we assume again 0–1 rationing in the labor market and thus, recalling that ℓ_t^d is the single firm's effective demand of labor and $\lambda_t^d \in [0, 1]$ is the probability that the firm is not rationed on the labor market, it follows that $E\ell_t = \lambda_t^d \ell_t^d$. On the goods market the rationing rule is the same proposed in the base model (see (2)), where the firm's effective supply becomes now $y_t^s = f\left(\ell_t^d\right) + s_t$.

From the maximization of expected profits, $\gamma_t^s\left[f\left(\ell_t^d\right) + s_t\right] - \alpha_t \ell_t^d$ we get each firm's effective labor demand as

$$\ell_t^d = \ell^d\left(\gamma_t^s; \alpha_t\right) = \left(\frac{\gamma_t^s ab}{\alpha_t}\right)^{\frac{1}{1-b}}, \tag{6}$$

which is independent of s_t. The aggregate labor demand then is $L_t^d = n'\ell^d\left(\gamma_t^s; \alpha_t\right) \equiv L^d\left(\gamma_t^s; \alpha_t\right)$ and, because only a fraction λ_t^d of firms can hire workers, the aggregate supply of goods is

$$Y_t^s = \lambda_t^d n' f\left(\ell^d\left(\gamma_t^s; \alpha_t\right)\right) + S_t \equiv Y^s\left(\lambda_t^d, \gamma_t^s; \alpha_t, S_t\right). \tag{7}$$

4.1 Temporary Equilibrium Allocations

For any t, the definition of a temporary equilibrium with rationing is now described by the following

Definition 2. *Given a real wage α_t, a real profit level π_t, real money balances m_t, inventories S_t, a level of public expenditure G, a tax rate tax and an expectation type $\tau \in \{h, 1\}$, a list of rationing coefficients $\left(\gamma_t^d, \gamma_t^s, \lambda_t^d, \lambda_t^s, \delta_t, \varepsilon_t\right) \in [0, 1]^6$ and an aggregate allocation $\left(\overline{L}_t, \overline{Y}_t\right)$ constitute a* temporary equilibrium with rationing *if the following conditions are fulfilled:*

(C1) $\overline{L}_t = \lambda_t^s L^s = \lambda_t^d L^d\left(\gamma_t^s; \alpha_t\right);$

(C2) $\overline{Y}_t = \gamma_t^s Y^s\left(\lambda_t^d, \gamma_t^s; \alpha_t, S_t\right)$

$= \gamma_t^d X^d\left(\lambda_t^s; \alpha_t, (1 - tax)\,\pi_t, \tau\right) + \delta_t m_t + \varepsilon_t G;$

(C3) $\left(1-\lambda_t^s\right)\left(1-\lambda_t^d\right)=0; \left(1-\gamma_t^s\right)\left(1-\gamma_t^d\right)=0;$

(C4) $\gamma_t^d\left(1-\delta_t\right)=0; \delta_t\left(1-\varepsilon_t\right)=0.$

The four conditions in the above definition have exactly the same interpretation as in the base model of Sect. 2, from which they differ just for the explicit consideration of the role of expectations and inventories. Colombo and Weinrich (2008, Proposition 1) show the existence of a unique temporary equilibrium allocation given by $\left(\overline{L}_t, \overline{Y}_t\right)=\left(\mathcal{L}\left(\alpha_t, \pi_t, m_t, S_t, G, tax, \tau\right), \mathcal{Y}\left(\alpha_t, \pi_t, m_t, S_t, G, tax, \tau\right)\right)$.

4.2 Dynamics

In this extended framework with expectations and inventories, the link between successive periods is given by the adjustment of prices, by the changes in the stock of money and in profits and by possible changes in the expectation type. For given τ, the adjustment of prices and wages is again such that the price rises (falls) whenever an excess of demand (supply) is observed. More precisely, our simulations are based on the following adjustment mechanisms:[9]

$$p_{t+1}=\begin{cases}\left[1-\mu_1\left(1-\gamma_t^s\right)\right]p_t & \text{if } \gamma_t^s<1,\\ \left[1+\mu_2\left(1-\dfrac{\gamma_t^d+\delta_t+\varepsilon_t}{3}\right)\right]p_t & \text{if } \gamma_t^d<1,\end{cases} \tag{8}$$

$$w_{t+1}=\begin{cases}\left[1-\nu_1\left(1-\lambda_t^s\right)\right]w_t & \text{if } \lambda_t^s<1,\\ \left[1+\nu_2\left(1-\lambda_t^d\right)\right]w_t & \text{if } \lambda_t^d<1,\end{cases} \tag{9}$$

where $\mu_1, \mu_2, \nu_1, \nu_2 \in [0,1]$ can be interpreted as the degree of flexibility of adjustment. The dynamics of the real wage is then described by the following equations:

$$\alpha_{t+1}=\begin{cases}\dfrac{1-\nu_1\left(1-\lambda_t^s\right)}{1-\mu_1\left(1-\gamma_t^s\right)}\alpha_t & \text{if } \left(\overline{L}_t, \overline{Y}_t\right)\in K,\\ \dfrac{1-\nu_1\left(1-\lambda_t^s\right)}{1+\mu_2\left(1-\dfrac{\gamma_t^d+\delta_t+\varepsilon_t}{3}\right)}\alpha_t & \text{if } \left(\overline{L}_t, \overline{Y}_t\right)\in C,\\ \dfrac{1+\nu_2\left(1-\lambda_t^d\right)}{1+\mu_2\left(1-\dfrac{\gamma_t^d+\delta_t+\varepsilon_t}{3}\right)}\alpha_t & \text{if } \left(\overline{L}_t, \overline{Y}_t\right)\in I,\\ \dfrac{1+\nu_2\left(1-\lambda_t^d\right)}{1-\mu_1\left(1-\gamma_t^s\right)}\alpha_t & \text{if } \left(\overline{L}_t, \overline{Y}_t\right)\in U,\end{cases} \tag{10}$$

[9] Differently than in the base model of Sect. 2, we consider now linear adjustment rules, to explicitly show that our results do not depend on the non-linearity of the adjustment mechanisms.

whereas the inflation factor $\theta_t = p_{t+1}/p_t$ is given by

$$\theta_t = \begin{cases} 1 - \mu_1 \left(1 - \gamma_t^s\right) & \text{if } \left(\overline{L}_t, \overline{Y}_t\right) \in K \cup U, \\ 1 + \mu_2 \left(1 - \frac{\gamma_t^d + \delta_t + \varepsilon_t}{3}\right) & \text{if } \left(\overline{L}_t, \overline{Y}_t\right) \in C \cup I. \end{cases} \quad (11)$$

The dynamics of profits, money and inventories follow from the definition of these variables and (10)–(11), i.e.,

$$\pi_{t+1} = \frac{\overline{Y}_t - \alpha_t \overline{L}_t}{\theta_t},$$

$$\begin{aligned} m_{t+1} &= \frac{M_{t+1}}{p_{t+1}} = \frac{1}{p_{t+1}} \left[(1 - tax)\, \Pi_{t-1} + w_t \overline{L}_t - p_t \overline{Y}_t + \delta_t M_t + \varepsilon_t p_t G\right] \\ &= \frac{1}{\theta_t} \left[(1 - tax)\, \pi_t + \alpha_t \overline{L}_t - \overline{Y}_t + \delta_t m_t + \varepsilon_t G\right] \\ &= \frac{1}{\theta_t} \left[\delta_t m_t + \varepsilon_t G + (1 - tax)\, \pi_t\right] - \pi_{t+1} \end{aligned}$$

and

$$S_{t+1} = Y^s \left(\lambda_t^d, \gamma_t^s; \alpha_t, S_t\right) - \overline{Y}_t = \lambda_t^d n' a \left(\frac{\gamma_t^s ab}{\alpha_t}\right)^{\frac{b}{1-b}} + S_t - \overline{Y}_t\,,$$

where

$$\overline{Y}_t = \mathcal{Y}\left(\alpha_t, \pi_t, m_t, S_t, G, tax, \tau\right) \quad \text{and} \quad \overline{L}_t = \mathcal{L}\left(\alpha_t, \pi_t, m_t, S_t, G, tax, \tau\right).$$

It follows that the dynamics of the model is then given by the sequence $\{(\alpha_t, m_t, \pi_t, S_t)\}_{t=1}^{\infty}$.

We now need to consider the possibility of expectation switching, which should occur whenever it is required in order to keep expectations correct along a trajectory of the system. To illustrate the point, consider the case in which consumers have deflationary expectations in period t ($\theta_t^e \leq 1$ or, equivalently, $\tau = h$) but the equilibrium in period t is such that there is excess demand on the goods market and thus $p_{t+1} > p_t$. The assumption $\tau = h$ in period t has then been incorrect and we substitute it by $\tau = 1$, i.e., $\theta_t^e > 1$. Obviously then a different equilibrium arises in period t but we claim that the type of equilibrium is still such that there is excess demand on the goods market. Thus, expectations have been adjusted so as to become correct. Analogously we correct the expectations in case $\theta_t^e > 1$ but the equilibrium in period t involves excess supply on the goods market. The rationale for doing this is given by the following lemma, that is proven in Colombo and Weinrich (2008).

Lemma 1. *Assume that for $\tau = h$ in period t an equilibrium with $\gamma_t^d < 1$ occurs. Then this inequality is preserved when switching in period t to $\tau = 1$. Conversely, assume that for $\tau = 1$ in period t an equilibrium with $\gamma_t^s < 1$ occurs. Then this inequality is preserved when switching in period t to $\tau = h$.*

Taking expectations switching into account, a trajectory of the dynamic system is given by a sequence $\{(\alpha_t, m_t, \pi_t, S_t, \tau_t)\}_{t=1}^{\infty}$. It is important to stress that these state variables are perfectly foreseen by economic agents in any period t, so that our dynamic system is a truly forward looking one.[10]

5 Numerical Analysis: The Study of Deflationary Recessions

The non-linear dynamic system representing the economy outlined in the previous section is substantially more complicated than the three-dimensional one investigated in Sect. 2. In particular, we now have to deal with a five-dimensional system – with state variables $\alpha_t, m_t, \pi_t, S_t$ and τ_t – composed of four non-degenerate subsystems each of which may become effective through endogenous regime switching. In analyzing the dynamics of the model, we therefore need to resort to numerical simulations. For this purpose, we consider the same benchmark parameter set used for the base model, that is (3), corresponding to the stationary Walrasian equilibrium, with the addition of $S_0 = 0$ and $\tau_0 = h$, with trading levels $L^* = Y^* = 100$.

Several factors are shown to complement the role of price and wage stickiness (flexibility) – that we illustrated in Sect. 3 – in determining the dynamic effects of a shock. The model outlined in Sect. 4, by considering explicitly firms' inventories and consumers' expectations, adds further propagation mechanisms for shocks and helps explaining their persistence.

Consumers' expectations, in particular, have played an important role in the policy debate. For instance, the 2008 Nobel Laureate Paul Krugman, in a series of influential articles on the Japanese deflationary recession (see Krugman, 1998), stressed the importance of creating inflationary expectations to overcome the liquidity trap in which the country got stuck. Although our setup does not offer insights on how to create inflationary expectations in the first place, it theoretically supports this claim. To see why, let us start from our benchmark parameter set (3) and consider a restrictive monetary shock determined by a reduction in the initial money stock to $m_0 = 40$, keeping all other parameters and initial values at their Walrasian levels. Letting $\nu_1 = 0.025$ and $\nu_2 = 0.1$ be the downward and upward adjustment

[10] In most cases of dynamic systems in economics, they are given by a system of implicit difference equations, in which case an explicit solution in the sense of a (local) flow of mappings cannot be computed analytically. On the other hand, models that avoid this problem – giving rise to truly forward looking dynamic systems – typically are not compatible with perfect foresight outside the stationary state. For a systematic discussion of this issue see, e.g., Böhm and Wenzelburger (1999).

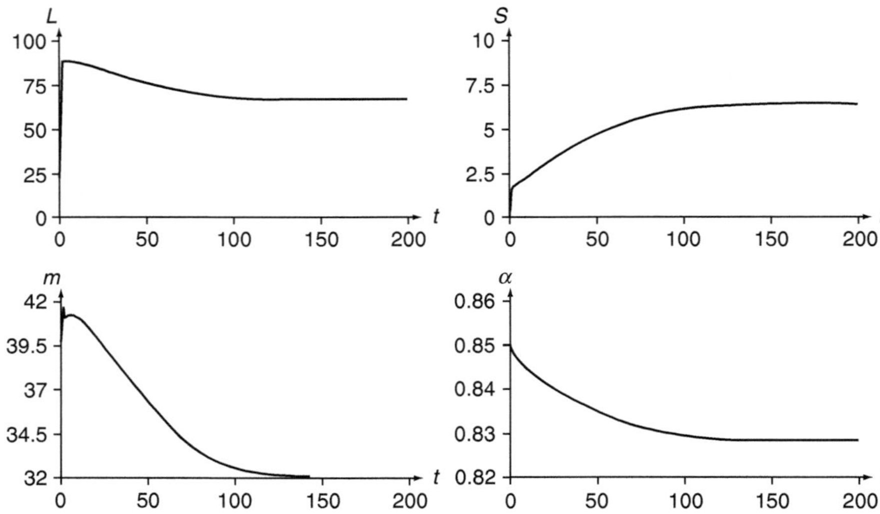

Fig. 3 Time series when $\nu_1 = 0.025$ and $m_0 = 40$

speeds of wages, respectively, and $\mu_1 = \mu_2 = 0.1$ the adjustment speeds of prices out of the Walrasian equilibrium, Fig. 3 shows the convergence of the dynamic system to a deflationary recessionary quasi-stationary state $(\overline{\alpha}, \overline{m}, \overline{\pi}, \overline{S}, \overline{\tau})$ entailing a permanent decrease in employment and output.[11]

Consider now a change in consumers' expectations from $\tau = h$ to $\tau = 1$, that is $(\alpha_0, m_0, \pi_0, S_0, \tau_0) = (\overline{\alpha}, \overline{m}, \overline{\pi}, \overline{S}, 1)$. Expecting inflation for the next period, young consumers demand all their planned life-time consumption in the first period, which boosts aggregate demand and can potentially lead the economy out of the recession. This is shown in Fig. 4 for $\nu_1 = 0.025$ and $\nu_2 = 0.1$, where the economy returns immediately to full employment and the inflationary expectations are confirmed. More precisely, the percentage price inflation, $\phi_t = 100\,(p_{t+1} - p_t)\,/\,p_t$, remains positive over a prolonged period of time, meaning that the economy finds itself each period in a state of repressed inflation, which confirms the validity of Krugman's policy suggestion.

It is also interesting to note that the degree of stickiness of the nominal wage (and price) seems to affect the ability of inflationary expectations to restore full employment. Figure 5 shows that, when a higher downward flexibility of the wage rate is assumed ($\nu_1 = 0.06$ in the figure), the increase in aggregate demand due to anticipated purchases by consumers is not sufficient to overcome the recession, and expectations return from inflationary to deflationary after one period. This confirms

[11] The downward speed of the wage adjustment plays a crucial role. Until approximately $\nu_1 = 0.018$ the economy is able to return to full employment after the monetary shock, whereas for speeds of wage adjustment larger than this it gets trapped in the underemployment situation shown in Fig. 3 (see Colombo and Weinrich, 2008 for a detailed bifurcation analysis).

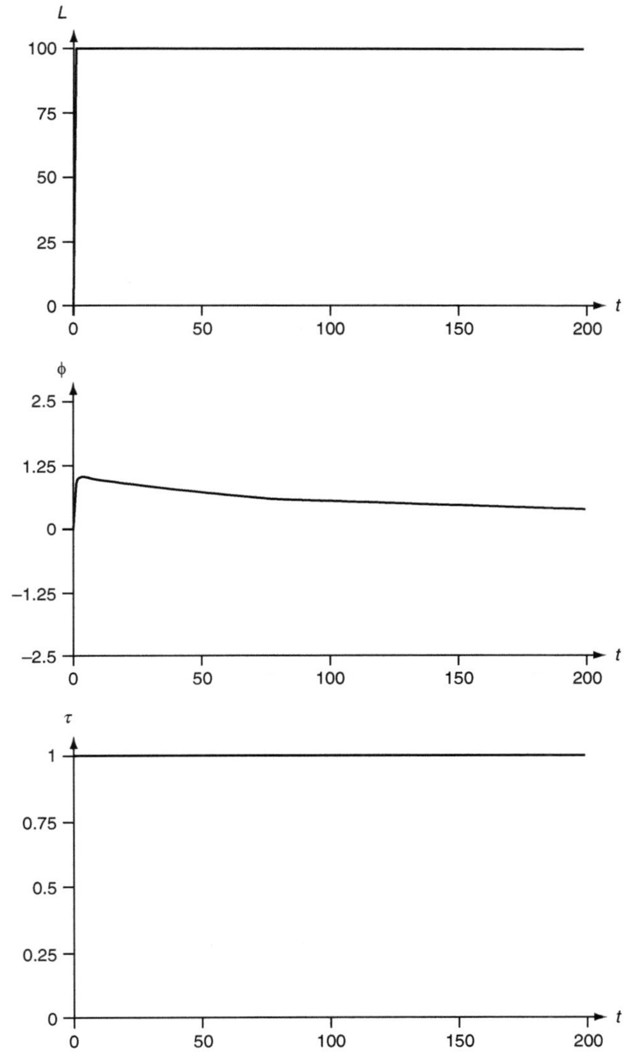

Fig. 4 Effect of inflationary expectations if $\nu_1 = 0.025$

that inflationary expectations need to be sustained over time to help escaping a deflationary recession; something in the spirit of the 'irresponsible' monetary policy – i.e., a monetary policy remaining expansionary even when prices start rising – advocated by Krugman (1998).

It is finally worth noting that, for intermediate values of downward nominal wage flexibility, multiple equilibria with self-confirming expectations will emerge (see Colombo and Weinrich, 2008), which suggests that imposing downward wage rigidity may be a useful measure to overcome a recession when inflationary expectations alone are not enough to do so.

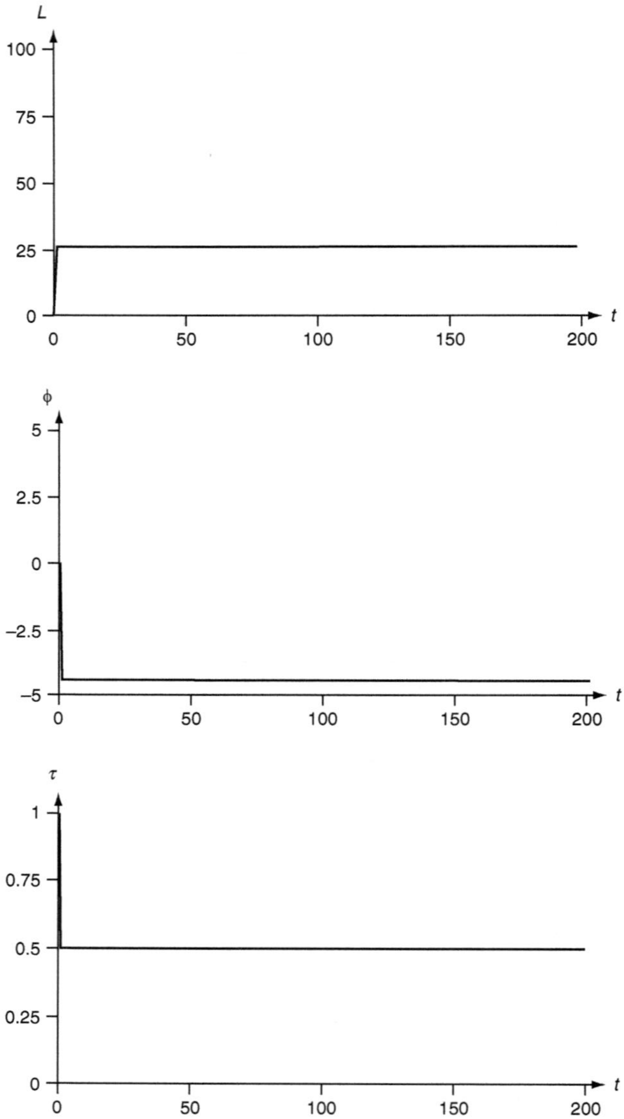

Fig. 5 Effect of inflationary expectations if $\nu_1 = 0.06$

Further measures that can be efficiently analyzed in our framework and complement the role of inflationary expectations are based on expansionary fiscal and monetary policies. Specifically, one such measure that received considerable attention in the policy debate has been proposed in 2003 by Ben Bernanke for the Japanese recession. It consists in a tax reduction accompanied by a transfer of funds from the Central Bank to the government to compensate for the loss in tax revenue.

In Colombo and Weinrich (2003b) we formally analyze this policy and show under which conditions it can effectively help to overcome the crisis.

By focusing on consumers' expectations, in the above analysis we left in the background the effects of firms' inventories. Although a full exploration of the role of inventories is behind the scope of this paper (we refer the reader to Colombo and Weinrich, 2006), it is easily seen that inventories amplify the importance of the spillover effects among markets. More precisely, following a restrictive monetary shock as the one considered above and provided wages are flexible downward, the nominal wage diminishes, which (if the decrease in the nominal wage is large enough) implies a reduction of the real wage as well. The presence of inventories, by increasing the fall of labor demand that in turn depresses labor income and aggregate demand, crucially reinforces this reduction. Eventually the economy converges to a quasi-stationary state with permanent unemployment, a constant low real wage and permanent deflation of the nominal variables.[12]

6 Concluding Remarks

In this paper, we outlined a truly forward looking dynamic framework that generates a wide variety of dynamic behaviors, ranging from cycles of different orders and complex/chaotic behavior to convergence towards quasi-stationary states where the economy lies persistently far away from its Walrasian equilibrium.

Being capable to endogenously determine the emergence of disequilibrium situations, our setup provides an ideal framework to represent and explain the causes of prolonged economic crises, such as the recent Japanese deflationary recession, and to evaluate the impact of alternative economic policies aimed at resolving them. In this sense, our modelling may also prove useful in the analysis of the effects and consequences of the remedies that are being proposed to face the current global recession determined by the breakdown of the financial system.

Several extensions are possible that would extend the reach of our analysis. Three of them stand at the center of our current research agenda in this field. The first deals with the formation of expectations. In the current framework, we focus on the role of agents' expectations without investigating the process from which they arise. Accounting endogenously for the mechanism of expectations formation may

[12] The dynamic behavior of the economy following a restrictive shock becomes completely different whenever the nominal wage is rigid downward. In this case, the real wage and the real money stock increase up to the point in which there start to be excess demand on the goods market and excess supply on the labor market. At this point the goods price starts increasing again, which implies a reduction of the real wage and of the real money stock until the economy converges back to the Walrasian equilibrium. Therefore, unlike in the case where nominal wages are flexible downward, imposing downward nominal wage rigidity may be a measure limiting the effects of deflationary recessions.

enrich the scope of our analysis, by further highlighting the transmission channels of shocks and the effectiveness of alternative economic policies.

A second feature of the model that would benefit from an in-depth examination deals with inventories. In this paper, we focus essentially on their role as a propagation mechanism of shocks and as a source of spillovers between markets. At the beginning of each period, the stock of inventories carried by a firm is simply what remains unsold at the end of the previous period. In this capacity, inventories are a passive element in the decision problem of firms, while it would be more satisfactory to consider them as strategic choice variables as suggested in the literature (see, e.g., Blinder and Fischer, 1981 and Blinder, 1982, where firms have a target inventory level and want to keep a certain inventories-to-sale ratio).

Finally, a third aspect of our modelling that could be refined has to do with the adjustment mechanism of prices and wages. As we discussed in the Introduction, the mechanism by which prices and wages are adjusted between periods is exogenously given in the current formulation of the model. On the one hand, this adds to the flexibility of the approach, by making it consistent with a wide variety of possible explanations for rigidities. On the other hand, in many circumstances it may be interesting to focus on specific sources of stickiness, and endogenize them, to better explore their implications in terms of the feedback and spillover effects arising in the economy.

To explore the implications of these extensions and to generalize the scope of our approach lies at the core of our current research.

References

Akerlof, G. A., & Yellen, J. L. (1985). A near-rational model of the business cycle with wage and price inertia. *Quarterly Journal of Economics, 100*, 823–838.

Ball, L., & Romer, D. (1990). Real rigidities and the non-neutrality of money. *Review of Economic Studies, 57*, 183–203.

Ball, L., & Romer, D. (1991). Sticky prices as coordination failure. *American Economic Review, 81*, 539–553.

Bernake, B., & Gertler, M. (1995). Inside the black box: The credit channel of monetary policy transmission. *Journal of Economic Perspectives, 9*, 27–48.

Bignami, F., Colombo, L., & Weinrich, G. (2004). Complex business cycles and recurrent unemployment in a non-Walrasian macroeconomic model. *Journal of Economic Behavior and Organization, 53*, 173–191.

Blanchard, O. J. (1990). Why does money affect output? A survey. In B. Friedman & F. Hahn (Eds.), *Handbook of monetary economics*. Amsterdam: North Holland.

Blanchard, O. J. (2000). What do we know about macroeconomics that Fisher and Wicksell did not? *Quarterly Journal of Economics, 115*, 1375–1409.

Blinder, A. S. (1982). Inventories and sticky prices. *American Economic Review, 72*, 334–348.

Blinder, A. S., & Fischer, S. (1981). Inventories, rational expectations, and the business cycle. *Journal of Monetary Economics, 8*, 277–304.

Böhm, V. (1989). *Disequilibrium and macroeconomics*. New York: Basil Blackwell.

Böhm, V., & Wenzelburger, J. (1999). Expectations, forecasting and perfect foresight – A dynamical systems approach. *Macroeconomic Dynamics 3*, 167–186.

Calvo, G. (1983). Staggered prices in a utility-maximizing framework. *Journal of Monetary Economics, 12*, 383–398.
Colombo, L., & Weinrich, G. (2003a). The Phillips curve as a long-run phenomenon in a macroeconomic model with complex dynamics. *Journal of Economic Dynamics and Control, 28*, 1–26.
Colombo, L., & Weinrich, G. (2003b). *Unemployment and inventories in the business cycle* (Contributi di ricerca in Econometria e Matematica n. 69). Università Cattolica, Milan. Retrieved from http://www.unicatt.it/docenti/weinrich/invent.pdf.
Colombo, L., & Weinrich, G. (2006). Deflationary recessions in a general equilibrium framework. In A. Namatame, T. Kaisouji, & Y. Aruka (Eds.), *Lecture notes in economics and mathematical systems*. Berlin: Springer.
Colombo, L., & Weinrich, G. (2008). The role of expectations and wages in deflationary recessions (mimeo). Retrieved from http://www.unicatt.it/docenti/weinrich/expdefrec.pdf.
Gale, D. (1979). Large economies with trading uncertainty. *Review of Economic Studies, 46*, 319–338.
Gale, D. (1981). Large economies with trading uncertainty: A correction. *Review of Economic Studies, 48*, 363–364.
Green, J. (1980). On the theory of effective demand. *Economic Journal, 90*, 341–353.
Kiyotaki, N., & Moore, J. (1997). Credit cycles. *Journal of Political Economy, 105*, 211–248.
Krugman, P. (1998). It's Baaack: Japan's slump and the return of the liquidity trap. *Brookings Papers on Economic Activity, 2*, 137–187.
Mankiw, N. G. (1985). Small menu costs and large business cycles. *Quarterly Journal of Economics, 100*, 529–537.
Rotemberg, J. J., & Woodford, M. (1991). Markups and the business cycle. In: O. Blanchard & S. Fischer (Eds.), *NBER macroeconomics annual*. Cambridge: MIT press.
Shapiro, C., & Stiglitz, J. (1984). Equilibrium Unemployment as a worker discipline device. *American Economic Review, 74*, 433–444.
Svensson, L. E. O. (1980). Effective demand and stochastic rationing. *Review of Economic Studies, 47*, 339–355.
Weinrich, G. (1982). On the theory of effective demand. *Economic Journal, 92*, 174–175.
Weinrich, G. (1984). On the theory of effective demand under stochastic rationing. *Journal of Economic Theory, 34*, 95–115.
Weinrich, G. (1988). On the foundations of stochastic non-price rationing and the adjustment of prices. *Decisions in Economics and Finance, 11*, 107–131.

On the Transition Dynamics in Endogenous Recombinant Growth Models

Fabio Privileggi

1 Introduction

Tsur and Zemel (2007) developed an endogenous growth model in which balanced long-run growth is obtained by assuming that the stock of knowledge evolves according to Weitzman's (1998) recombinant expansion process and is used, together with physical capital, as input factor by competitive firms in order to produce a unique physical good. At each instant new knowledge is produced by an independent R&D sector directly controlled by a "regulator" who aims at maximizing the discounted utility of a representative consumer over an infinite horizon. The optimal resources required for new knowledge production are obtained by the regulator in the form of a tax levied on the consumers. The economy, thus, envisages two sectors, a competitive one devoted to the production of the unique physical good, and a regulated R&D sector in which the public good "knowledge" is being directly financed by the regulator and produced according to Weitzman's production function.

In such framework Tsur and Zemel provide conditions under which the economy performs sustained constant balanced growth in the long run; moreover, when balanced growth occurs, they also characterize the asymptotic optimal tax rate and the common growth rate of all variables. Hence, by endogenizing the optimal choice for investing in knowledge production, their result generalizes Weitzman's (1998) endogenous growth model in which the investment in knowledge production was assumed to be constant and exogenously determined.

In this paper we further extend the Tsur and Zemel results by studying more accurately the transition dynamics along a characteristic turnpike curve in the knowledge-capital state space already discussed in Tsur and Zemel (2007). For a specific parametrization of the model and when the conditions allowing sustained

F. Privileggi
Dip. di Politiche Pubbliche e Scelte Collettive Polis, Università del Piemonte Orientale, Via Cavour 84, 15100 Alessandria, Italy
and
IEL - Collegio Carlo Alberto, Via Real Collegio, 30 10024 Moncalieri (Torino), Italy,
e-mail: fabio.privileggi@sp.unipimn.it

G.I. Bischi et al. (eds.), *Nonlinear Dynamics in Economics, Finance and the Social Sciences*, DOI 10.1007/978-3-642-04023-8_14,

long-run growth are met, we are able to (numerically) compute the optimal policy – in terms of optimal consumption – and thus the optimal time-path trajectories of the stock of knowledge, capital, output and consumption – as well as their transition growth rates – while the economy is being headed along the turnpike curve toward its long-run constant balanced growth behavior.

Our method is based on the standard technique of transforming the state and control variables of the Hamiltonian describing the optimal dynamics of (a slightly generalized version of) the Tsur and Zemel model – all diverging in the long-run – into "detrended" state-like and control-like variables, both converging to a saddle-path stable steady state in the appropriate space as time elapses. To study such detrended system we apply the time-elimination method introduced by Mulligan and Sala-i-Martin (1991) (see also Mulligan and Sala-i-Martin, 1993 and Barro and Sala-i-Martin, 2004, pp. 593–596) so that the optimal detrended consumption policy can be calculated by means of numerical methods for ODEs; then, substituting such policy in the ODE of the state-like variable and solving it – again numerically – with respect to time, the optimal time-path trajectories of both state-like and control-like variables are obtained. Eventually, these trajectories are reconverted into time-path trajectories for the original model, thus allowing for a detailed analysis of the transition dynamics of all relevant variables.

Two technical difficulties had to be overcome: (1) finding a proper probability function for the Weitzman's recombinant process suitable for the change of variables in the construction of the detrended system of ODEs, and (2) the exploitation of a singular point – other than the saddle-path steady state – which can be used as initial condition for calculating specific solutions for the ODE describing the policy. Due to the high instability of the system of ODEs characterizing the detrended variables, we have been able to fully solve the model only for a set of values of the parameters; more precisely, our approach works satisfactory only on a manifold of dimension one in the parameters' space (see Remark 2 at the end of Sect. 4).

Section 2 discusses the original contribution by Weitzman (1998) on the production of new knowledge by combining existing ideas and its generalization to the endogenous recombinant growth framework provided by Tsur and Zemel (2007). The central contribution of this paper is contained in Sect. 3, where, under a suitable choice for the functions of the model – in particular, for the probability of success in matching pairs of ideas – we are able to transform the original diverging dynamics into an equivalent system of two ODEs in two "detrended" variables converging asymptotically to a steady state in the appropriate space. This allows for numeric computation of the optimal policy of both the detrended system and the original diverging dynamics, which is implemented in Sect. 4 for a specific set of parameters' values. Finally, after using the optimal policy obtained so far to numerically trace out the optimal time-path trajectories, Sect. 5 is dedicated to a qualitative discussion of the transition dynamics thus obtained, while Sect. 6 reports some concluding remarks and topics for future research.

2 Endogenous Recombinant Growth

Weitzman's (1998) knowledge production device postulates that originally unprocessed (*seed*) ideas are blended with all other ideas available in order to generate new *hybrid* seed ideas; a costly selection process permits in turn to extract from those a subset of *fertile* seed ideas that are again recombined with all the existent fertile ideas to produce yet new hybrids. This process occurs indefinitely, generating knowledge growth. The hybridization is based on matching m ideas together and then checking whether the resulting new idea is fertile (i.e., successful). If $A(t)$ is the stock of knowledge available at time t (measured as the total number of fertile ideas), let $C_m[A(t)]$ denote the number of different combinations of m elements (hybrids) of $A(t)$; i.e.: $C_m[A(t)] = A(t)!/\{m![A(t) - m]!\}$ [e.g., $C_2(A) = A(A - 1)/2$]. Therefore, at time t the number of hybrid seed ideas is given by

$$H(t) = C_m[A(t)] - C_m[A(t-1)]. \tag{1}$$

If π is the probability of obtaining a successful idea from each matching, the number of new successful ideas at time t is given by (Weitzman, 1998, eqn. (2) on p. 337):

$$\Delta A(t) = A(t+1) - A(t) = \pi H(t) = \pi \{C_m[A(t)] - C_m[A(t-1)]\}, \tag{2}$$

which, in a discrete time framework, defines a *recombinant expansion process* of second order representing the potential knowledge production path. Therefore, the stock of knowledge A has the potential of growing at an increasing rate of growth (Weitzman, 1998, Lemma on p. 338). However, potentially explosive growth is actually precluded by scarcity of resources employed in the matching process; as a matter of fact, Weitzman (1998) reconciles his theory with standard endogenous growth models (see, e.g., Romer, 1996, Aghion and Howitt, 1999, or Barro and Sala-i-Martin, 2004) by showing that knowledge growth – as well as the growth rate of GNP in real economies – is actually bounded. Accordingly, the knowledge generation mechanism envisaged by Weitzman uses two inputs: hybrid seed ideas, H, and physical resources, J. The latter affects the probability π of producing successful ideas by increasing it with larger J for each given H, while J becomes less productive for larger H. To summarize, π results to be increasing in the ratio J/H.

Thus, the *production function for new knowledge* ΔA is

$$\Delta A = W(J, H) = H\pi(J/H), \tag{3}$$

corresponding to Weitzman (1998, (28) on p. 346). Note that W in (3) is homogeneous of degree 1. In the sequel we shall assume the following.

Assumption 1. *The function* $\pi : \mathbb{R}_+ \to [0,1]$ *is independent of time and is such that* $\pi' > 0$, $\pi'' < 0$, $\pi(0) = 0$ *and* $\pi(\infty) \leq 1$; *moreover,*[1] $\lim_{x \to 0+} \pi'(x) < +\infty$.

Provided that J is a constant fraction of the total output y, $J = sy$, Weitzman (1998) establishes that in the long run the asymptotic growth rate is a positive constant depending on the exogenously determined saving rate s.

2.1 The Framework

Tsur and Zemel (2007), made an important refinement of Weitzman's analysis by endogenizing the (optimal) resources J employed in the production of new knowledge.[2] Their model features a "regulator" who has the task of choosing the optimal amount J to be employed in production of new knowledge – which, in turn, is being assigned to all firms producing the amount y of a unique (physical) output – in order to maximize the discounted utility of a representative consumer over an infinite horizon. Output producing firms operate in a competitive environment, while the regulator has the power to levy the exact amount J as a tax on the representative consumer, through which, given all the H hybrid seed ideas freely available, new useful knowledge is being directly generated according to (3), and is immediately and freely passed to the output producing firms.

The difficulty in dealing with the second-order dynamic (2) is overcome by switching from the Weitzman's discrete time formulation into a continuous time model. This allows the authors to rewrite (1) as

$$H(t) = C_m' [A(t)] \dot{A}(t), \tag{4}$$

where $\dot{A}(t)$ is the derivative of the stock of knowledge with respect to time. By replacing $\Delta A(t)$ with $\dot{A}(t)$ in (3) we obtain the analogous of (3) in continuous time:

$$\dot{A}(t) = H(t)\, \pi [J(t)/H(t)], \tag{5}$$

where the probability of generating a new fertile idea π still satisfies Assumption 1.

By combining (4) and (5) the law of motion for the stock of knowledge $A(t)$ is

$$\dot{A}(t) = J(t)/\varphi [A(t)], \tag{6}$$

[1] For simplicity, in the sequel $\lim_{x \to 0+} \pi'(x)$ will be denoted by $\pi'(0)$.

[2] Our analysis slightly departs from that of Tsur and Zemel by allowing J to be any amount of physical capital available in the economy, while the authors constrain such resources to be only a fraction $0 \leq s \leq 1$ of the total output y. In other words, in our economy the regulator has the power to extract resources also from existing physical capital, in addition to the whole total output y.

where

$$\varphi(A) = C'_m(A)\,\pi^{-1}\left[1/C'_m(A)\right] \tag{7}$$

is the *expected unit cost of knowledge production*. Note that $\varphi(\cdot)$ is decreasing and, as knowledge keeps spreading, it converges to

$$\lim_{A\to\infty} \varphi(A) = 1/\pi'(0) > 0, \tag{8}$$

where $1/\pi'(0)$ is strictly positive by Assumption 1.

With no loss of generality, we shall assume that labour is constant and normalized to one:[3] $L \equiv 1$. The output producing firms use a neoclassical production function,

$$y(t) = F\left[k(t), A(t)\right], \tag{9}$$

depending on aggregate capital and knowledge-augmented labour $A(t)L$, for $L = 1$.

Assumption 2. $F : \mathbb{R}^2_+ \to \mathbb{R}_+$ *exhibits constant returns to scale and is such that* $F_k > 0$, $F_A > 0$, $F_{kk} < 0$, $F_{AA} < 0$, $F_{kA} > 0$, *and satisfies* $\lim_{k\to 0^+} F(k, A) = +\infty$ *for all* $A > 0$.

Each firm i maximizes instantaneous profit by renting capital k_i and hiring labour $L_i \leq 1$ from the households, taking as given the capital rental rate r, the labour wage w and the stock of knowledge A. Since all firms use the same technology and operate in a competitive market, and all households are the same, the subscript i can be dropped and (9) can be rewritten as $y = Af(k/A)$, where

$$f(x) = F(x, 1). \tag{10}$$

Since firms act competitively, in equilibrium their profit is zero, that is, households earn $y = Af(k/A) = rk + w$; moreover, the amount of capital demanded, k, satisfies

$$f'(k/A) = r. \tag{11}$$

Given that, at each instant t, a fraction $J(t)$ of the whole endowment of the economy, $k(t) + y(t)$, is being employed to finance R&D firms, and a fraction $c(t)$ is being consumed, capital evolves through time according to

$$\dot{k}(t) = y(t) - J(t) - c(t), \tag{12}$$

where it is assumed that capital does not depreciate. Since the upper bound[4] for $J(t)$ and $c(t)$ is jointly given by $J(t) + c(t) \leq k(t) + y(t)$, $\dot{k}(t)$ in (12) may be negative.

[3] Tsur and Zemel (2007) assume that the amount of labour is L, constant through time even if not necessarily equals to one. As stationarity with respect to time of L is the strong assumption here, normalizing labour to $L \equiv 1$ has the advantage of simplifying notation at no cost.

[4] See footnote 2.

Assuming that all households enjoy an instantaneous utility $u[c(t)]$, with $u: \mathbb{R}_+ \to \mathbb{R}_+$ increasing and strictly concave, the "regulator" solves

$$\max_{\{c(t),J(t)\}} \int_0^\infty u\left[c\left(t\right)\right] e^{-\rho t} dt \tag{13}$$

$$\text{subject to } \begin{cases} \dot{A}\left(t\right) = J\left(t\right) / \varphi\left[A\left(t\right)\right] \\ \dot{k}\left(t\right) = F\left[k\left(t\right), A\left(t\right)\right] - J\left(t\right) - c\left(t\right) \\ J\left(t\right) + c\left(t\right) \leq k\left(t\right) + F\left[k\left(t\right), A\left(t\right)\right] \\ k\left(t\right) \geq 0,\ J\left(t\right) \geq 0,\ c\left(t\right) \geq 0 \\ k\left(0\right) = k_0 > 0,\ A\left(0\right) = A_0 > 0, \end{cases}$$

where $\rho > 0$ is the (constant) discount rate. Equation (13) may be interpreted as a maximum welfare problem, where k and A are the state variables and c and J are the controls. Suppressing the time argument, the current-value Hamiltonian associated to (13) is

$$H\left(A, k, J, c, \vartheta_1, \vartheta_2\right) = u\left(c\right) + \vartheta_1\left[F\left(k, A\right) - J - c\right] + \vartheta_2 J / \varphi\left(A\right), \tag{14}$$

where ϑ_1, ϑ_2 are the costates of k and A respectively. Necessary conditions are:

$$u'\left(c\right) = \vartheta_1 \tag{15}$$

$$J = \begin{cases} 0 & \text{if } \vartheta_2 / \varphi\left(A\right) < \vartheta_1 \\ \tilde{J} & \text{if } \vartheta_2 / \varphi\left(A\right) = \vartheta_1 \\ k + F\left(k, A\right) - c & \text{if } \vartheta_2 / \varphi\left(A\right) > \vartheta_1 \end{cases} \tag{16}$$

$$\dot{\vartheta}_1 = \rho\vartheta_1 - \vartheta_1 F_k\left(k, A\right) \tag{17}$$

$$\dot{\vartheta}_2 = \rho\vartheta_2 - \vartheta_1 F_A\left(k, A\right) + \vartheta_2 J \varphi'\left(A\right) / \left[\varphi\left(A\right)\right]^2 \tag{18}$$

$$\lim_{t\to\infty} H\left(t\right) e^{-\rho t} = 0, \tag{19}$$

where $\tilde{J}$ in (16) will be defined in (22). The case $J = k + F(k, A) - c$ when $\vartheta_2/\varphi(A) > \vartheta_1$ in (16) can be ruled out by the Inada condition of Assumption 2.

Taking time derivative of $\vartheta_1 = \vartheta_2/\varphi(A)$ in (16) and using (17) and (18) gives

$$F_k\left(k, A\right) - F_A\left(k, A\right) / \varphi\left(A\right) = 0, \tag{20}$$

defining the locus on the space (A, k) on which the marginal product of capital equals that of knowledge per unit cost. Equation (20) can be rewritten as $z(k/A) = \varphi(A)$, where $z(x) = f(x)/f'(x) - x$, with f defined in (10), is increasing in x; thus, (20) can be expressed as a function of the only variable A:

$$\tilde{k}\left(A\right) = z^{-1}\left[\varphi\left(A\right)\right] A, \tag{21}$$

where z^{-1} is the inverse of $z(x)$.

Differentiating $\tilde{k}(A)$ with respect to time and using (12) and (6) yields

$$\tilde{J}(t) = [\tilde{y}(t) - c(t)]\,\varphi[A(t)] / \left\{\tilde{k}'[A(t)] + \varphi[A(t)]\right\}, \tag{22}$$

where $\tilde{y}(t) = F\{\tilde{k}[A(t)], A(t)\}$, expressing the optimal investment in R&D, $\tilde{J}(t)$, as a function of the optimal consumption $c(t)$, when the economy grows along the curve $\tilde{k}(A)$ defined in (21); that is, when $\vartheta_2(t)/\varphi[A(t)] = \vartheta_1(t)$ in (16).

We consider the limit of (21) for large A, which becomes linear, by defining:

$$\tilde{k}_\infty(A) = \tilde{\eta}A + q, \tag{23}$$

where, by (8), $\tilde{\eta} = z^{-1}[1/\pi'(0)]$ and q is a non-negative constant. Note that $\tilde{k}(A)$ lies above $\tilde{k}_\infty(A)$ for all $A < \infty$, approaching $\tilde{k}_\infty(A)$ as A increases. The intercept q depends on the number of ideas m being matched at each instant t in (4).

Proposition 1. *The intercept q in (23) is zero whenever $m > 2$, while $q > 0$ for $m = 2$.*

Proof. Since $\tilde{k}_\infty(A) = \tilde{\eta}A + q$ is the asymptote of $\tilde{k}(A)$,

$$q = \lim_{A\to+\infty}\left[\tilde{k}(A) - \tilde{\eta}A\right] = \lim_{A\to+\infty}\left\{z^{-1}[\varphi(A)] - z^{-1}\left[1/\pi'(0)\right]\right\}A. \tag{24}$$

As $\varphi(A)$ is decreasing and, under Assumption 1, bounded away from zero [specifically, $0 < 1/\pi'(0) \le \varphi(A) \le \varphi(A_0)$], by Assumption 2 $z^{-1}[\varphi(A)] - z^{-1}[1/\pi'(0)]$ in (24) is $o[\varphi(A)]$. Thus, since, by (7), $O[\varphi(A)] = O[C'_m(A)] = O(A^{m-1})$ [i.e., $C'_m(A) \sim A^{m-1}$ for large A], if $m > 2$ the limit in (24) is zero, while, if $m = 2$, such limit must be nonzero; as $z^{-1}[\varphi(A)] - z^{-1}[1/\pi'(0)] > 0$ for all $A < +\infty$, $q > 0$ holds whenever $m = 2$. □

Another locus will be considered, that on which the marginal product of capital equals the individual discount rate, $f'(k/A) = \rho$, which, by (11), implies $r = \rho$. As $f'(\cdot)$ is decreasing, also such curve can be expressed as a function of A:

$$\hat{k}(A) = \hat{\eta}A, \tag{25}$$

with $\hat{\eta} = (f')^{-1}(\rho)$; that is, $\hat{k}(A)$ is the linear function with slope $\hat{\eta} > 0$.

The curves $\tilde{k}(A)$, $\tilde{k}_\infty(A)$ and $\hat{k}(A)$ defined in (21), (23) and (25) will be labeled *turnpike*, *asymptotic turnpike* and *stagnation line* respectively. The optimal investment in R&D along the turnpike $\tilde{k}(A)$ defined in (22), $\tilde{J}(t)$, will be referred as the *singular policy*. We shall assume the following.

Assumption 3. *The instantaneous utility is CIES: $u(c) = (c^{1-\sigma} - 1)/(1-\sigma)$, with $\sigma \ge 1$.*

Proposition 2 (Tsur and Zemel, 2007).

(i) A necessary condition for the economy to sustain long-run growth is

$$\hat{\eta} > \tilde{\eta}; \tag{26}$$

conversely, if $\hat{\eta} \leq \tilde{\eta}$ the economy eventually reaches a steady (stagnation) point on the line $\hat{k}(A)$ corresponding to zero growth.

(ii) Under (26), for any given initial knowledge stock A_0 there is a corresponding threshold capital stock $k^{sk}(A_0) \geq 0$ such that whenever $k_0 \geq k^{sk}(A_0)$ the economy – possibly after an initial transition outside the turnpike – first reaches the turnpike $\tilde{k}(A)$ in a finite time, and then continues to grow along it as time elapses until the asymptotic turnpike $\tilde{k}_\infty(A)$ is reached in the long-run. Along $\tilde{k}_\infty(A)$ the economy follows a balanced growth path characterized by a common constant growth rate of output, knowledge, capital and consumption given by

$$\gamma = (r_\infty - \rho)/\sigma > 0, \tag{27}$$

where $r_\infty = \lim_{A\to\infty} f'[\tilde{k}_\infty(A)/A] = f'(\tilde{\eta})$ defines the long-run capital rental rate.[5] *Moreover, $\tilde{J}(t) < \tilde{y}(t)$ for large t, and the income shares devoted to investments in knowledge and capital are constant and given respectively by*

$$s_\infty = \gamma / \left\{r_\infty \left[1 + \tilde{\eta}\pi'(0)\right]\right\} \quad \text{and} \quad s_\infty^k = \gamma\tilde{\eta}\pi'(0) / \left\{r_\infty \left[1 + \tilde{\eta}\pi'(0)\right]\right\}. \tag{28}$$

If $k_0 < k^{sk}(A_0)$ the economy eventually stagnates.

Proposition 2, whose proof can be found in Tsur and Zemel (2007), establishes that if (26) holds and k_0 is sufficiently high with respect to initial knowledge stock A_0, the economy grows along a turnpike path which, in the long run, converges to a balanced growth path with knowledge and capital growing at the same constant rate and with constant saving rate, thus confirming Weitzman's result in a more general setting.

As the case $\vartheta_2/\varphi(A) > \vartheta_1$ in (16) is ruled out, two optimal regimes are possible:

1. Zero R&D, corresponding to $J \equiv 0$, which, if maintained forever, eventually leads the economy to some steady state (stagnation point) on the line $\hat{k}(A)$.
2. A path along the turnpike $\tilde{k}(A)$ – maybe started after a finite period of transition outside the turnpike – corresponding to the singular policy $\tilde{J}$ in (22), which envisages growth for all variables as time elapses and, if maintained forever, eventually lead to a balanced growth path along the asymptotic turnpike $\tilde{k}_\infty(A)$.

Under (26) and if $k_0 \geq k^{sk}(A_0)$ it can be shown that the turnpike $\tilde{k}(A)$ is "trapping", i.e., the economy keeps growing along it after it is reached. Hence, there are two types of transitions: one driving the system toward the turnpike starting from

[5] Note that, under (26), $r_\infty = f'(\tilde{\eta}) > f'(\hat{\eta}) = f'[(f')^{-1}(\rho)] = \rho$.

outside it, and another characterizing the optimal path along $\tilde{k}(A)$ after it has been entered. We shall focus on the latter; specifically, we shall assume that (26) holds, implying that the stagnation line $\hat{k}(A)$ lies strictly above[6] the turnpike $\tilde{k}(A)$ for A sufficiently large, moreover, we shall restrict our attention to the case $k_0 = \tilde{k}(A_0)$. In this scenario $k_0 \geq k^{sk}(A_0)$ is certainly satisfied, as the turnpike $\tilde{k}(A)$ is trapping.

2.2 Dynamics Along the Turnpike

We now adapt the optimal conditions (15)–(19) to the system's behavior along the turnpike $\tilde{k}(A)$. All variables on the turnpike will be labeled with a "$\sim$" symbol.

Suppressing the time argument and using (22), (6) becomes

$$\dot{A} = [\tilde{y}(A) - \tilde{c}] / \left[\tilde{k}'(A) + \varphi(A)\right], \tag{29}$$

where $\tilde{y}(A) = F[\tilde{k}(A), A] = Af[\tilde{k}(A)/A]$ with $f(\cdot)$ defined in (10). Equation (29) is the unique dynamic constraint as $\dot{\tilde{k}} = \tilde{k}'(A)\dot{A} = \tilde{k}'(A)[\tilde{y}(A) - \tilde{c}]/[\tilde{k}'(A) + \varphi(A)]$; therefore, now the unique state variable is A, and, by (22), the unique control is $\tilde{c}$.

Thus, the "regulator" solves

$$\max_{\{\tilde{c}(t)\}} \int_0^{\infty} u[\tilde{c}(t)] e^{-\rho t} dt \tag{30}$$

$$\text{subject to} \begin{cases} \dot{A}(t) = \{\tilde{y}[A(t)] - \tilde{c}(t)\} / \left\{\tilde{k}'[A(t)] + \varphi[A(t)]\right\} \\ 0 \leq \tilde{c}(t) \leq \tilde{k}[A(t)] + \tilde{y}[A(t)] \\ A(0) = A_0 > 0. \end{cases}$$

The current-value Hamiltonian for problem (30) is

$$\tilde{H}(A, \tilde{c}, \vartheta) = u(\tilde{c}) + \vartheta [\tilde{y}(A) - \tilde{c}] / \left[\tilde{k}'(A) + \varphi(A)\right], \tag{31}$$

where ϑ is the costate variable associated with A. Necessary conditions are:

$$\vartheta = u'(\tilde{c}) \left[\tilde{k}'(A) + \varphi(A)\right] \tag{32}$$

$$\dot{\vartheta} = \left\{\rho - \left[\tilde{y}'(A) - \left(\tilde{k}''(A) + \varphi'(A)\right)\dot{A}\right] / \left[\tilde{k}'(A) + \varphi(A)\right]\right\} \vartheta \tag{33}$$

$$\lim_{t \to \infty} \tilde{H}(t) e^{-\rho t} = 0, \tag{34}$$

where $\dot{A}$ in (33) is given by (29).

---

[6] This holds for all $A > 0$ when $m > 2$, while for A large enough if $m = 2$.

Since, by (20), $F_A[\tilde{k}(A), A] = F_k[\tilde{k}(A), A]\varphi(A)$ along the turnpike and, by (11), $\tilde{r}(A) = F_k[\tilde{k}(A), A]$, where $\tilde{r}(A)$ is the capital rental rate on the turnpike, $\tilde{y}'(A) = \tilde{r}(A)[\varphi(A) + \tilde{k}'(A)]$. Hence, dividing by ϑ, (33) can be rewritten as

$$\dot{\vartheta}/\vartheta = \rho - \tilde{r}(A) + \dot{A}\left[\tilde{k}''(A) + \varphi'(A)\right] / \left[\tilde{k}'(A) + \varphi(A)\right]. \tag{35}$$

Taking time derivative of (32), dividing by ϑ and coupling with (35), under Assumption 3 we get

$$\dot{\tilde{c}}/\tilde{c} = [\tilde{r}(A) - \rho]/\sigma = \left\{f'\left[\tilde{k}(A)/A\right] - \rho\right\}/\sigma, \tag{36}$$

where in the second equality (11) and (10) have been used.

From (29) and (36) we obtain the following system of ODEs defining the optimal dynamics for $A(t)$ and $\tilde{c}(t)$ along the turnpike under Assumption 3:

$$\begin{cases} \dot{A} = \left\{f\left[\tilde{k}(A)/A\right]A - \tilde{c}\right\} / \left[\tilde{k}'(A) + \varphi(A)\right] \\ \dot{\tilde{c}} = \tilde{c}\left\{f'\left[\tilde{k}(A)/A\right] - \rho\right\}/\sigma. \end{cases} \tag{37}$$

Proposition 2(ii) states that in the long run the ratios $\dot{A}/A$ and $\dot{\tilde{c}}/\tilde{c}$ obtained from (37) converge to the balanced growth rate $\gamma = (r_\infty - \rho)/\sigma$.

3 Model Specification and Analysis

We now suitably restrict the class of models under investigation.

Assumption 4. *In addition to Assumption 3, the followings hold:*

(i) Only pairs of ideas will be matched in the recombinant process: $m = 2$.
(ii) The probability function $\pi : \mathbb{R}_+ \to [0, 1]$ *of the recombinant process is*

$$\pi(x) = \beta x / (\beta x + 1), \qquad \beta > 0. \tag{38}$$

(iii) The production function has the Cobb–Douglas form: $F(k, A) = \theta k^\alpha A^{1-\alpha} = \theta A(k/A)^\alpha$, *with* $\theta > 0$ *and* $0 < \alpha < 1$.

Clearly, $\pi(\cdot)$ in (38) satisfies Assumption 1; parameter β measures the degree of efficiency of the Weitzman matching process, the larger β the higher probability of obtaining a new successful idea out of each (pairwise) matching of seed ideas.

Since, when $m = 2$, $C_2'(A) = (2A - 1)/2$, and from (38) we get $\pi^{-1}(y) = y/[\beta(1 - y)]$, substituting both in (7) yields the following explicit form for $\varphi(A)$:

$$\varphi(A) = (2A - 1)/[\beta(2A - 3)] = (1/\beta)[1 + 2/(2A - 3)]. \tag{39}$$

As $\pi'(0) = \beta$, Assumption 4(iii) and (39) yields:

$$\tilde{k}(A) = [\alpha/(1-\alpha)]\,\varphi(A)\,A = \{\alpha/[\beta(1-\alpha)]\}\,[1 + 2/(2A-3)]\,A \quad (40)$$
$$\tilde{k}_{\infty}(A) = \{\alpha/[\beta(1-\alpha)]\}\,(A+1) \qquad (\text{i.e., } \tilde{\eta} = q = \alpha/[\beta(1-\alpha)]) \quad (41)$$
$$\hat{k}(A) = (\theta\alpha/\rho)^{1/(1-\alpha)}\,A \qquad \left(\text{i.e., } \hat{\eta} = (\theta\alpha/\rho)^{1/(1-\alpha)}\right), \quad (42)$$

and the growth condition (26) becomes

$$\rho < \theta\alpha\,[\beta(1-\alpha)/\alpha]^{1-\alpha}\,. \quad (43)$$

It is seen from (40) that the initial condition A_0 must be in the open interval $(3/2, +\infty)$, and the graph of $\tilde{k}(A)$ is a U-shaped curve on it. Since the stock of knowledge A cannot be depleted and the economy is bound to follow the optimal investment in R&D policy $\tilde{J} > 0$ defined in (22), along the turnpike A must grow: $\dot{A}(t) > 0$ for all $t \geq 0$. Therefore, a U-shaped $\tilde{k}(A)$ means that capital $\tilde{k}(t)$ decreases $[\dot{\tilde{k}}(t) < 0]$ when t is small and increases $[\dot{\tilde{k}}(t) > 0]$ for larger t, envisaging that in early times it is optimal to take away some physical capital from the output-producing sector and invest it in R&D, so that the stock of knowledge A can take-off. Moreover, $\dot{A} > 0$ in (29) – and thus in (37) – has important implications.

Proposition 3. *Under Assumption 4, the* optimal policy *along the turnpike, $\tilde{c}(A)$, satisfies*

$$\begin{cases} \tilde{c}(A) > \tilde{y}(A) \text{ for } 3/2 < A < A^s \\ \tilde{c}(A^s) = \tilde{y}(A^s) \\ \tilde{c}(A) < \tilde{y}(A) \text{ for } A > A^s, \end{cases} \quad (44)$$

where

$$A^s = 1 + (1/2)\left(\alpha + \sqrt{1 + 4\alpha + \alpha^2}\right). \quad (45)$$

Moreover, $\tilde{c}'(A) \leq 0$ in a neighborhood of A^s.

Proof. By differentiating $\tilde{k}(A)$ in (40) it is easily seen that the denominator of (29), $\tilde{k}'(A) + \varphi(A)$, vanishes on the unique point A^s defined in (45), which belongs to the domain $(3/2, +\infty)$ as $A^s > 3/2$ for all $0 < \alpha < 1$; moreover, $\tilde{k}'(A) + \varphi(A) < 0$ for $3/2 < A < A^s$ and $\tilde{k}'(A) + \varphi(A) > 0$ for $A > A^s$. Therefore, $\dot{A}(t) > 0$ for all $t \geq 0$ in (29) implies (44). Since it can be checked that A^s is also the unique (minimum) stationary point for the optimal output $\tilde{y}(A)$ – i.e., $\tilde{y}'(A^s) = 0$ – and (44) states that the graph of the optimal policy $\tilde{c}(A)$ must intersect the graph of the optimal output $\tilde{y}(A)$ from above on $A = A^s$, $\tilde{c}'(A) \leq 0$ must hold in a neighborhood of A^s. □

Proposition 3 will be useful in handling the point corresponding to $(A^s, \tilde{c}(A^s))$ in the "detrended" system.

3.1 State-Like and Control-Like Variables

When the economy performs sustained growth in the long run, there are no steady states toward which the system eventually converges. Thus, we transform the state variable A and the control $\tilde{c}$ in a state-like variable, μ, and a control-like variable, χ, respectively, so that $\mu(t)$ and $\chi(t)$ converge to some fixed points μ^* and χ^* in the space (μ, χ) as time elapses. We choose the following transformations:

$$\mu = \tilde{k}(A)/A = [\alpha/(1-\alpha)]\,\varphi(A) = \{\alpha/[\beta(1-\alpha)]\}\,[1 + 2/(2A-3)] \quad (46)$$

$$\chi = \tilde{c}/A, \quad (47)$$

where in (46) we used (40) and (39). Hence, A is related to μ as follows:

$$A = \alpha/[\beta(1-\alpha)\mu - \alpha] + 3/2. \quad (48)$$

Given the "detrended" optimal policy $\chi(\mu)$, the optimal policy of (30) is

$$\tilde{c}(A) = \chi[(\alpha/(1-\alpha))\,\varphi(A)]\,A. \quad (49)$$

Under Assumption 4(iii), from (37) we obtain the following ratios:

$$\dot{A}/A = \left\{\theta\left[\tilde{k}(A)/A\right]^{\alpha} - \tilde{c}/A\right\} / \left[\tilde{k}'(A) + \varphi(A)\right] \quad (50)$$

$$\dot{\tilde{c}}/\tilde{c} = \left\{\theta\alpha\left[\tilde{k}(A)/A\right]^{\alpha-1} - \rho\right\}/\sigma. \quad (51)$$

The growth rate of μ in (46) is $\dot{\mu}/\mu = \tilde{k}'(A)\dot{A}/\tilde{k}(A) - \dot{A}/A$; therefore, $\dot{\mu} = [\tilde{k}'(A) - \mu]\dot{A}/A$, which, coupled with (50) and using (47), yields

$$\dot{\mu} = \left[\tilde{k}'(A) - \mu\right](\theta\mu^{\alpha} - \chi)/\left[\tilde{k}'(A) + \varphi(A)\right]. \quad (52)$$

As (39) equals to $2/(2A-3) = \beta\varphi(A) - 1$ and $\varphi'(A) = -4/[\beta(2A-3)^2]$, φ' is a function of φ: $\varphi'(A) = -(1/\beta)[2/(2A-3)]^2 = -(1/\beta)[\beta\varphi(A) - 1]^2$; moreover, (39) may also be rewritten as $A = 1/[\beta\varphi(A) - 1] + 3/2$, while (46) is equivalent to $\varphi(A) = [(1-\alpha)/\alpha]\mu$. Hence, By differentiating (40) and substituting these expressions for $\varphi'(A)$, A and $\varphi(A)$, after a fair amount of algebra $\tilde{k}'(A)$ in (52) becomes

$$\tilde{k}'(A) = \{\alpha/[2\beta(1-\alpha)]\}\left\{6\beta[(1-\alpha)/\alpha]\mu - 3\beta^2[(1-\alpha)/\alpha]^2\mu^2 - 1\right\}. \quad (53)$$

We can now rewrite (52) only in terms of variables μ and χ:

$$\dot{\mu} = \left[1 - \frac{2\beta(1-\alpha)\mu}{2\beta(1-\alpha)(1+2\alpha)\mu - 3\beta^2(1-\alpha)^2\mu^2 - \alpha^2}\right](\theta\mu^{\alpha} - \chi). \quad (54)$$

By (50), (51) and (46), the growth rate of χ in (47) is $\dot{\chi}/\chi = (\theta\alpha\mu^{\alpha-1} - \rho)/\sigma - (\theta\mu^{\alpha} - \chi)/[\tilde{k}'(A) + \varphi(A)]$, which, by replacing $\tilde{k}'(A)$ as in (53) and $\varphi(A) = [(1-\alpha)/\alpha]\mu$, yields the following ODE for the control-like variable χ:

$$\dot{\chi} = \left[\frac{\theta\alpha\mu^{\alpha-1} - \rho}{\sigma} - \frac{2\alpha\beta\,(1-\alpha)\,(\theta\mu^{\alpha} - \chi)}{2\beta\,(1-\alpha)\,(1+2\alpha)\,\mu - 3\beta^2\,(1-\alpha)^2\,\mu^2 - \alpha^2}\right]\chi. \quad (55)$$

Hence, we must study the following system of ODEs:

$$\begin{cases} \dot{\mu} = [1 - 2\beta\,(1-\alpha)\,\mu/Q\,(\mu)]\,(\theta\mu^{\alpha} - \chi) \\ \dot{\chi} = \left[\left(\theta\alpha\mu^{\alpha-1} - \rho\right)/\sigma - 2\alpha\beta\,(1-\alpha)\,(\theta\mu^{\alpha} - \chi)\,/\,Q\,(\mu)\right]\chi, \end{cases} \quad (56)$$

where

$$Q\,(\mu) = -3\beta^2\,(1-\alpha)^2\,\mu^2 + 2\beta\,(1-\alpha)\,(1+2\alpha)\,\mu - \alpha^2. \quad (57)$$

3.2 Fixed Points and Phase Diagram

Since $A > 3/2$, from (46) one immediately obtains the range $(\mu^*, +\infty)$, with

$$\mu^* = \alpha/\left[\beta\,(1-\alpha)\right], \quad (58)$$

for the state-like variable μ, with endpoints corresponding to $A \to +\infty$ and $A \to (3/2)^+$ respectively. In other words, μ^* in (58) is the *steady value* for variable μ corresponding to long-run behavior of the economy along the asymptotic turnpike $\tilde{k}_\infty(A)$ [μ^* is the slope of $\tilde{k}_\infty(A)$, as seen in (41)]. The feasible set for the detrended variables (μ, χ) therefore is $S = [\mu^*, +\infty) \times \mathbb{R}_{++}$, where we added the boundary μ^* corresponding to the asymptotic dynamics ($A = +\infty$) of the original model.

From the first equation in (56), two loci on which $\dot{\mu} = 0$ are found in S: the curve

$$\chi = \theta\mu^{\alpha} \quad (59)$$

and the vertical line $\mu \equiv \mu^*$, with μ^* as in (58). Equation (59) vanishes the second factor in the RHS of the first equation in (56), while μ^* is the largest (and only feasible) solution of $Q(\mu) - 2\beta(1-\alpha)\mu = 0$, with $Q(\mu)$ defined in (57), vanishing the first factor in the RHS of the same equation.

From the second equation in (56), the unique locus on which $\dot{\chi} = 0$ is given by

$$\chi = \theta\mu^{\alpha} - Q\,(\mu)\left(\theta\alpha\mu^{\alpha-1} - \rho\right)/\left[2\alpha\beta\sigma\,(1-\alpha)\right]. \quad (60)$$

$Q(\mu)$ turns out to have a unique (admissible) root, call it μ^s, satisfying

$$Q\,(\mu) = -3\beta^2\,(1-\alpha)^2\,\mu^2 + 2\beta\,(1-\alpha)\,(1+2\alpha)\,\mu - \alpha^2 = 0, \quad (61)$$

with $Q(\mu) > 0$ for $\mu^* \leq \mu < \mu^s$ and $Q(\mu) < 0$ for $\mu > \mu^s$. Thus, whether the locus (60) lies above or below the locus (59) depends on whether $\mu^* \leq \mu < \mu^s$ or $\mu > \mu^s$, and on the sign of $(\theta\alpha\mu^{\alpha-1} - \rho)$. On $\mu = \mu^s$, however, they intersect, and this yields our *first steady state*: (μ^s, χ^s), with $\chi^s = \theta(\mu^s)^\alpha$, which happens to correspond to the point $(A^s, \tilde{c}(A^s))$ discussed in Proposition 3 for the original dynamic (37). To see this, recall that, from (44), $\tilde{c}(A^s) = \tilde{y}(A^s)$ must hold on the critical point A^s defined in (45); by replacing A^s in (46) and (47), we get,

$$\mu^s = \left(1 + 2\alpha + \sqrt{1 + 4\alpha + \alpha^2}\right) / \left[3\beta\left(1 - \alpha\right)\right], \qquad \chi^s = \theta\left(\mu^s\right)^\alpha, \tag{62}$$

where μ^s coincides with the largest (and only admissible) solution of (61).

It is immediately seen that $\mu^* < \mu^s$ for all feasible values of parameters α and β, which means that $Q(\mu^*) > 0$ must hold; moreover, using (58), the necessary condition for growth (43) can be rewritten as $[\theta\alpha(\mu^*)^{\alpha-1} - \rho] > 0$. Therefore, we conclude that the locus (60) intersects the locus $\mu \equiv \mu^*$ strictly below locus (59). Since along such vertical line $\dot{\mu} = 0$, we have found the *second steady state* of system (56): (μ^*, χ^*), where χ^* is (60) evaluated at $\mu = \mu^*$, specifically,

$$\chi^* = \theta\left\{\alpha / \left[\beta\left(1 - \alpha\right)\right]\right\}^\alpha \left(1 - 1/\sigma\right) + \rho / \left[\beta\sigma\left(1 - \alpha\right)\right]. \tag{63}$$

Clearly, under Assumption 4 $\chi^* > 0$. As $\theta\mu^\alpha$ in (59) is increasing in μ and $\chi^* < \theta(\mu^*)^\alpha$, it follows that (μ^*, χ^*) lies south-west of (μ^s, χ^s). We shall see in short that (μ^*, χ^*) is the saddle-path stable steady state to which system (56) converges in the long-run. Hence, χ^* is the asymptotic slope of the optimal consumption $\tilde{c}(A)$ steadily growing at the constant rate γ in the original model.

As condition (43) states that $\rho < \theta\alpha(\mu^*)^{\alpha-1}$ must hold and, as $0 < \alpha < 1$, $\theta\alpha\mu^{\alpha-1}$ is a decreasing function of μ, there must be a unique value $\hat{\mu} > \mu^*$ such that $[\theta\alpha(\hat{\mu})^{\alpha-1} - \rho] = 0$. It is clear from the last factor in the second term in the RHS of (60) that the two loci (60) and (59) must intersect in $\mu = \hat{\mu}$; hence $(\hat{\mu}, \hat{\chi})$, with

$$\hat{\mu} = \left(\theta\alpha/\rho\right)^{\frac{1}{1-\alpha}}, \qquad \hat{\chi} = \theta\left(\theta\alpha/\rho\right)^{\frac{\alpha}{1-\alpha}}, \tag{64}$$

is the *third* (and last) *steady state* associated to (56). From (42), $\hat{\mu}$ in (64) corresponds to the value $\hat{A}$ at which $\tilde{k}(A)$ intersects $\hat{k}(A)$ in the original model. By equating (40) and (42) [or by substituting $\hat{\mu}$ as in (64) into (48)], $\hat{A}$ turns out to be

$$\hat{A} = \alpha / \left[\beta\left(1 - \alpha\right)\left(\theta\alpha/\rho\right)^{\frac{1}{1-\alpha}} - \alpha\right] + 3/2, \tag{65}$$

which in turn, if replaced in (49) and using $\hat{\chi}$ as in (64), yields the value of the optimal policy at the intersection point $\hat{A}$, $\tilde{c}(\hat{A}) = \hat{\chi}\hat{A}$, of the original model.

The position of the last steady state, $(\hat{\mu}, \hat{\chi})$, depends on how large the discount factor ρ is with respect to the parameters α, θ and β. Since $\mu^* < \mu^s$ implies $\theta\alpha(\mu^s)^{\alpha-1} < \theta\alpha(\mu^*)^{\alpha-1}$, three scenarios may occur, all satisfying condition (43):

1. If $\rho < \theta\alpha(\mu^s)^{\alpha-1}$, $\mu^s < \hat{\mu}$ and $(\hat{\mu}, \hat{\chi})$ lies north-east of (μ^s, χ^s).

2. If $\rho = \theta\alpha(\mu^s)^{\alpha-1}$, $\mu^s = \hat{\mu}$ and the two steady states collapse: $(\hat{\mu}, \hat{\chi}) = (\mu^s, \chi^s)$.
3. If $\theta\alpha(\mu^s)^{\alpha-1} < \rho < \theta\alpha(\mu^*)^{\alpha-1}$, $\mu^* < \hat{\mu} < \mu^s$ and $(\hat{\mu}, \hat{\chi})$ lies north-east of (μ^*, χ^*) and south-west of (μ^s, χ^s).

We shall focus on the third case, corresponding to a scenario in which A^s lies on the left of $\hat{A}$, on which the turnpike $\tilde{k}(A)$ intersects the stagnation line $\hat{k}(A)$.

Proposition 4. *Under Assumption 4 and provided that* $\theta\alpha(\mu^s)^{\alpha-1} < \rho < \theta\alpha(\mu^*)^{\alpha-1}$ *holds, the two fixed points* (μ^*, χ^*) *and* $(\hat{\mu}, \hat{\chi})$ *can be classified as follows:*

1. (μ^*, χ^*), *with coordinates defined in (58) and (63), is* saddle-path stable, *with the stable arm converging to it from north-east whenever the initial values* $(\mu(t_0), \chi(t_0))$ *are suitably chosen.*
2. $(\hat{\mu}, \hat{\chi})$, *with coordinates defined in (64), is an unstable clockwise-rotating spiral.*

Proof. Above the locus (59) the term $(\theta\mu^\alpha - \chi)$ in the first equation of (56) is negative, while it is positive below. $Q(\mu)$ in (57) is such that $Q(\mu) > 0$ for $\mu^* < \mu < \mu^s$, while $Q(\mu) < 0$ for $\mu > \mu^s$; therefore, $[1 - 2\beta(1-\alpha)\mu/Q(\mu)]$ is negative for $\mu^* < \mu < \mu^s$ and positive for $\mu > \mu^s$. Hence: if $\mu^* < \mu < \mu^s$, $\dot{\mu} > 0$ above locus (59) and $\dot{\mu} < 0$ below; while, if $\mu > \mu^s$, $\dot{\mu} < 0$ above locus (59) and $\dot{\mu} > 0$ below.

Since $\chi > 0$, the sign of $\dot{\chi}$ in the second equation of (56) depends on the sign of the term in square brackets in the RHS. From the sign of $Q(\mu)$ we infer that for $\mu^* < \mu < \mu^s$ such term is positive above locus (60) and it is negative below, while the converse holds for $\mu > \mu^s$. Thus, when $\mu^* < \mu < \mu^s$, $\dot{\chi} > 0$ above locus (60) and $\dot{\chi} < 0$ below; conversely, if $\mu > \mu^s$, $\dot{\chi} < 0$ above locus (60) and $\dot{\chi} > 0$ below.

The analysis above is sufficient to trace out the phase diagram for the case $\theta\alpha(\mu^s)^{\alpha-1} < \rho < \theta\alpha(\mu^*)^{\alpha-1}$, i.e., when $\mu^* < \hat{\mu} < \mu^s$, which is reported in Fig. 1. Clearly, (μ^*, χ^*) is *saddle-path stable*; it can be guessed that its stable arm is increasing and lying below locus (60) on the interval $[\mu^*, \mu^s)$. To check its saddle-path stability, consider the Jacobian of (56) evaluated at (μ^*, χ^*):

$$J\left(\mu^*, \chi^*\right) = \begin{bmatrix} \frac{\rho-\beta\theta(1-\alpha)(\mu^*)^\alpha}{\sigma} & 0 \\ -\frac{(1-\alpha)\left[c_1(\mu^*)^{2\alpha}+c_2(\mu^*)^\alpha+\rho^2\right]}{\alpha\sigma^2} & \frac{\rho+\beta\theta(1-\alpha)(\sigma-1)(\mu^*)^\alpha}{\sigma} \end{bmatrix}, \tag{66}$$

where $c_1 = (\beta\theta)^2\alpha\sigma(1-\alpha)(\sigma-1)$, $c_2 = \beta\theta\rho(\alpha+\sigma-1)$. By (43) the terms on the diagonal have opposite signs; hence, $\det[J(\mu^*, \chi^*)] < 0$ and (μ^*, χ^*) is a saddle.

As (μ^*, χ^*) lies strictly below locus (59) and the unique intersection point between the loci (60) and (59) on the interval $[\mu^*, \mu^s)$ is the fixed point $(\hat{\mu}, \hat{\chi})$, it must be the case that (60) crosses (59) from below on $(\hat{\mu}, \hat{\chi})$. Therefore, $(\hat{\mu}, \hat{\chi})$ is a clockwise rotating *spiral* and the eigenvalues of the Jacobian of (56) evaluated at $(\hat{\mu}, \hat{\chi})$ are complex. To establish instability we need to show that their real part is positive, or, equivalently, that $\mathrm{tr}[J(\hat{\mu}, \hat{\chi})] > 0$. The Jacobian is

$$J\left(\hat{\mu}, \hat{\chi}\right) = \frac{1}{Q\left(\hat{\mu}\right)} \begin{bmatrix} \left[Q\left(\hat{\mu}\right) - 2\beta\left(1-\alpha\right)\hat{\mu}\right]\rho & 2\beta\left(1-\alpha\right)\hat{\mu} - Q\left(\hat{\mu}\right) \\ -\frac{(1-\alpha)\left[\rho Q(\hat{\mu})+2\sigma\alpha^2\beta\theta(\hat{\mu})^\alpha+\rho^2\right]\theta(\hat{\mu})^{\alpha-1}}{\sigma} & 2\alpha\beta\left(1-\alpha\right)\theta\left(\hat{\mu}\right)^\alpha \end{bmatrix},$$

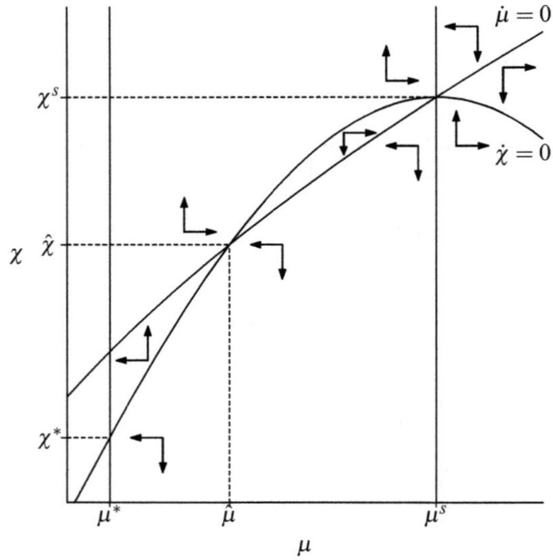

Fig. 1 Phase diagram of system (56) when $\theta\alpha(\mu^s)^{\alpha-1} < \rho < \theta\alpha(\mu^*)^{\alpha-1}$

with $Q(\hat{\mu}) > 0$, as $\mu^* < \hat{\mu} < \mu^s$. Since $[\alpha\theta(\hat{\mu})^{\alpha-1} - \rho] = 0$ on $(\hat{\mu}, \hat{\chi})$, it is immediately seen that $\mathrm{tr}[J(\hat{\mu}, \hat{\chi})] = \rho Q(\hat{\mu}) > 0$, and the proof is complete. □

Remark 1. The critical point (μ^s, χ^s), with coordinates defined in (62), cannot be classified analytically, as the Jacobian matrix of (56) evaluated at (μ^s, χ^s) has some elements diverging either to $-\infty$ or to $+\infty$ as (μ, χ) approaches (μ^s, χ^s), the sign of infinity depending on the direction along which $(\mu, \chi) \to (\mu^s, \chi^s)$.

We have seen in Sect. 2.1 that $\tilde{k}(A) > \tilde{k}_\infty(A)$ for all A (and thus for all t); this is consistent with $\mu(t) > \mu^*$ for all t. Hence, the stable trajectory must approach (μ^*, χ^*) from the right. We denote by $\chi(\mu)$ such trajectory, which is the *optimal policy expressed in terms of state-like and control-like variables*. Its slope on (μ^*, χ^*) is the slope of the eigenvector associated to the negative eigenvalue of (66) (see Barro and Sala-i-Martin, 2004, p. 596), that is,

$$\chi'(\mu^*) = \frac{\beta\theta\alpha\sigma(1-\alpha)(\sigma-1)(\mu^*)^{2\alpha} + \rho(\alpha+\sigma-1)(\mu^*)^{\alpha} + [\rho^2/(\beta\theta)]}{\alpha\sigma^2(\mu^*)^{\alpha}}, \tag{67}$$

which is clearly positive. Hence, $\chi(\mu)$ approaches (μ^*, χ^*) from north-east in a (right) neighborhood of μ^*; consequently, along the turnpike both ratios $\tilde{k}(A)/A$ and $\tilde{c}/A$ must decline in time when they are approaching the asymptotic turnpike.

Under the assumption that $\theta\alpha(\mu^s)^{\alpha-1} < \rho < \theta\alpha(\mu^*)^{\alpha-1}$, $\mu^* < \hat{\mu} < \mu^s$; by translating $\hat{\mu}$ into $\hat{A}$ through (65), it follows that the intersection point between $\tilde{k}(A)$ and $\hat{k}(A)$ lies on the right of the singular point A^s defined in (45). Therefore, by condition (44) of Proposition 3, $\tilde{c}(\hat{A}) < \tilde{y}(\hat{A})$, which is equivalent to $\chi(\hat{\mu}) < \theta(\hat{\mu})^{\alpha} = \hat{\chi}$. Hence, the optimal trajectory $\chi(\mu)$ keeps well below the (unstable) steady state

$(\hat{\mu}, \hat{\chi})$, which thus happens to be harmless for our analysis, at least for the case[7] $\theta\alpha(\mu^s)^{\alpha-1} < \rho < \theta\alpha(\mu^*)^{\alpha-1}$.

Conversely, the steady state (μ^s, χ^s) is the most problematic as on one hand its stability cannot be checked analytically, while on the other hand the optimal policy $\chi(\mu)$ must actually cross it.[8] However, since in our scenario $(A^s, \tilde{k}(A^s)) \neq (\hat{A}, \tilde{k}(\hat{A}))$, the system in the original model is not on the stagnation line when it hits $(A^s, \tilde{k}(A^s))$ and thus cannot stop over it; accordingly, the detrended system cannot stop over (μ^s, χ^s). All these "singularities" attached to (μ^s, χ^s) led us to opt for a qualitative approach based on information gathered on a neighborhood of (μ^s, χ^s). Condition (44) of Proposition 3 for $A \neq A^s$ translates into

$$\begin{cases} \chi(\mu) < \theta(\mu)^\alpha & \text{for } \mu^* < \mu < \mu^s \\ \chi(\mu) > \theta(\mu)^\alpha & \text{for } \mu > \mu^s, \end{cases} \tag{68}$$

which, in turn, means that the optimal policy must lie below the locus (59) when $\mu^* < \mu < \mu^s$ and above it when $\mu > \mu^s$. A closer inspection of a neighborhood of (μ^s, χ^s) in Fig. 1 shows that it is attractive on the area above the locus (59) (above $\chi = \theta\mu^\alpha$) and on the right of the vertical line $\mu \equiv \mu^s$, while it is repulsive below $\chi = \theta\mu^\alpha$ and on the left of $\mu \equiv \mu^s$. As $\theta\mu^\alpha$ is increasing in μ, this suggests that the optimal policy $\chi(\mu)$ must be increasing on (μ^s, χ^s) and the optimal trajectory $(\mu(t), \chi(t))$ must cross (μ^s, χ^s) from north-east to south-west as time elapses.

3.3 *Time Elimination, Policy Function and Initial Conditions*

In order to study the policy function $\chi(\mu)$ – which is the conjugate of $\tilde{c}(A)$ in the original model – we apply the technique developed by Mulligan and Sala-i-Martin (1991) and tackle the unique ODE given by the ratio between the equations in (56):

$$\chi'(\mu) = \frac{\left[\left(\alpha\theta\mu^{\alpha-1} - \rho\right)/\sigma\right] Q(\mu) - 2\alpha\beta(1-\alpha)\left[\theta\mu^\alpha - \chi(\mu)\right]}{\left[Q(\mu) - 2\beta(1-\alpha)\mu\right]\left[\theta\mu^\alpha - \chi(\mu)\right]}\chi(\mu), \tag{69}$$

where $Q(\mu)$ is defined in (57).

The natural choice for the initial condition of (69) is the saddle-path stable steady state (μ^*, χ^*), while the value of $\chi'(\mu^*)$ in (67) will be used to select the stable arm outside (μ^*, χ^*). The previous analysis, however, has endowed us with another

[7] A similar situation occurs when $\rho < \theta\alpha(\mu^s)^{\alpha-1}$, in which case $\tilde{c}(\hat{A}) > \tilde{y}(\hat{A})$, and thus $\chi(\hat{\mu}) > \theta(\hat{\mu})^\alpha = \hat{\chi}$. Only when $\rho = \theta\alpha(\mu^s)^{\alpha-1}$, and the two points $\hat{A}$ and A^s collapse, the optimal trajectory necessarily must cross the (unstable) steady state $(\hat{\mu}, \hat{\chi})$; in this case, however, the point $(\hat{\mu}, \hat{\chi}) = (\mu^s, \chi^s)$ inherits the peculiar singularity properties of (μ^s, χ^s), thus becoming a "supersingular" point to be handled with circumspection.

[8] Condition (44) of Proposition 3 states that $\tilde{c}(A^s) = \tilde{y}(A^s)$, which implies $\chi(\mu^s) = \theta(\mu^s)^\alpha = \chi^s$.

reference point, the singular point (μ^s, χ^s), which may be exploited as initial condition as well. Although the Jacobian of (56) evaluated on (μ^s, χ^s) is intractable, we are able to compute the slope of the policy at $\mu = \mu^s$ by applying l'Hôpital rule to the RHS of (69) evaluated at $\mu = \mu^s$. Since $Q(\mu^s) = 0$ and $[\theta(\mu^s)^\alpha - \chi(\mu^s)] = 0$, we obtain the following quadratic equation in $\chi'(\mu^s)$:

$$\begin{aligned} &2\beta\sigma(1-\alpha)\mu^s[\chi'(\mu^s)]^2 - 4\alpha\beta\sigma(1-\alpha)\chi^s\chi'(\mu^s) \\ &- \left\{\left[\alpha\theta(\mu^s)^{\alpha-1} - \rho\right]Q'(\mu^s) - 2\alpha^2\beta\sigma\theta(1-\alpha)(\mu^s)^{\alpha-1}\right\}\chi^s = 0. \end{aligned} \tag{70}$$

Substituting μ^s and χ^s as in (62) and $Q'(\mu^s) = -2\beta(1-\alpha)\sqrt{1+4\alpha+\alpha^2}$ into (70) two positive real solutions appear, the largest being larger than the slope of the locus (59) at $\mu = \mu^s$. However, this happens only when $\theta\alpha(\mu^s)^{\alpha-1} < \rho < \theta\alpha(\mu^*)^{\alpha-1}$; this is why we chose to confine our numerical approach to such scenario.

4 Numeric Simulation of the Optimal Policy

By applying the *Fehlberg fourth-fifth order Runge–Kutta method with degree four interpolant* method (see, e.g., Shampine and Corless, 2000) implemented through Maple 12.02 to ODE (69), we were able to find satisfactory result only for single sets of parameters values. We chose values for parameters α, ρ, σ and θ which are often assumed in the macroeconomic literature (see, e.g., Mulligan and Sala-i-Martin, 1993): $\alpha = 0.5$, $\rho = 0.04$ and $\theta = \sigma = 1$. Note that $\sigma = 1$ implies logarithmic instantaneous utility. For such parameters' values, β must satisfy the necessary growth condition (43), which turns out to be $\beta > 0.0064$.

We plan to exploit the steady state (μ^*, χ^*) and the singular point (μ^s, χ^s) [see (58), (63) and (62)] as initial conditions in order to trace out two different curves as numeric solutions of (69) through Maple 12.02. Both curves provide an approximation for the same (unique) trajectory representing the optimal policy[9] $\chi(\mu)$ for $\mu \geq \mu^*$. For the chosen parameters' values, such two curves happen to be sufficiently close to each other for a reasonably large range of μ values only for a unique admissible value of the technological parameter: $\beta = 0.0124$. Since, for $\alpha = 0.5$, $\rho = 0.04$, $\theta = \sigma = 1$ and $\beta = 0.0124$, each curve provides a reliable approximation of $\chi(\mu)$ around its own initial condition and both match on most of the open interval (μ^*, μ^s), our idea is to approximate the whole $\chi(\mu)$ by using the first curve

[9] Such trajectory is the unique true solution of (69) corresponding to the stable arm of the saddle point (μ^*, χ^*) and, at the same time, crossing the singular point (μ^s, χ^s). Other solutions of (69) may cross at most one of the two points, like, for example, the trajectory corresponding to the unstable arm of (μ^*, χ^*), or other unknown trajectories possibly crossing the singular point (μ^s, χ^s). We owe such clarification to an anonymous referee.

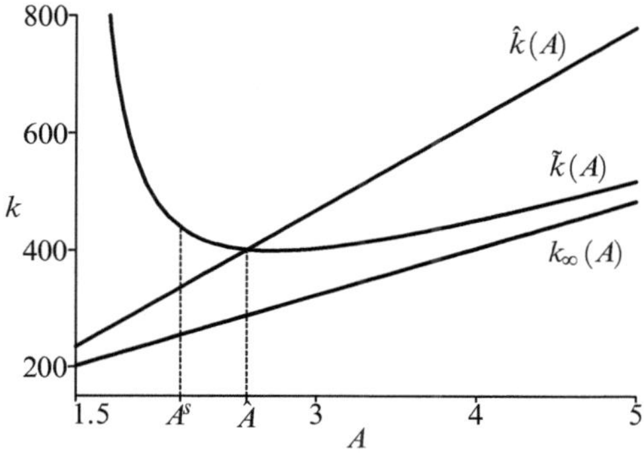

Fig. 2 The turnpike $\tilde{k}(A)$, the asymptotic turnpike $k_\infty(A)$ and the stagnation line $\hat{k}(A)$ for $\alpha = 0.5$, $\rho = 0.04$, $\theta = \sigma = 1$ and $\beta = 0.0124$

for μ close to μ^* and the second one for μ close to (and larger than) μ^s, while "joining"them together on some "intermediate" value on which they almost match.

For our parameters' values, (62) yields $\mu^s = 204.4503$, which implies $\rho = 0.04 > 0.035 = \theta\alpha(\mu^s)^{\alpha-1}$, corresponding to the third scenario of Sect. 3.2, in which $A^s < \hat{A}$. Figure 2 portraits the turnpike $\tilde{k}(A)$, the asymptotic turnpike $\tilde{k}_\infty(A)$ and the stagnation line $\hat{k}(A)$ as in (40), (41) and (42); as expected, $A^s = 2.1514 < 2.567 = \hat{A}$.

In view of Proposition 2, the long-run capital rental rate is $r_\infty = f'(\tilde{\eta}) = 0.0557$, the long-run common constant growth rate is $\gamma = 0.0157$, while the long-run income shares invested in knowledge and capital are the same: $s_\infty = s_\infty^k = 0.1408$.

The steady states are $(\mu^*, \chi^*) = (80.6452, 6.4516)$, $(\hat{\mu}, \hat{\chi}) = (156.25, 12.5)$ and $(\mu^s, \chi^s) = (204.4503, 14.2986)$. Figure 3 shows the loci $\dot{\mu} = 0$ and $\dot{\chi} = 0$ in slim black, while the thick curves are the result of the numeric solution of (69) representing the policy $\chi(\mu)$: the black one uses (μ^*, χ^*) as initial condition and (67), $\chi'(\mu^*) = 0.0687$, for the selection of the stable arm; the grey one has (μ^s, χ^s) as initial condition and slope given by the largest solution of (70) on $\mu = \mu^s$, $\chi'(\mu^s) = 0.0602$. The two approximate trajectories will be labeled $\chi^*(\mu)$ and $\chi^s(\mu)$ respectively.

Even for our choice of parameters' values the Maple 12.02 algorithm is capable of computing the trajectory $\chi^*(\mu)$ only up to a point: it actually stops at $\bar{\mu} \simeq 197 < 204.4503 = \mu^s$, falling short of the singular point, (μ^s, χ^s). On the other hand, as it is clear from Fig. 3, trajectory $\chi^s(\mu)$ heavily underestimates the policy for values of μ approaching μ^* (i.e., far away from μ^s). The two curves, however, seem sufficiently close to each other on most of the interval (μ^*, μ^s), thus suggesting that the numeric approach actually works satisfactorily for these values of parameters.

In order to estimate the whole policy $\chi(\mu)$, for all $\mu \geq \mu^*$, we shall use $\chi^*(\mu)$ for μ values close to μ^*, and $\chi^s(\mu)$ for μ values closer to μ^s. Since from Fig. 3 it is clear that $\chi^*(\hat{\mu}) \approx \chi^s(\hat{\mu})$, we shall define the approximated policy as a piecewise function by joining the two trajectories at the point $\hat{\mu} = 156.25 \in (\mu^*, \mu^s)$:

$$\chi(\mu) = \begin{cases} \chi^*(\mu) & \text{for } \mu^* \leq \mu \leq \hat{\mu} \\ \chi^s(\mu) & \text{for } \mu \geq \hat{\mu}. \end{cases} \tag{71}$$

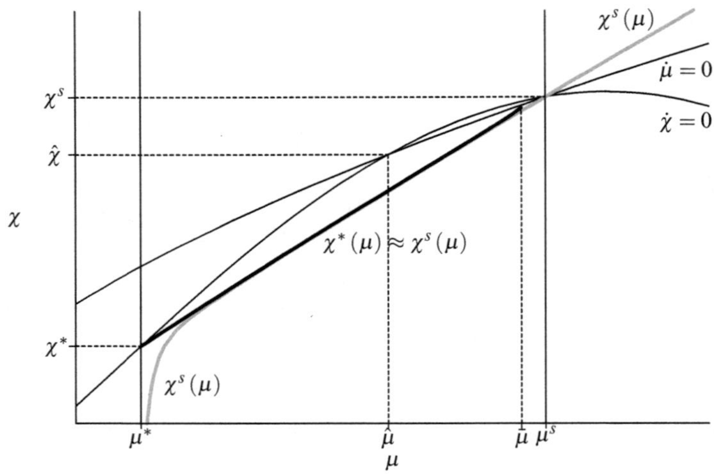

Fig. 3 Loci $\dot{\mu}=0$ and $\dot{\chi}=0$ (*slim black curves*) and approximate trajectories $\chi^*(\mu)$ and $\chi^s(\mu)$ (*black and grey thick curves* respectively) for $\alpha=0.5$, $\rho=0.04$, $\theta=\sigma=1$ and $\beta=0.0124$

Surprisingly, already for $\beta=0.0123$, or $\beta=0.0125$, while keeping fixed all other parameters, the curves $\chi^*(\mu)$ and $\chi^s(\mu)$ in Fig. 3 split apart, while the range of μ for which the numeric algorithm is able to perform starts to shrink dramatically; this is why we take as reliable only the solution obtained for $\beta=0.0124$.

Remark 2. We tried different values for the parameters α, ρ, σ and θ; for all feasible set of values for such parameters we found a scenario similar to that described above, at least under condition $\theta\alpha(\mu^s)^{\alpha-1}<\rho<\theta\alpha(\mu^*)^{\alpha-1}$: only for one specific value of parameter β, related to the choice of α, ρ, σ and θ, the two numerical solutions – $\chi^*(\mu)$ with initial condition (μ^*,χ^*) and $\chi^s(\mu)$ with initial condition (μ^s,χ^s) – turned out to be sufficiently close to each other on a large part of the interval (μ^*,μ^s). We conclude, thus, that the numeric approach works satisfactory only on a manifold of dimension one in the parameters' space.

5 Discussion

To get the approximated time-path trajectory of μ we substitute the optimal policy $\chi(\mu)$ as in (71) into the first equation of (56), yielding the following ODE in t,

$$\dot{\mu}(t)=\{1-2\beta(1-\alpha)\mu(t)/Q[\mu(t)]\}\{\theta[\mu(t)]^{\alpha}-\chi[\mu(t)]\}, \quad (72)$$

with $Q(\cdot)$ defined in (57), which can be numerically solved. Since $\chi(\mu)$ in (71) is defined piecewise, we need to choose an instant $\hat{t}>0$ on which the trajectory has the (common) value $\hat{\mu}=156.25$; then, the initial value $\mu_0=\mu(0)$ will be given

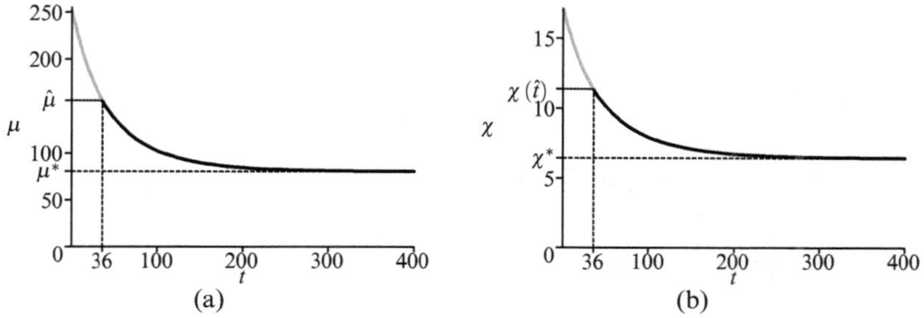

Fig. 4 (**a**) $\mu(t)$ and (**b**) $\chi(t)$ for $\alpha = 0.5$, $\rho = 0.04$, $\theta = \sigma = 1$ and $\beta = 0.0124$

by evaluating in $t = 0$ the solution of (72) with $\chi(\cdot) = \chi^s(\cdot)$ and $\mu(\hat{t}) = \hat{\mu}$ as initial condition. For different $\hat{t}$ we can consider any initial value $\mu_0 = \mu(0) > \hat{\mu}$.

In our example we assume $\hat{t} = 36$, corresponding to $\mu_0 = 251.977$ in $t = 0$. According to (71), we define $\mu(t)$ as the solution of (72) with $\chi(\cdot) = \chi^s(\cdot)$ for $0 \leq t \leq \hat{t}$ [corresponding to $\hat{\mu} \leq \mu(t) \leq \mu_0$], and as the solution of (72) with $\chi(\cdot) = \chi^*(\cdot)$ for $t \geq \hat{t}$ [corresponding to $\mu^* \leq \mu(t) \leq \hat{\mu}$]. Figure 4a plots $\mu(t)$ for $0 \leq t \leq 400$ by distinguishing the part (in grey) obtained through $\chi^s(\cdot)$ for $0 \leq t \leq \hat{t} = 36$ from the part eventually converging to μ^* (in black) obtained by means of $\chi^*(\cdot)$ for $t \geq 36$.

The time-path trajectory $\chi(t)$ is then computed by letting $\chi(t) = \chi[\mu(t)]$ in (71), with $\mu(t)$ just obtained, for all $0 \leq t \leq 400$. Figure 4b reports the result, again by emphasizing in grey the part for $0 \leq t \leq \hat{t} = 36$. In $t = 0$, $\chi(0) = \chi_0 = 17.1194$, corresponding to $\mu_0 = 251.977$, while in $t = \hat{t} = 36$, $\chi(36) = 11.3688$; clearly, $\chi(\hat{t}) = 11.3688 < 12.5 = \hat{\chi}$, as expected.

With $\mu(t)$ and $\chi(t)$ at hand, we can compute the optimal consumption $\tilde{c}(A)$ and output $\tilde{y}(A)$ along the turnpike $\tilde{k}(A)$ in the original model as functions of A. By (48) we find the initial stock of knowledge $A_0 = 1.9707$ in $t = 0$, corresponding to μ_0. To A_0 corresponds an initial capital $k_0 = \tilde{k}(A_0) = 496.57$ in $t = 0$. $\tilde{c}(A)$ is then obtained through (49), with $\chi(\cdot)$ defined in (71): $\chi^s(\cdot)$ for $A_0 \leq A \leq \hat{A}$ (corresponding to $\hat{\mu} \leq \mu \leq \mu_0$), and $\chi^*(\cdot)$ for $A \geq \hat{A}$ (corresponding to $\mu^* \leq \mu \leq \hat{\mu}$). Figure 5a reports $\tilde{k}(A)$, $\tilde{y}(A)$ and $\tilde{c}(A)$ just evaluated on a scale larger than in Fig. 2. Figure 5b magnifies the intersection point between $\tilde{y}(A)$ and the $\tilde{c}(A)$ occurring on A^s, close to A_0 and to the left of $\hat{A}$. Since on $[A_0, \hat{A}]$ $\tilde{c}(A)$ is being built through $\chi^s(\cdot)$ in (72), this portion of its graph is emphasized in grey, as we did in previous figures.

The time-path trajectory of the stock of knowledge $A(t)$ is obtained by evaluating (48) at $\mu(t)$ for all t, while time-path trajectories $\tilde{k}(t)$ and $\tilde{y}(t)$ follow by construction. The consumption time-path trajectory $\tilde{c}(t)$ is computed by evaluating (49) at $A(t)$ for all t. These trajectories are drawn in Fig. 6a, while Fig. 6b reports the time path-trajectory of the capital rental rate r; once again, their dependence on the $\chi^s(\cdot)$ arm of the policy in (71) for $0 \leq t \leq \hat{t} = 36$ is emphasized in grey.

From Figs. 2, 5a and 6a, emerges that the dynamics along the turnpike are characterized by a much larger amount of physical capital than any other variable. A large

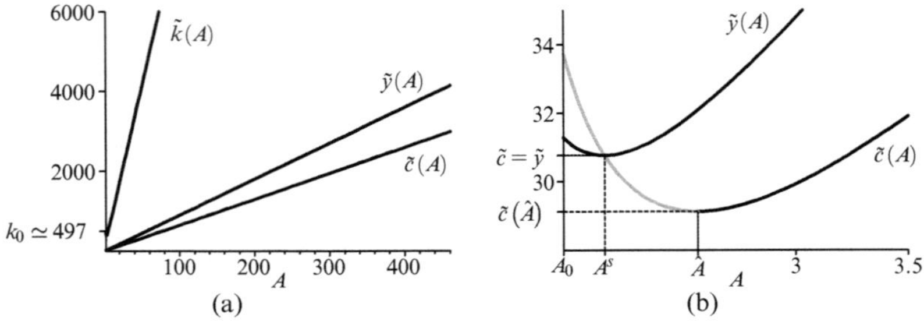

Fig. 5 (**a**) $\tilde{c}$, $\tilde{y}$ and $\tilde{k}$ as functions of A along the turnpike; (**b**) $\tilde{c}$ and $\tilde{y}$ close to $A_0 = 1.9707$

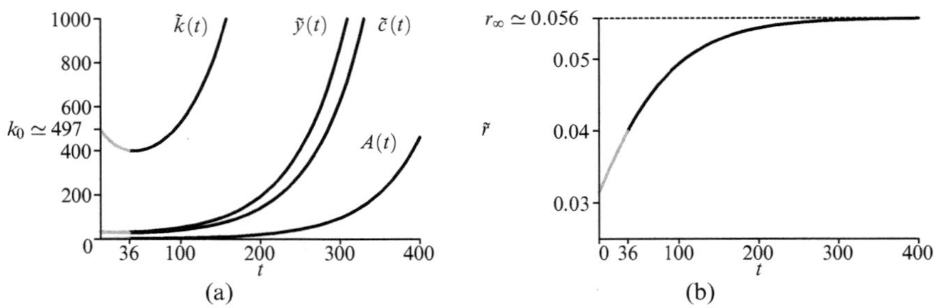

Fig. 6 (**a**) Time-path trajectories of A, $\tilde{k}$, $\tilde{y}$ and $\tilde{c}$; (**b**) time-path trajectory for $\tilde{r}$

initial capital, $k_0 = 496.57$, compared to very few initial ideas, $A_0 = 1.9707$, is required to let the recombinant process to take-off. Such amount, even if for a short time, is partially being "eaten up" by both consumption [$\tilde{c}(A) > \tilde{y}(A)$ for $A_0 \leq A \leq A^s$] and investment in R&D, thus envisaging an initial period of decline for capital $\tilde{k}$. Figure 5b shows that output and consumption decrease for a short time as well; specifically, output declines until $\tilde{c}(A)$ hits $\tilde{y}(A)$ at $A = A^s$, and consumption decreases until the turnpike crosses the stagnation line on $A = \hat{A}$ (see Fig. 2) at $\hat{t} = 36$. For larger t all variables start to increase, with a much higher $\tilde{k}$ with respect to all others, especially to A. For example, when $A \simeq 73$, $\tilde{k} \simeq 6000$ in Fig. 5a.

In our example, thus, sustained growth requires a large exploitation of physical resources, at least relatively to knowledge, even under a "balanced" ($\alpha = 0.5$) Cobb–Douglas technology. Such "asymmetry" is explained by the ratio between the (low) price of capital – numéraire – and the relatively high unit cost of knowledge production: for $\beta = 0.0124$ $\varphi(A)$ turns out to be significantly larger than 1, as $\varphi(A) > \lim_{A \to \infty} \varphi(A) = 1/\pi'(0) = 1/\beta = 80.6452$ (see also, Figs. 8a and 8b).

Figure 6a exhibits a system which actually takes some time to take-off. Provided that our economy starts with very few ideas ($A_0 = 1.9707$) and sufficiently large capital ($k_0 = 496.57$), the initial transient dynamics happen to last quite long; especially $A(t)$ takes no less than 200 periods before becoming significant [note that in the meantime $\tilde{k}(t)$ already started to "blow up"]. For example, it takes around 282 periods to reach the stock $A \simeq 73$, corresponding to $\tilde{k} \simeq 6,000$. Similarly, the

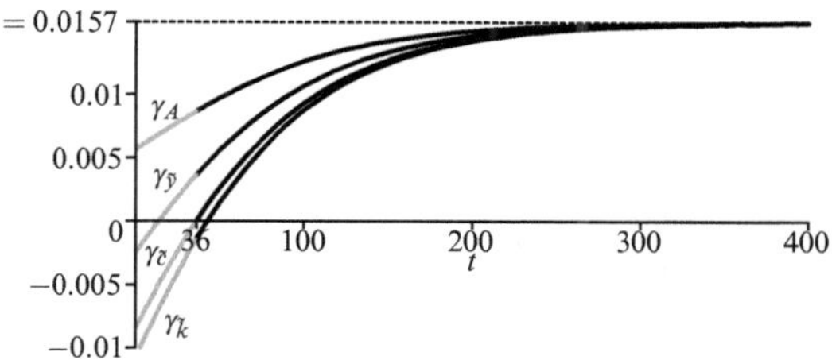

Fig. 7 Growth rates γ_A, $\gamma_{\tilde{k}}$, $\gamma_{\tilde{y}}$ and $\gamma_{\tilde{c}}$, of A, $\tilde{k}$, $\tilde{y}$ and $\tilde{c}$ as functions of time

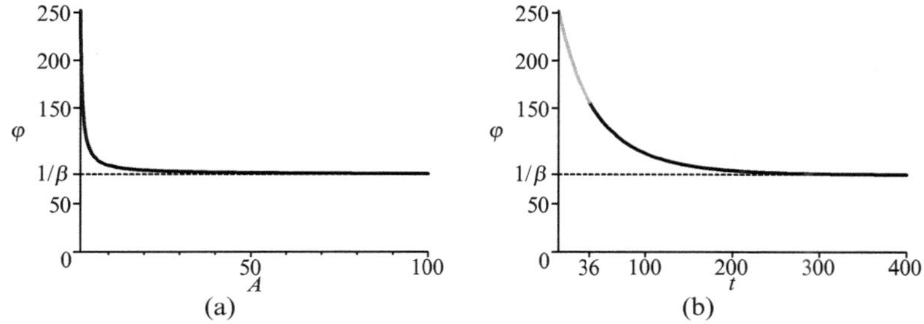

Fig. 8 (**a**) Unit cost of knowledge production, φ, as a function of A; (**b**) its time-path trajectory

constant ratio $\tilde{c}(A)/\tilde{y}(A)$ visible in Fig. 5a – due to almost linearity of $\tilde{c}(A)$ and $\tilde{y}(A)$ and which can be checked to be close to the asymptotic ratio 0.07184, corresponding to the saving rate $s_\infty + s^k_\infty = 0.2816$ – is actually not reached before at least 300 periods. To conclude, Figs. 2 and 5a should be read carefully when one introduces time: of course the economy grows along the turnpike $\tilde{k}(A)$, but at a very slow pace in early times, while keeps accelerating until it "explodes" along $\tilde{k}_\infty(A)$.

Figure 6b adds more information to the analysis: even if $\tilde{k}$ is always (much) larger than A, its productivity keeps rising in time, as confirmed by its increasing rental rate, $\tilde{r}$, until it reaches its asymptotic value, $r_\infty = 0.0557$.

Figure 7 confirms everything in terms of rates of growth. By construction, $A(t)$ is the only variable with rate of growth $\gamma_A = \dot{A}/A$ always positive, while $\tilde{k}(t)$, $\tilde{y}(t)$ and $\tilde{c}(t)$, all experience negative growth at early times, where $\gamma_{\tilde{k}} = \dot{\tilde{k}}/\tilde{k}$, $\gamma_{\tilde{y}} = \dot{\tilde{y}}/\tilde{y}$ and $\gamma_{\tilde{c}} = \dot{\tilde{c}}/\tilde{c}$ are negative. Interestingly, it can be observed that $\tilde{c}(t)$ reaches its absolute minimum in $\hat{t} = 36$ [corresponding to $\tilde{c}(\hat{A})$, as confirmed by Fig. 5b].

The striking feature of recombinant growth is evident in Fig. 7: all growth rates are increasing in time while approaching their asymptotic common value $\gamma = 0.0157$. This reflects the original Weitzman's (1998) hypothesis: in early times ideas are scarce and thus have the potential of growing at increasing rates, in the long-run limited physical resources to be invested in R&D – with respect to the exploding number of ideas – cools down growth to the more realistic case of constant rates.

Figure 8a shows the graph of the unit cost of knowledge production $\varphi(A)$ as in (39), which is sharply decreasing in A for A close to A_0. Such jump, however, is to be diluted when time is considered, as shown in Fig. 8b where φ is plotted as a function of t, since A starts to grow significantly only after some time (see Fig. 6a).

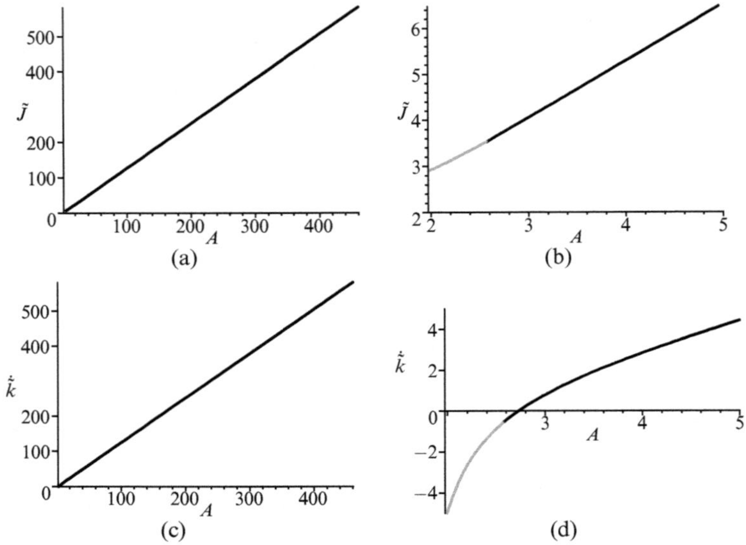

Fig. 9 (**a**) $\tilde{J}(A)$, (**b**) its detail for A close to A_0; (c) $\dot{\tilde{k}}(A)$, (d) its detail for A close to A_0

Investment in R&D $\tilde{J}$ and investment in capital $\dot{\tilde{k}}$ as functions of A are plotted in Fig. 9; $\tilde{J}$ is computed by using $\tilde{c}(A)$ and $\tilde{y}(A)$, $\varphi(A)$ and $\tilde{k}'(A)$ – obtained by differentiating (40) with respect to A – in (22). From Figs. 9a and 9c, where a large range of A values is considered, we learn that both look linear in A and have the same magnitude, implying that they become the same well before reaching their asymptotic (common) constant share $s_\infty = J_\infty / y_\infty = s^k_\infty = \dot{k}_\infty / y_\infty = 0.1408$. Only for A close to A_0 their behavior differ, as magnified by Figs. 9a and 9d.

It is interesting to compare the magnitude of $\tilde{J}(A)$ and $\dot{\tilde{k}}(A)$ in Figs. 9a and 9d with that of $\tilde{c}(A)$ and $\tilde{y}(A)$ in Figs. 5a and 5b: for all A – also close to A_0 – the optimal dynamics postulate relatively small investment in both factors with respect to consumption and output. Figures 10a and 10b confirm this in terms of investment shares, $\tilde{s} = \tilde{J}/\tilde{y}$ and $\tilde{s}^k = \dot{\tilde{k}}/\tilde{y}$. Both are increasing in A and reach their asymptotic value $s_\infty = s^k_\infty = 0.1408$ quite rapidly, although $\tilde{s}^k < 0$ for small A. Such quick jumps to their asymptotic value is consistent with the linearity exhibited by $\tilde{J}(A)$ and $\dot{\tilde{k}}(A)$ in Figs. 9a and 9c.

Also the dynamics of $\tilde{J}$ (or $\tilde{s}$) confirm Weitzman's (1998) evolution of knowledge: when A – and thus seed ideas H – is scarce function (5) exhibits low productivity; accordingly, only few resources are employed in R&D, while they increase as A – and H – become more abundant. In the long-run are the physical resources that become scarce with respect to knowledge – they grow slower than what (potentially) could do knowledge – and bound the rate of investment $\tilde{s}$ to its asymptotic value s_∞.

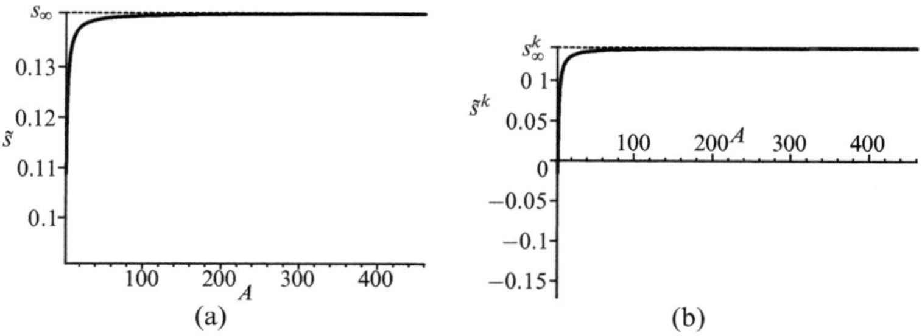

Fig. 10 (**a**) $\tilde{s} = \tilde{J}/\tilde{y}$ as a function of A; (**b**) $\tilde{s}^k = \dot{\tilde{k}}/\tilde{y}$ as a function of A

The graphs of new (successful) knowledge production, $\dot{A}$, and seed ideas, $\tilde{H}$, as functions of A are reported in Fig. 11; the former is given by (29), while the latter is computed from (4) using $\dot{A}$ and $C_2'(A) = A - 1/2$. Strict convexity of $\tilde{H}$ in Figs. 11c and 11d, associated to linearity of $\dot{A}$ (also for A close to A_0) in Figs. 11a and 11b, is consistent to formula (4), which implies quadratic growth for $\tilde{H}$ when $\dot{A}$ grows linearly. It is worth noting the difference in magnitudes between seed ideas $\tilde{H}$ and the actual successful ideas $\dot{A}$ produced out of $\tilde{H}$: such low returns are justified by the choice of a very small value for the efficiency parameter, $\beta = 0.0124$, in (38), requiring abundant seed ideas to guarantee sustained growth of knowledge.

To conclude, Fig. 12 shows time-path trajectories of $\tilde{J}$, $\dot{\tilde{k}}$, $\tilde{s}$, $\tilde{s}^k$, $\dot{A}$ and $\tilde{H}$. Due to slow growth of $A(t)$ in early times, linearity of investments $\tilde{J}$ and $\dot{\tilde{k}}$, and of new knowledge $\dot{A}$, evident in Figs. 9a, 9c and 9a, correspond to convex time-path trajectories, as shown in Figs. 12a, 12b and 12e. For the same reason, convexity of $\tilde{H}$ in Fig. 11c becomes more accentuated in Fig. 12f; similarly, the sudden jumps to their asymptotic value of $\tilde{s}$ and $\tilde{s}^k$ in Figs. 10a and 10b is being smoothed in Figs. 12c and 12d. Specifically, both need at least 200 periods before approaching their long-run (common) constant value s_∞.

6 Conclusions

The exercise performed in this paper is a very preliminary attempt to tackle the transition dynamics in the recombinant growth model introduced by Tsur and Zemel (2007). For CIES instantaneous utility and Cobb–Douglas production in the output sector, we chose a suitable function for the Weitzman's (1998) probability of obtaining a successful idea from pairwise matchings of seed ideas, so that the original optimal dynamics along the turnpike, which is diverging in the long-run, can be "detrended" to an equivalent system converging to a steady state. In the space of the detrended variables we exploit the asymptotic steady state plus a singular point, across which the optimal policy must get through at some early instant, in order to numerically compute two trajectories which, for a specific choice for the

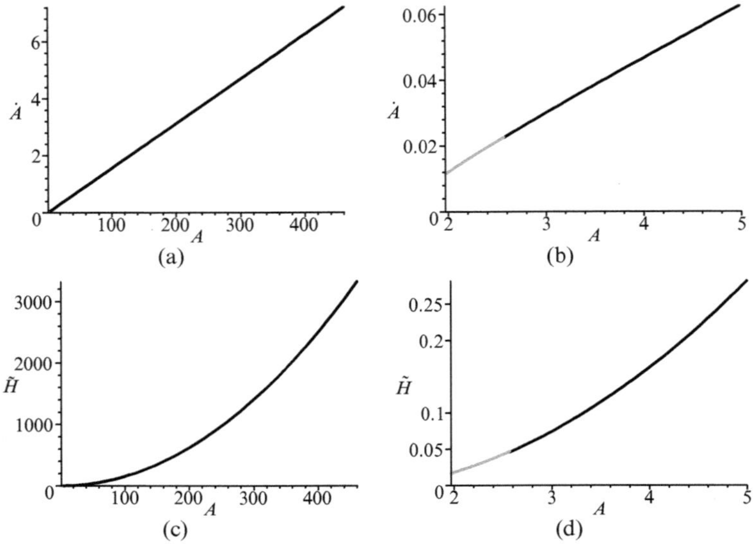

Fig. 11 (**a**) and (**b**) $\dot{A}$ as a function of A; (**c**) and (**d**) $\widetilde{H}$ as a function of A

parameters' values, happen to be sufficiently close to each other on a large range between such two points. By joining together these trajectories at an intermediate point, we build an approximation of the optimal policy which must be reasonably close to the true policy on all variables' domain. By converting such trajectory into the original state variable (stock of knowledge) and control variable (consumption) trajectories, we obtain a good approximation of the optimal consumption, which in turn, again by solving numerically an ODE, yields the transition optimal time-path trajectories of the stock of knowledge, physical capital, output and consumption – as well as their transition growth rates – along the turnpike.

We believe that our main technical contribution is the appropriate form chosen for the Weitzman's probability function defined in Assumption 4(ii), which allows for "detrending" the original system (37) into the equivalent system (56).

If, on one hand the optimal policy obtained in Sect. 4, and used to build time-path trajectories in Sect. 5, may clearly be of interest per se, on the other hand it is insufficient for studying how the system's transitional behavior is being affected by changes in the technological parameter β of the probability function π of Assumption 4(ii), while keeping fixed all other parameters' values. In order to further investigate this topic one needs either to improve the numerical computation of system (56) so that the matching of the two aforementioned trajectories in the detrended space is maintained at least on a nontrivial interval of values for parameter β, or trying a completely different approach on either system (37) or system (56) by means of analytical tools in order to explicitly find the true form of the optimal trajectories. One may tackle the latter by looking for some special function that may prove useful in solving one of (37) or (56); see, e.g., Boucekkine and Ruiz-Tamarit (2008) for

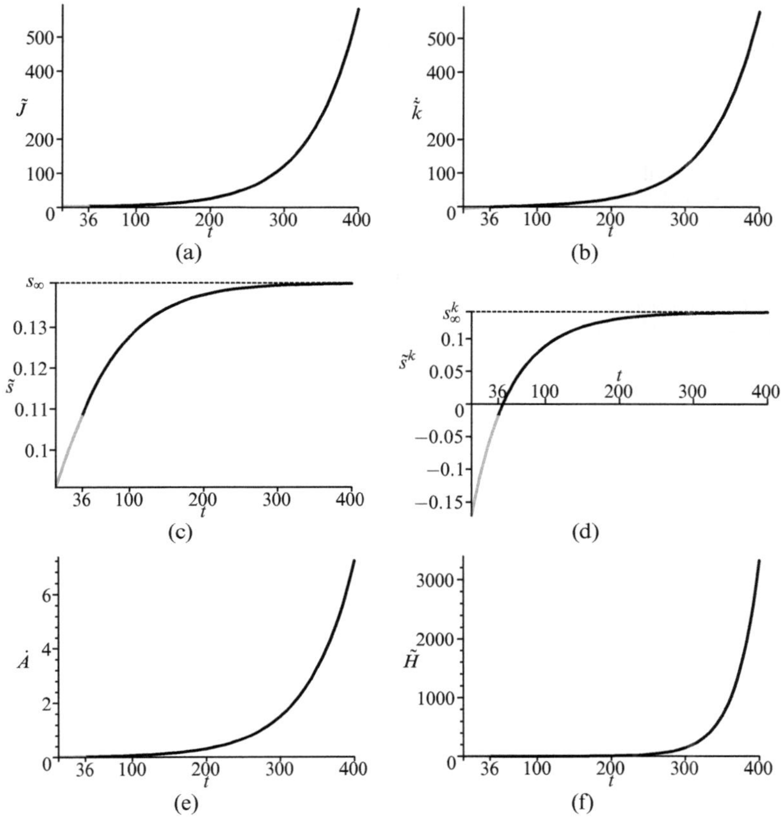

Fig. 12 Time-path trajectories of (**a**) $\tilde{J}$, (**b**) $\dot{\tilde{k}}$, (**c**) $\tilde{s} = \tilde{J}/\tilde{y}$, (**d**) $\tilde{s}^k = \dot{\tilde{k}}/\tilde{y}$, (**e**) $\dot{A}$, (**f**) $\tilde{H}$

a recent application of the Gaussian hypergeometric functions to the Lucas–Uzawa model. Both approaches will be investigated in future research projects.

Acknowledgements We thank Giovanni Ramello for bringing our attention to recombinant growth models, Raouf Boucekkine for precious (and critical) technical suggestions and Mauro Sodini, met at the MDEF 2008, for encouragement at a time when we were nearly giving up looking for a suitable "detrendization" of the model. We are also grateful to Carla Marchese, for her help in the economic interpretation of the results. All remaining errors are, of course, ours.

References

Aghion, P., & Howitt, P. (1999). *Endogenous growth theory*. Cambridge, MA: The MIT Press.

Barro, R. J., & Sala-i-Martin, X. (2004). *Economic growth* (2nd edn.). Cambridge, MA: The MIT Press.

Boucekkine, R., & Ruiz-Tamarit, J. R. (2008). Special functions for the study of economic dynamics: The case of the Lucas–Uzawa model. *Journal of Mathematical Economics, 44*, 33–54.
Mulligan, C. B., & Sala-i-Martin, X. (1991). *A note on the time-elimination method for solving recursive dynamic economic models* (NBER Working Paper Series 116).
Mulligan, C. B., & Sala-i-Martin, X. (1993). Transitional dynamics in two-sector models of endogenous growth. *The Quarterly Journal of Economics, 108*, 737–773.
Romer, D. (1996). *Advanced macroeconomics*. New York: McGraw-Hill.
Shampine, L. F., & Corless, R. M. (2000). Initial value problems for ODEs in problem solving environments. *Journal of Computational and Applied Mathematics, 125*, 31–40.
Tsur, Y., & Zemel, A. (2007). Towards endogenous recombinant growth. *Journal of Economic Dynamics and Control, 31*, 3459–3477.
Weitzman, M. L. (1998). Recombinant growth. *The Quarterly Journal of Economics, 113*, 331–360.

Political Accountability: A Stochastic Control Approach

Michele Longo and Alessandra Mainini

1 Introduction

In a democracy elections are the primary mechanism to discipline politicians. Indeed, policy-makers care for being in office and this affects their policy choices; for instance, they refrain from rent-extraction in order to be re-elected and benefit from future rents. Hence, elections provide implicit incentives that allow voters to align politicians' preferences with their own ones. This role is crucial because constitutions do not offer explicit incentive schemes (cf. Persson et al., 1997), that is, forms of compensation based on some performance measure as may happen in a relationship between employers and employees.

Early political agency models, such as Barro (1973) and Ferejohn (1986), describe the disciplining effect of elections assuming that voters are *backward-looking* (i.e., re-election is a reward for incumbent's past performance) and that the incumbent and the challenger are identical. This implies that even a small change in voters' preferences makes them leave their announced voting rule.

When heterogeneity is introduced in features that are beyond the control of the politician like in the case of competence, elections remain a device for keeping politicians accountable but in a different way. For example, in the so-called *career concerns* models, based on Holmström (1982) seminal work about manager's career, voters are *forward-looking*, that is, they are interested in politician's future performance instead of her past achievements. Citizens have beliefs about politician's competence and vote for the candidate that makes them better off in the post-election period. In these models, the incumbent's current performance provides a signal of her future competence: hence, politicians in office might opportunistically perform today by refraining from rent-extraction in order to improve their re-election probability.

M. Longo (✉)
Università Cattolica di Milano, Largo Gemelli, 1, 20123 Milano, Italy,
e-mail: michele.longo@unicatt.it

G.I. Bischi et al. (eds.), *Nonlinear Dynamics in Economics, Finance and the Social Sciences*, DOI 10.1007/978-3-642-04023-8_15,

In this paper, we study a career concerns model where politicians differ in their competence, which is not observed by voters nor by politicians themselves.[1] All agents are rational, expected utility maximizers, and learn symmetrically[2] about politician's ability by observing the economy wealth which is a noisy signal of politician's competence. Time is continuous and divided into two periods: a pre-election period and a post-election period. Citizens are risk neutral, get utility from each end-of-period economy wealth and, at the end of the first period, vote for the politician in office or a challenger randomly chosen among the population: elections are the sole means in voters' hands to control their utility. On the other hand, in each term in office, the incumbent chooses a rent seeking behaviour, which is unobserved by voters and negatively affects economy wealth, and the size of public sector. Politicians are risk averse and derive utility only from the rent.

The model is set in a stochastic framework where the value of the public sector is defined by an Itô diffusion process whose drift depends on politician's ability, which is modelled as a random variable with known prior distribution. Moreover, we assume the presence of a private sector with constant value. Then the economy wealth depends on politician's ability and results from the extent of public intervention and the amount of resources diverted by the incumbent.[3] In this framework, the incumbent politician maximizes her expected utility by choosing a rent-seeking behaviour and a size of the public sector. The first period maximization takes into account the end-of-period vote, whereas in the second and last one the problem is not conditioned by elections. In any case, the only information available to the incumbent is the one generated by the economy wealth (competence is not observed) and this makes the problem of incomplete information type.

The continuous-time choice has a threefold motivation: first, it enables the politician to build a reputation (i.e., beliefs about her future competence) through time during the first period; second, in this setting it is possible to describe the politician's optimal policies within the single period; third, continuous-time is amenable to mathematical techniques such as stochastic control and filtering.

From a mathematical point of view, the analysis entails using filtering techniques to re-formulate the problem within a complete information setting. Then, by using a suitable change of measure and relying on the dynamic programming principle we

---

[1] "One key issue is whether the politicians know their own types. In the case of competence, it seems less plausible to suppose that politicians know their own capacities completely and may be learning about this along with the voters" (Besley, 2006). "The assumption of symmetric learning considerably simplifies the analysis, as there is no possibility of signaling. For many aspects of politician quality, this is a reasonable assumption. For example, a new legislator is unlikely to know how good she will be at negotiating with lobbyists and party leaders" (Ashworth, 2005).

[2] Symmetric learning prevents from equilibria multiplicity typical of signaling models (cf. Rogoff, 1990).

[3] Such a model specification has been widely used in mathematical finance to study consumption/investment decisions in financial markets with unobservable returns (cf., inter alia, Lakner, 1995; Karatzas and Zhao, 2001; Rieder and Bäuerle, 2005; Björk et al. unpublished).

write the Hamilton–Jacobi–Bellman equation for the value function and, at least for the second period, we provide an explicit solution in stochastic form together with the optimal controls. A feature of the first period problem is that part of the functional to be maximized depends on a *zero-one* random variable (the vote outcome) that makes the problem similar to the *digital option* framework.

Our analysis finds support for the traditional idea that elections lead a politician to be opportunistically more aligned with voters' preferences. The presence of a re-election constraint modifies politicians' optimal policies because they are interested in being re-elected and not only in getting the rent. This is due to *career concerns* motivations.

The structure of the paper is as follows. Section 2 describes the model. Section 3 deals with the analysis of both the first and second period. In Sect. 4 we study a particular case. Section 5 concludes.

2 The Model

Time is continuous and divided into two periods of equal length $T > 0$: a pre-election period $[0, T]$ and a post-election period $[T, 2T]$. At the end of the first period the incumbent politician runs for re-election against a challenger randomly chosen among the population. We assume that the second period is the last: hence, the politician in office in this period has no electoral concerns.

Uncertainty is described by a complete probability space $(\Omega, \mathcal{F}, \mathbb{P})$ equipped with a filtration $\mathbb{F} := (\mathcal{F}_t)$, $0 \leq t \leq 2T$, satisfying the usual conditions of $\mathbb{P}$-null sets augmentation and right-continuity. On $(\Omega, \mathcal{F}, \mathbb{P})$ a standard one-dimensional $\mathbb{F}$-Brownian motion (W_t) and a random variable ε independent of (W_t) are defined. In this setting, we have a private sector with value constant and equal to one and a public sector whose value evolves over time according to

$$dG_t = G_t(\varepsilon dt + \sigma dW_t), \tag{1}$$

where σ is a positive constant and $\varepsilon \in \{0, 1\}$ denotes competence in managing the public sector. ε is assumed *unknown* to all agents (politicians as well as citizens), constant within each period, and with the following Bernoulli prior distribution:

$$\mathbb{P}[\varepsilon = 1] = p, \quad \mathbb{P}[\varepsilon = 0] = 1 - p. \tag{2}$$

Even though agents do not observe incumbent's competence, they can continuously observe the realizations of G and thus infer the true value of it. Let

$$\mathbb{F}^G := \left(\mathcal{F}_t^G\right)_{0 \leq t \leq 2T} \tag{3}$$

be the $\mathbb{P}$-augmented filtration generated by G and,[4] for each $0 \le t \le 2T$,

$$\hat{\varepsilon}_t := \mathbb{E}\left[\varepsilon \mid \mathcal{F}_t^G\right] \tag{4}$$

be the conditional expectation of the incumbent's ability at time t relative to the observable information $\mathcal{F}_t^G$. Notice that $\hat{\varepsilon}_t$ is a $\mathcal{F}_t^G$-measurable random variable with distribution $(\pi_t, 1-\pi_t)$, where, for all $0 \le t \le 2T$,

$$\pi_t := \mathbb{P}\left[\varepsilon = 1 \mid \mathcal{F}_t^G\right]. \tag{5}$$

Within the previous framework, which is common knowledge, the state variable $X(\varepsilon)$ denotes the economy wealth and the control variables u and k represent, respectively, the proportion of the public sector with respect to the economy size and the instantaneous rent the politician extracts during her office which is not observed by citizens. Therefore, if at time t the initial wealth is $x > 0$ and the politician chooses a policy pair (u_s, k_s) then the economy wealth evolves according to the SDE

$$dX_s(\varepsilon) = u_s X_s(\varepsilon)(\varepsilon ds + \sigma dW_s) - k_s ds, \quad X_t(\varepsilon) = x, \tag{6}$$

where either $0 \le t \le s \le T$ or $T \le t \le s \le 2T$.

Definition 1. The policy $(u, k) := (u_s, k_s)$, $0 \le t \le s \le 2T$, is called *admissible* at time t with initial wealth $x > 0$ if:

- It is progressively measurable with respect to $\mathbb{F}^G$.
- $0 \le u_s \le 1$ and $\mathbb{E}\left[\int_t^{2T} u_s^2 ds\right] < \infty$.
- $k_s \ge 0$ and $\mathbb{E}\left[\int_t^{2T} k_s ds\right] < \infty$.
- $X_s(\varepsilon) \ge 0$.

We denote by $\mathcal{A}_t(x)$ the set of all admissible policies at time t with initial wealth x and by

$$X_s^{t,x;u,k}(\varepsilon) \tag{7}$$

the unique strong solution of (6) with control $(u, k) \in \mathcal{A}_t(x)$.

Politicians have CRRA preferences and get utility from the rent they extract in each instant t. In the second period, the re-elected politician maximizes

$$\mathbb{E}\left[\int_T^{2T} \frac{k_s^\alpha}{\alpha} ds\right], \quad 0 < \alpha < 1, \tag{8}$$

[4] Notice that the observable filtration $\mathbb{F}^G$ is strictly included in the "true" filtration $\mathbb{F}$ because of the presence of the random variable ε in the drift coefficient of (1).

over all $(u, k) \in \mathcal{A}_T(x)$ and subject to

$$dX_s(\hat{\varepsilon}_T) = u_s X_s(\hat{\varepsilon}_T)(\hat{\varepsilon}_T ds + \sigma dW_s) - k_s ds, \quad X_T(\hat{\varepsilon}_T) = x, \tag{9}$$

where $\hat{\varepsilon}_T$ is defined by (4). A control pair $(u^*(\hat{\varepsilon}_T), k^*(\hat{\varepsilon}_T)) \in \mathcal{A}_T(x)$ is called optimal for the re-elected politician if $\mathbb{E}\left[\int_T^{2T}\left[(k_s^*(\hat{\varepsilon}_T))^\alpha/\alpha\right]ds\right] = v_2(T, x)$, where

$$v_2(T, x) := \sup_{(u,k)\in\mathcal{A}_T(x)} \mathbb{E}\left[\int_T^{2T} \frac{k_s^\alpha}{\alpha} ds\right] \tag{10}$$

is the second period politician's value function (or indirect utility). On the other hand, if in the second period a new politician is in office she maximizes (8) subject to (9) with ε in place of $\hat{\varepsilon}_T$. The optimal policies $(u^*(\varepsilon), k^*(\varepsilon))$ are defined analogously.

We assume that the first period incumbent politician gets utility in the second period only if re-elected, otherwise her utility is zero. If we denote by $\mathcal{R} : \Omega \to \{0, 1\}$ the $\mathcal{F}_T^G$-measurable *re-election function*, that is $\mathcal{R} = 1$ means voters re-elect the politician in office whereas $\mathcal{R} = 0$ indicates they vote for her opponent, then in the first period the incumbent maximizes

$$\mathbb{E}\left[\int_0^T \frac{k_s^\alpha}{\alpha} ds + \mathcal{R} v_2\left(T, X_T^{0,x;u,k}(\varepsilon)\right)\right], \tag{11}$$

where v_2 is defined by (10), over all $(u, k) \in \mathcal{A}_0(x)$ and subject to (6) with $t = 0$.

Voters are risk-neutral, get utility from the end-of-period wealth, and can affect their utility only through elections. They re-elect the incumbent politician at time T if and only if[5]

$$\mathbb{E}\left[X_{2T}^{T,x;u^*(\hat{\varepsilon}_T),k^*(\hat{\varepsilon}_T)}(\hat{\varepsilon}_T)\right] \geq \mathbb{E}\left[X_{2T}^{T,x;u^*(\varepsilon),k^*(\varepsilon)}(\varepsilon)\right], \tag{12}$$

where $x > 0$ is the economy wealth at time T.

3 The Analysis

As the model description suggests, the analysis will proceed backwards. From now on, we drop in the notation of the state variables the dependence on the control variables, for example, we denote the unique solution of (6) associated to $(u, k) \in \mathcal{A}_t(x)$ with $X_s^{t,x}(\varepsilon)$ instead of $X_s^{t,x;u,k}(\varepsilon)$.

[5] Cf., inter alia, Rogoff (1990, p. 25).

3.1 Post-election Period

After elections, the politician in office can be either the first period incumbent, whose updated probability of being competent is p^I, that is we assume

$$\pi_T = p^I, \quad 0 \le p^I \le 1, \tag{13}$$

or a new politician, whose probability of being skilled is p [see (2)]. Whoever the politician in office is, she solves the same stochastic control problem. Thus, we carry out the analysis for the new politician; to obtain the value function and the optimal feedback controls of the re-elected incumbent, it is sufficient to write p^I instead of p.

The time homogeneous structure of the model enables us to shift backwards the second period analysis. Indeed, if we define $\overline{W}_t := W_{t+T} - W_T$, $\overline{\mathcal{F}}_t := \mathcal{F}_{t+T}$, $\overline{u}_t := u_{t+T}$, and so on, then we can formally describe the second period exactly as before but with the time variable running from 0 to T. In this case t denotes the time elapsed from the beginning of period. For the sake of simplicity, we keep the same notation for both periods.

In order to use dynamic programming techniques, we define, for any given $(t,x) \in [0,T] \times \mathbb{R}_+$, the functional

$$J_2(t,x;u,k) := \mathbb{E}\left[\int_t^T \frac{k_s^\alpha}{\alpha} ds\right], \tag{14}$$

where $(u,k) \in \mathcal{A}_t(x)$ and $X_s^{t,x}(\varepsilon)$ is the unique solution of (6), and the value function

$$v_2(t,x) := \sup_{(u,k)\in\mathcal{A}_t(x)} J_2(t,x;u,k). \tag{15}$$

Since the politician has access only to the information contained in $\mathbb{F}^G$ and not to the full information in $\mathbb{F}$, problem (15) is not Markovian and the dynamic programming principle does not hold.[6]

Now, by means of a suitable change of measure,[7] we transform the partial information problem (15) into an equivalent complete information problem.[8] To this end, define the processes

$$Y_t := W_t + \left(\frac{\varepsilon}{\sigma}\right)t, \quad 0 \le t \le T, \tag{16}$$

---

[6] From a mathematical point of view, the issue is that the $\mathbb{F}$-Brownian motion (W_t) needs not to be a Brownian motion w.r.t. the "smaller" filtration $\mathbb{F}^G$.

[7] By now this is a standard technique in filtering theory (cf. Lakner, 1995; Karatzas and Zhao, 2001; Pham, 2009).

[8] In fact, under the new probability measure we can represent the processes G in (1) and X in (6) as a martingale and a semimartingale, respectively, w.r.t. the observable filtration $\mathbb{F}^G$.

with $Y_0 = 0$, and

$$Z_t(\varepsilon) := \exp\left(-\frac{\varepsilon}{\sigma}W_t - \frac{1}{2}\left(\frac{\varepsilon}{\sigma}\right)^2 t\right), \quad 0 \leq t \leq T. \tag{17}$$

Then $Z_t(\varepsilon)$ is a $(\mathbb{P}, \mathbb{F})$-martingale with $Z_0(\varepsilon) = 1$ and, by Girsanov Theorem, (Y_t) is an $\mathbb{F}$-Brownian motion under the probability measure $\tilde{\mathbb{P}}$ defined by the Radon–Nikodym derivative[9]

$$\frac{d\tilde{\mathbb{P}}}{d\mathbb{P}} = Z_T(\varepsilon). \tag{18}$$

Observe that $\tilde{\mathbb{P}}$ and $\mathbb{P}$ are equivalent and that $\tilde{\mathbb{P}}[\varepsilon = 1] = \mathbb{P}[\varepsilon = 1] = p$. Moreover, (Y_t) is also an $\mathbb{F}^G$-Brownian motion under $\tilde{\mathbb{P}}$ (see Lakner, 1995, Proposition 4.1, p. 255). Finally, since ε is $\mathcal{F}_0$-measurable and independent of (W_t), and (Y_t) has independent increments, (Y_t) and ε are independent under $\tilde{\mathbb{P}}$, so that

$$\tilde{\mathbb{P}}\left[\varepsilon = 1 \mid \mathcal{F}_t^G\right] = \tilde{\mathbb{P}}[\varepsilon = 1] = p. \tag{19}$$

In the new filtered probability space $\left(\Omega, \mathcal{F}, \tilde{\mathbb{P}}, \mathbb{F}^G\right)$ we can represent the processes (1) and (6) as Itô diffusions driven by the $\mathbb{F}^G$-Brownian motion (Y_t). Indeed, we have[10]

$$d\tilde{G}_s = \sigma \tilde{G}_s dY_s, \quad \tilde{G}_t = g \tag{20}$$

and

$$d\tilde{X}_s = \sigma u_s \tilde{X}_s dY_s - k_s ds, \quad \tilde{X}_t = x. \tag{21}$$

Clearly, since we are working on the probability space $\left(\Omega, \mathcal{F}, \tilde{\mathbb{P}}\right)$, we need to represent also the functional (14) as an expected value w.r.t. $\tilde{\mathbb{P}}$. By observing that

$$\mathbb{E}\left[\int_t^T \frac{k_s^\alpha}{\alpha} ds\right] = \tilde{\mathbb{E}}\left[\int_t^T Z_s^{-1}(\varepsilon) \frac{k_s^\alpha}{\alpha} ds\right], \tag{22}$$

where $Z_t^{-1}(\varepsilon)$ is the $\left(\tilde{\mathbb{P}}, \mathbb{F}\right)$-martingale

$$Z_t^{-1}(\varepsilon) = \exp\left(\frac{\varepsilon}{\sigma}Y_t - \frac{1}{2}\left(\frac{\varepsilon}{\sigma}\right)^2 t\right), \quad 0 \leq t \leq T, \tag{23}$$

[9] $\tilde{\mathbb{P}}$ is the *probability of reference* in filtering theory.

[10] Observe that no learning occurs in $\left(\Omega, \mathcal{F}, \tilde{\mathbb{P}}, \mathbb{F}^G\right)$ [see (19)] since the representation of the observable processes in this space with respect to (Y_t) removes the unobservable parts of the drifts.

it can be proved that[11]

$$\tilde{\mathbb{E}}\left[\int_t^T Z_s^{-1}(\varepsilon)\frac{k_s^\alpha}{\alpha}ds\right] = \tilde{\mathbb{E}}\left[\int_t^T (Q_s + 1 - p)\frac{k_s^\alpha}{\alpha}ds\right], \tag{24}$$

where $Q_t := pZ_t^{-1}(1)$ is driven by the Zakai equation

$$dQ_t = (1/\sigma)\, Q_t dY_t, \quad Q_0 = p. \tag{25}$$

Q_t is called the *unnormalized* conditional probability that $\varepsilon = 1$ given the information available at time t, $\mathcal{F}_t^G$, and an application of the Bayes' rule (see Karatzas and Shreve, 1988, Lemma 3.5.3, p. 193) yields the following relation with the conditional probability π_t in (5):

$$\pi_t = \mathbb{E}\left[\chi_{\{\varepsilon=1\}} \mid \mathcal{F}_t^G\right] = \frac{\tilde{\mathbb{E}}\left(\chi_{\{\varepsilon=1\}} Z_s^{-1}(\varepsilon) \mid \mathcal{F}_t^G\right)}{\tilde{\mathbb{E}}\left(Z_t^{-1}(\varepsilon) \mid \mathcal{F}_t^G\right)} = \frac{Q_t}{Q_t + 1 - p}, \tag{26}$$

where χ_A is the characteristic function of the set $A \subset \Omega$. Therefore, in the new filtered probability space $\left(\Omega, \mathcal{F}, \tilde{\mathbb{P}}, \mathbb{F}^G\right)$, the equivalent complete information problem is characterized by the two-dimensional Markov process

$$\begin{cases} d\tilde{X}_s = u_s\sigma\tilde{X}_s dY_s - k_s ds, & \tilde{X}_t = x, \\ dQ_s = (1/\sigma)\, Q_s dY_s, & Q_t = q, \end{cases} \tag{27}$$

and the value function is

$$\tilde{v}_2(t,x,q) := \sup_{(u,k)\in\mathcal{A}_t(x)} \tilde{\mathbb{E}}\left[\int_t^T \frac{k_s^\alpha}{\alpha}\left(Q_s^{t,x,q} + 1 - p\right)ds\right], \tag{28}$$

for all $(t,x,q) \in [0,T] \times \mathbb{R}_+ \times \mathbb{R}_+$. The Dynamic Programming Principle holds for this problem (see Fleming and Rishel, 1975, Chap. VI) and, under appropriate regularity conditions, yields the following Hamilton–Jacobi–Bellman (HJB) equation for $\tilde{v}_2$:

$$0 = w_t + \sup_{0\le u\le 1, k\ge 0}\left\{\mathbb{A}^{u,k}[w](x,q) + \frac{k^\alpha}{\alpha}(q + 1 - p)\right\}, \tag{29}$$

for all $(t,x,q) \in [0,T) \times \mathbb{R}_+ \times \mathbb{R}_+$, and boundary condition

$$w(T,x,q) = 0, \quad (x,q) \in \mathbb{R}_+ \times \mathbb{R}_+, \tag{30}$$

where

$$\mathbb{A}^{u,k}[w](x,q) = \frac{1}{2}\sigma^2u^2x^2w_{xx} + uxqw_{xq} + \frac{1}{2}\frac{q^2}{\sigma^2}w_{qq} - kw_x \tag{31}$$

[11] See for example Décamps et al. (2005) and Borkar (2005).

is the differential operator associated to the state dynamics (27). Assuming $w_{xx} < 0$,[12] the maximization in the RHS of (29) gives the following optimal values for u and k:

$$u_2^*(t,x,q) = -\frac{xqw_{xq}}{\sigma^2 x^2 w_{xx}}, \quad k_2^*(t,x,q) = \left(\frac{q+1-p}{w_x}\right)^{1/(1-\alpha)}. \tag{32}$$

Direct substitution in (29) yields

$$0 = w_t - \frac{1}{2}\frac{q^2 w_{xq}^2}{\sigma^2 w_{xx}} + \frac{1}{2}\frac{q^2}{\sigma^2} w_{qq} + \frac{1-\alpha}{\alpha}(q+1-p)^{1/(1-\alpha)}(w_x)^{-\alpha/(1-\alpha)}. \tag{33}$$

Standard homogeneity arguments and a power transformation (cf. Zariphopoulou, 2001) suggest a solution of the HJB equation (33) of the form

$$w(t,x,q) = \frac{x^\alpha}{\alpha}[h(t,q)]^{1-\alpha}, \tag{34}$$

where h solves

$$0 = h_t + \frac{1}{2}\frac{q^2}{\sigma^2} h_{qq} + (q+1-p)^{1/(1-\alpha)}, \tag{35}$$

for all $(t,q) \in [0,T) \times \mathbb{R}_+$, with boundary condition

$$h(T,q) = 0, \quad q \in \mathbb{R}_+. \tag{36}$$

Under appropriate regularity and growth conditions, this non-homogeneous linear parabolic equation admits, through the *Feynman–Kac* formula, the following solution in stochastic form:

$$\tilde{w}_2(t,q) := \tilde{\mathbb{E}}\left[\int_t^T \left(\Phi_s^{t,q} + 1 - p\right)^{1/(1-\alpha)} ds\right], \tag{37}$$

where the process $\Phi_s^{t,q}$, $t \le s \le T$, solves

$$d\Phi_s = (1/\sigma)\,\Phi_s dB_s, \quad \Phi_t = q,$$

with (B_s) a standard Brownian motion on $\left(\Omega, \mathcal{F}, \tilde{\mathbb{P}}, \mathbb{F}^G\right)$. Finally, a standard verification argument (see Fleming and Rishel, 1975, Proposition 4.1, p. 159) will prove that

$$\tilde{v}_2(t,x,q) = \frac{x^\alpha}{\alpha}[\tilde{w}_2(t,q)]^{1-\alpha}, \tag{38}$$

[12] This will be checked later.

and

$$u_2^*(t,x,q) = \frac{q\,(\tilde{w}_2)_q}{\sigma^2 \tilde{w}_2}, \quad k_2^*(t,x,q) = x\frac{(q+1-p)^{1/(1-\alpha)}}{\tilde{w}_2}, \tag{39}$$

are the optimal feedback control functions, where $(\tilde{w}_2)_q = \partial\tilde{w}_2/\partial q$. Observe that $\tilde{v}_2$ is strictly concave w.r.t. x, $k_2^* > 0$, and the optimal state variable $\hat{X}$ is positive since its dynamics becomes geometric once we substitute u_2^* and k_2^* into (27).

3.2 A Re-election Rule

In order to better describe the voters' strategy we shift forwards the optimal trajectories computed in the previous subsection. Hence, under the new probability measure $\tilde{\mathbb{P}}$, voters' expected utility in the second period is

$$\tilde{\mathbb{E}}\left[\hat{X}_{2T}^{T,x,\eta}\left(\hat{Q}_{2T}^{T,x,\eta} + 1 - \eta\right)\right], \tag{40}$$

where $\eta \in \{p, p^I\}$ [see (2) and (13)], and $\hat{X}^{T,x,\eta}$ and $\hat{Q}^{T,x,\eta}$ are the second period optimal state dynamics with initial condition $\hat{X}_T^{T,x,\eta} = x$ and $\hat{Q}_T^{T,x,\eta} = \eta$. Voters are rational and choose whether to re-elect the incumbent or vote for a challenger by comparing the expected utility they get under each choice. Define the set $R \subset \mathbb{R}_+$ (*re-election set*) as follows:

$$\begin{aligned} R := \Big\{p' \in \mathbb{R}_+ \mid \tilde{\mathbb{E}}\left[\hat{X}_{2T}^{T,x,p'}\left(\hat{Q}_{2T}^{T,x,p'} + 1 - p'\right)\right] \\ \geq \tilde{\mathbb{E}}\left[\hat{X}_{2T}^{T,x,p}\left(\hat{Q}_{2T}^{T,x,p} + 1 - p\right)\right]\Big\}. \end{aligned} \tag{41}$$

Observe that R is non empty ($p \in R$) and does not depend on x. Indeed, once we substitute the feedback controls (39) in (27), the dynamics for $\tilde{X}$ becomes geometric and, as a result, for any $\lambda \in \mathbb{R}$, $\tilde{X}_{2T}^{t,\lambda x,q} = \lambda \tilde{X}_{2T}^{t,x,q}$. This means that voters' election rule does not depend on the economic wealth at the time of elections but only on the incumbent's reputation of being competent. Hence, whenever R is an $\mathcal{F}_T^G$-measurable set, the re-election function takes the form

$$\mathcal{R} = \chi_{\left\{Q_T^{0,x,p} \in R\right\}}. \tag{42}$$

Therefore, in the complete information setting, the first period maximization problem (11) becomes

$$\tilde{\mathbb{E}}\left[\int_0^T \frac{k_s^\alpha}{\alpha}\left(Q_s^{0,x,p} + 1 - p\right)ds + \chi_{\left\{Q_T^{0,x,p} \in R\right\}}\tilde{v}_2\left(T, \tilde{X}_T^{0,x,p}, Q_T^{0,x,p}\right)\right], \tag{43}$$

subject to (27) with $t = 0$ and p in place of q, where $\tilde{v}_2$ is given by (28).

Remark 1. If voters' expected utility (40) were increasing in η, then the re-election rule (42) would be

$$\mathcal{R} = \chi_{\left\{Q_T^{0,x,p} \geq p\right\}} \quad \text{or, equivalently,} \quad \mathcal{R} = \chi_{\{\pi_T \geq p\}}.$$

In other words, voters re-appoint the incumbent politician if and only if the conditional probability that she is competent is higher than the ex-ante probability p. That is, citizens vote for the more competent between the incumbent politician and the challenger.

3.3 Pre-election Period

In this section, we see how elections make politicians accountable. Indeed, we will show the existence of an opportunistic behaviour on the incumbent side.

In the first period the incumbent politician deals with the same problem as in period two, except that at time T she runs for re-election against an opponent randomly chosen among the population. Then, under the assumption of rationality, her policies must be optimal given the voters' re-election rule. We suppose citizens always believe they are on the equilibrium path, that is, they believe that the incumbent extracts the equilibrium rent: hence, incumbent's beliefs coincide with voters' beliefs. Consequently, for any given $(t, x, q) \in [0, T] \times \mathbb{R}_+ \times \mathbb{R}_+$, the politician maximizes

$$\begin{aligned} &\tilde{J}_1\left(t, x, q; u, k\right) \\ &:= \tilde{\mathbb{E}}\left[\int_t^T \frac{k_s^\alpha}{\alpha}(Q_s^{t,x,q} + 1 - p)ds + \chi_{\{Q_T^{t,x,q} \in R\}} \tilde{v}_2(0, \tilde{X}_T^{t,x,q}, Q_T^{t,x,q})\right] \end{aligned} \tag{44}$$

over all admissible control pairs $(u, k) \in \mathcal{A}_t\left(x\right)$ and subject to (27), where $\tilde{v}_2$ and R are respectively defined in (28) and (41). This means that the politician obtains the second period value function if and only if she is re-appointed at $t = T$. The first period value function is

$$\tilde{v}_1(t, x, q) := \sup_{(u,k) \in \mathcal{A}_t(x)} \tilde{J}_1\left(t, x, q; u, k\right). \tag{45}$$

Now, if $\tilde{v}_1$ is smooth enough in $[0, T) \times \mathbb{R}_+ \times \mathbb{R}_+$, we expect it satisfies the HJB equation

$$0 = w_t + \sup_{0 \leq u \leq 1, k \geq 0} \left\{\mathbb{A}^{u,k}\left[w\right]\left(x, q\right) + \frac{k^\alpha}{\alpha}\left(q + 1 - p\right)\right\}, \tag{46}$$

where $\mathbb{A}^{u,k}[w]$ is as in (31), together with the boundary condition

$$w(T,x,q) = \begin{cases} \tilde{v}_2(0,x,q), & (x,q) \in \mathbb{R}_+ \times R, \\ 0, & (x,q) \in \mathbb{R}_+ \times (\mathbb{R}_+ \backslash R). \end{cases} \tag{47}$$

Observe that (46) and (29) have the same differential part but different boundary conditions. Along the lines of the previous analysis we expect a solution of the form

$$w(t,x,q) = \frac{x^\alpha}{\alpha}[h(t,q)]^{1-\alpha}, \tag{48}$$

where h solves

$$0 = h_t + \frac{1}{2}\frac{q^2}{\sigma^2}h_{qq} + (q+1-p)^{1/(1-\alpha)}, \tag{49}$$

for all $(t,q) \in [0,T) \times \mathbb{R}_+$, with boundary condition

$$h(T,q) = \begin{cases} \tilde{w}_2(0,q), & q \in R, \\ 0, & q \in \mathbb{R}_+ \backslash R, \end{cases} \tag{50}$$

where $\tilde{w}_2$ is defined by (37). The optimal first period policies are

$$u_1^*(t,x,q) = \frac{qh_q}{\sigma^2 h}, \quad k_1^*(t,x,q) = x\frac{(q+1-p)^{1/(1-\alpha)}}{h}. \tag{51}$$

Proposition 1. *Assume that the boundary value problem (49)–(50) admits a smooth solution $\tilde{w}_1(t,q)$ on $[0,T) \times \mathbb{R}_+$ such that*

$$\tilde{v}_1(t,x,q) = \frac{x^\alpha}{\alpha}\tilde{w}_1(t,q), \tag{52}$$

for all $(t,x,q) \in [0,T] \times \mathbb{R}_+ \times \mathbb{R}_+$. Then,

$$k_1^*(t,x,q) \leq k_2^*(t,x,q), \tag{53}$$

for all $(t,x,q) \in [0,T] \times \mathbb{R}_+ \times \mathbb{R}_+$, where k_1^* and k_2^* are respectively defined by (51) and (39).

Proof. Observe that $\tilde{v}_2 \geq 0$ [see (28)]. Hence, from the definitions of $\tilde{v}_1$ and $\tilde{v}_2$ [see (45) and (28)] it follows

$$\tilde{v}_1(t,x,q) \geq \tilde{v}_2(t,x,q),$$

for all $(t,x,q) \in [0,T] \times \mathbb{R}_+ \times \mathbb{R}_+$. Then, the characterizations (38) and (52) imply

$$\tilde{w}_1(t,q) \geq \tilde{w}_2(t,q),$$

for all $(t, q) \in [0, T] \times \mathbb{R}_+$. Finally,

$$k_1^*(t, x, q) = \frac{x\,(q + 1 - p)^{1/(1-\alpha)}}{\tilde{w}_1\,(t, q)} \leq \frac{x\,(q + 1 - p)^{1/(1-\alpha)}}{\tilde{w}_2\,(t, q)} = k_2^*\,(t, x, q)\,.$$

□

According to traditional accountability literature, relation (53) shows that the incumbent politician diverts less resources under electoral pressure, all other things the same. That is, politicians interested in holding office refrain from rent-extraction in the pre-election period in order to benefit from post-election rents. Rational voters, by maximizing their expected utility, give the politician implicit incentives which make her more aligned with their preferences.

We conclude this section with a technical observation. The above argument relies on the fact that the value function is smooth enough, at least in the interior of its domain, and satisfy the HJB equation (46)–(47). These features are far from straightforward to establish due to the lack of continuity along the boundary at $t = T$, and certainly deserve attention for future research.

4 Special Case

For $\alpha = 1/2$ we explicitly solve the boundary value problem (35)–(36), which enables us to recover the second period value function $\tilde{v}_2$ [see (38)] and study the dependence of the optimal policies on the variables t, x, and q. The solution of (35)–(36) takes the form (cf. Liu, 2007)

$$\tilde{w}_2\,(t, q) = \tilde{a}\,(t)\,q^2 + \tilde{b}\,(t)\,q + \tilde{c}\,(t)\,, \tag{54}$$

where $\tilde{a}\,(t)$, $\tilde{b}\,(t)$, and $\tilde{c}\,(t)$ are the solution of the system of ODEs

$$\begin{cases} a'\,(t) + \left(1/\sigma^2\right) a\,(t) + 1 = 0 \\ b'\,(t) + 2\,(1 - p) = 0 \\ c'\,(t) + (1 - p)^2 = 0 \end{cases} \tag{55}$$

with endpoint conditions

$$a\,(T) = 0, \quad b\,(T) = 0, \quad c\,(T) = 0. \tag{56}$$

That is

$$\begin{cases} \tilde{a}(t) = \sigma^2 \left(\exp\left((T-t)/\sigma^2\right) - 1\right), \\ \tilde{b}(t) = 2(1-p)(T-t), \\ \tilde{c}(t) = (1-p)^2 (T-t). \end{cases} \tag{57}$$

The following observations are in order:

1. k_2^* is increasing in t. That is, politician subtracts more resources in proximity of her end of office (thus reducing the citizens' expected utility), all other things the same. Hence, we should observe a decreasing level of the economic wealth caused by this egoistic behaviour.
2. k_2^* is decreasing in q. Hence, more competent politicians extract lower rents, all other things the same.
3. u_2^*, the proportion between private and public sector, is independent of the economy size. Hence, our model specification would suggest the existence of an optimal level of public intervention into the economy.
4. u_2^* is increasing in q. A competent politician invests more in public sector than a low ability one.

5 Conclusions

We analyze a *career concerns* political agency model in a two-period continuous time stochastic framework where politician's competence is unobserved. Standard filtering techniques permit us to transform the politician's partial information problem into a complete information one. Then, relying on the dynamic programming principle we write the Hamilton–Jacobi–Bellman equation for the value function and, at least for the second period, we provide a solution in stochastic form together with the optimal policies. By comparing the optimal rent before and after elections, we find support for the traditional idea that elections lead a politician to be opportunistically more aligned with voters' preferences. The presence of a re-election constraint modifies politician's optimal policies because of *career concerns* motivations. Moreover, for certain parameter values we solve explicitly the second period part of the model and study the optimal policies. In particular, in absence of an electoral constraint the politician subtracts more resources in proximity of her end of office.

We conclude the paper with some proposals for future work. First, we believe that the assumption of an economy ending at a fixed time is very restrictive although widely used in literature. One way to overcome this undesirable assumption is to adopt an infinite-horizon economy with an overlapping generation structure of short-lived politicians and long-lived voters. Second, it is not clear whether the citizens vote for the candidate expected to be more competent or not (see Remark 1), and this should become even more obscure in presence of risk-averse voters. We think that a more detailed analysis of the re-election rule $\mathcal{R}$ and, in particular, the shape and parameters' dependence of the set R [see (41)] in case of both risk-neutral and risk-averse voters is worthy of attention. All these are subjects for future research.

Acknowledgements We are very grateful to an anonymous referee for suggesting several improvements. We also thank the seminar participants at the MDEF 2008. All errors are solely our responsibility. Financial support from the *Università Cattolica di Milano* is gratefully acknowledged.

References

Ashworth, S. (2005). Reputational dynamics and political careers. *The Journal of Law, Economics and Organization, 21*, 441–466.

Barro, R. J. (1973). The control of politicians: An economic model. *Public Choice, 14*, 19–42.

Besley, T. (2006). *Principled agents? The political economy of good government*. Oxford: Oxford University Press.

Borkar, V. S. (2005). Controlled diffusion processes. *Probability Surveys, 2*, 213–244.

Décamps, J. P., Mariotti, T., & Villeneuve, S. (2005). Investment timing under incomplete information. *Mathematics of Operations Research, 30*, 472–500.

Ferejohn, J. (1986). Incumbent performance and electoral control. *Public Choice, 50*, 5–25.

Fleming, W. H., & Rishel, R. (1975). *Deterministic and stochastic optimal control*. New York: Springer.

Holmström, B. (1982). Managerial incentive problems: A dynamic perspective. In *Essays in economics and management in honor of Lars Wahlbeck*. Helsinki: Swedish School of Economics. Reprinted in *Review of Economic Studies, 66*, 169–182 (1999).

Karatzas, I., & Shreve, S. E. (1988). *Brownian motion and stochastic calculus*. New York: Springer.

Karatzas, I., & Zhao, X. (2001). Bayesian adaptive portfolio optimization. In E. Jouini, J. Cvitanić, & M. Musiela (Eds.), *Option pricing, interest rates and risk management* (pp. 632–669). Cambridge: Cambridge University Press.

Lakner, P. (1995). Utility maximization with partial information. *Stochastic Processes and Their Applications, 56*, 247–273.

Liu, J. (2007). Portfolio selection in stochastic environments. *The Review of Financial Studies, 20*, 1–39.

Persson, T., Roland, G., & Tabellini, G. (1997). Separations of powers and accountability. *The Quarterly Journal of Economics, 112*, 1163–1202.

Pham, H. (2009). Portfolio optimization under partial observation: Theoretical and numerical aspects. In D. Crisan & B. Rozovsky (Eds.), *Handbook of nonlinear filtering*. Oxford: Oxford University Press.

Rieder, U., & Bäuerle, N. (2005). Portfolio optimization with unobservable Markov-modulated drift process. *Journal of Applied Probability, 42*, 362–378.

Rogoff, K. (1990). Equilibrium political budget cycles. *American Economic Review, 80*, 21–36.

Zariphopoulou, T. (2001). A solution approach to valuation with unhedgeable risks. *Finance and Stochastics, 5*, 61–82.

Behavioral Portfolio Choice and Disappointment Aversion: An Analytical Solution with "Small" Risks

Enrico Saltari and Giuseppe Travaglini

1 Introduction

The standard portfolio model based on expected utility (EU) theory predicts a large equity position for most households. Empirical analysis demonstrates, however, that the composition of household's wealth is characterized by a small proportion of risky assets. A consolidated empirical literature provides measures of these financial phenomena (Brandolini et al., 2004; Faiella and Neri, 2004; Giannetti and Koskinen, 2009; Heaton and Lucas, 1996; Mankiw and Zeldes, 1991; Guiso and Zingales unpublished). For instance, in Italy over the period 1965–2006 the percentage of stocks held by households has been on average 9% of the total wealth. Similar proportions can be found in the portfolio of families in United States, France, Germany and Great Britain.

The puzzling aspect of these data is that the excess return on equities – a measure of the risk premium – has been often positive and even large. Dimson et al. (2002) illustrate that during the twentieth century it was around 6% in United States, Germany and Great Britain. This return was even higher and close to 7% in Italy and France. Similar rates of return are computed by Mehra and Prescott (1985, 2003) and Campbell (2003) over the same period for the main industrialized countries.

Loosely speaking the puzzle is the following. Given that equities yield such a high risk premium, why do households buy so few stocks? Almost all calibrated version of dynamic portfolio choice models with standard preferences (even when augmented with other important ingredients like transaction costs or borrowing constraints) fail in replicating the previous basic facts. Indeed, given plausible estimated stochastic processes for stock market returns, an implausibly high risk aversion is needed to keep the investors away from stocks.

Obviously, the evolution of the excess return is also characterized by its volatility. Dimson et al. (2002) provide some interesting data also on this aspect. The standard deviation of the risk premium is relatively high in Italy (32%) and Germany (35%),

E. Saltari (✉)
Department of Public Economics, University of Rome, "La Sapienza",
via del Castro Laurenziano 9, 00161 Rome, Italy,
e-mail: Enrico.Saltari@uniroma1.it

G.I. Bischi et al. (eds.), *Nonlinear Dynamics in Economics, Finance and the Social Sciences*, DOI 10.1007/978-3-642-04023-8_16,

whereas it is smaller and approximately equal to 20% in United States, France and Great Britain. Thus, there is evidence to suggest that the undersized proportion of equities in the household's portfolio depends on how a risk-averse agent perceives the trade-off between expected returns and riskiness.

Behavioral finance is a new approach to financial markets that has emerged in response to the difficulties of the traditional approach. The main innovation has been to look more deeply into preferences. Bernartzi and Thaler (1995) managed to reconcile theory and data, by assuming that agents exhibit "myopic loss aversion" based on the prospect theory by Kanheman and Tversky (1979). More recently Ang et al. (2005) adopted a different, theoretically more rigorous approach to the problem. They study portfolio allocation under preferences that show Disappointment Aversion (DA) as in Gul (1997). By calibrating stochastic processes for stocks and bonds returns to actual US data, they simulate their model to generate reasonable portfolio allocations even for moderate values of the disappointment aversion.

The intuition for this result is simple. Under DA the agent gives higher weight to losses and therefore she is less attracted by risky assets. However, in contrast to the prospect theory, the DA utility is an axiomatic theory. Perhaps, its most interesting difference with respect to the more traditional behavioral approach is that the reference point is updated endogenously by the agent without having to make any arbitrary exogenous assumption.

Just because of the endogeneity of the reference point in the value function, DA, although promising, does not deliver closed form solution to the optimal portfolio choice. Thus, numerical solutions are the standard tool for studying DA preferences. The drawback to numerical solutions is, however, that it is often difficult to determine why results come out the way they do, and this disadvantage may tend to obscure the underlying economics.

In this paper we argue that under DA preferences it is possible to obtain an analytical solution. Basically, we present a solution to the problem of the optimal share of risky asset. We study the case of "small" risks in a static model, delivering explicit values for portfolio weights. The fundamental economic principles of DA preferences are particularly clear in this setting. Under DA the percentage of risky assets in the portfolio is proportional to the ratio between the mean and the variance of the excess return, where the coefficient of proportionality is the reciprocal of the risk aversion. However, the mean and the variance of the excess returns do not depend on the original probability distribution. Rather, they depend critically on a new probability distribution which is affected by the degree of disappointment aversion.

What is the added value of this outcome in comparison with the traditional portfolio model? In the standard model of portfolio choice the Arrow–Pratt approximation implies that the risk yields a second-order effect on welfare compared to the effect of the mean of the lottery. So, when the risk is "small" the optimal portfolio choice for a risk averse agent is unaffected by risk. DA utility solves this paradox. Employing our analytical solution, we show that the share of the risky asset is coherent with the evidence. One of our interesting result is that when the risk is "small" the DA preferences provide a share of the risky asset which is an order of magnitude less than the corresponding share under the EU theory.

The paper is organized as follows. The next section describes the second-order risk aversion property in the context of the EU theory. Section 3 explores the potential of changing the utility function using the DA preferences. In this section we develop the basic model and compare the optimal shares under DA with those of the EU theory. Section 4 extends this result to continuous random returns. Section 5 concludes.

2 The Traditional Portfolio Choice

Consider an agent with initial wealth W. His endowment can be invested in a portfolio with risky and risk-free assets. The return of the risky asset is the random variable x_0. The riskless interest rate is equal to r. The problem of the agent is to determine the optimal composition $(W - \alpha, \alpha)$ of the portfolio where α is the amount of wealth invested in the risky asset. The end period value of the portfolio is

$$(W - \alpha)(1 + r) + \alpha (1 + x_0) = W (1 + r) + \alpha (x_0 - r) = w_0 + \alpha x,$$

where $\tilde{x} = \tilde{x}_0 - r$ is the excess return and $w_0 = W (1 + r)$.

The aim of the agent is to choose α so as to maximize expected utility $U(\alpha)$

$$\max U(\alpha) = E u (w_0 + \alpha x).$$

Let us assume that $u(.)$ is at least twice differentiable with $u' > 0$ and $u'' < 0$. The first-order condition is

$$U'(\alpha^*) = E\left[\tilde{x} u'\left(w_0 + \alpha^* x\right)\right] = 0,$$

where α^* is the optimal choice.

This standard model of the portfolio choice has an interesting but somewhat counterintuitive property. To focus on what concerns us here, note that under risk aversion the agent will purchase the risky asset if and only if $E\tilde{x} > 0$. To see this, suppose indeed that $\alpha^* = 0$. In such a case the first-order condition (which is also sufficient because of risk aversion) reduces to

$$U'(0) = u'(w_0) E(x) = 0.$$

Since $u'(w_0)$ is positive, $U'(0) = 0$ if and only if $E(x) \leq 0$. In other words, the risk averse agent will prefer the risky asset if and only if the expected excess return is positive. This is the well-known Local Risk Neutrality Theorem of Arrow (Arrow, 1965). To use the words of Arrow, if the expected excess return is positive, we are in presence of a favorable gamble and "the risk averter [...] always takes some part of a favorable gamble" (p. 155). If instead the excess return is a zero-mean risk, the agent will not hold the risky security.

The upshot is that for a risk averse agent the portfolio choice depends only on the expected excess return, but not on the volatility. This is because when the utility function is differentiable the risk premium tends to zero as the *square* of the size of the risk.[1] This property was called second order risk aversion by Segal and Spivak (1990): it implies that risk yields a second order effect on utility compared to the effect of the mean of the corresponding lottery. Hence, when agent purchases a small amount of risk, the expected return is of the first order, while the risk is of the second order. This latter claim has an important role in the following discussion.

2.1 "Small" Risks and the Traditional Portfolio Choice

An analytical solution to the portfolio problem is not in general available. Suppose we want to approximate the solution by using a first order Taylor expansion around $\alpha = 0$. The problem with this approach is that the size of the risk is endogenous. A way-out of this difficulty is suggested by Gollier (2001) by studying the case of a "small" risk. Following Gollier, define the excess return as

$$x(k) = k\mu + y \quad \text{and} \quad E(y) = 0, \tag{1}$$

where $\mu > 0$. In (1) when k tends to zero the excess return becomes a zero-mean risk y so that, because of Arrow Theorem seen above, α too tends to zero. In this sense we speak of small risks.

When $k > 0$, we obtain a solution for $\alpha^*(k)$ from the first-order condition

$$Ex(k)\,u'\left(w_0 + \alpha^*(k)\,x(k)\right) = 0. \tag{2}$$

Using a first-order expansion around the point $k = 0$ and the fact that $x(0) = y$ so that $\alpha^*(0) = 0$, we get (see Appendix 1 for the mathematical details)

$$\mu u'(w_0) + Ey^2 u''(w_0)\,\alpha^{*\prime}(0) = 0. \tag{3}$$

Using the approximation of $\alpha^*(k)$ around $k = 0$

$$\alpha^*(k) = \alpha^*(0) + k\alpha^{*\prime}(0)$$

[1] To see this most simply, remember the Arrow–Pratt approximation by which the risk premium π is proportional to the variance of the risk, σ^2 :

$$\pi \simeq \frac{1}{2}\sigma^2 A(w_0),$$

where A is the coefficient of absolute risk aversion. If we reduce the size of the risk, this will affect the risk premium as the square of the size of the risk because of the variance.

we can write the first-order condition as

$$k\mu u'(w_0) + Ey^2 u''(w_0)\,\alpha^*(k) = 0,$$

that is

$$\alpha^*(k) = -\frac{k\mu}{Ey^2}\frac{u'(w_0)}{u''(w_0)} = \frac{E(x)}{var(x)}\frac{1}{A(w_0)}, \tag{4}$$

where $A(w_0)$ is the Arrow–Pratt coefficient of absolute risk aversion. The optimal amount invested in the risky asset is proportional to the ratio between the mean and the variance and inversely related to the absolute risk aversion. This is the well-known result obtained in a dynamic framework by Samuelson (1969) and Merton (1990): the optimal portfolio choice is multiplicatively separable in risk aversion and the market price of risk.

3 Disappointment Aversion Preferences and Portfolio Choice

In this section we propose a behavioral approach to portfolio choice by adopting the axiomatic theory of disappointment aversion (DA) of preferences of Gul (1997). We start employing a basic model with only two states with outcomes z_1 and z_2. The reason why it is useful to use another name for the excess return will be clear in a moment. Define the *disappointment-averse expected utility* $V(\alpha)$ as

$$V(\alpha) = p_1 u(w_0 + \alpha z_1) + p_2 u(w_0 + \alpha z_2) - \beta p_2 [V(\alpha) - u(w_0 + \alpha z_2)], \tag{5}$$

where $z_1 > 0 > z_2$. The last term in the expression (5) captures the effect of the disappointment. When the "bad" outcome z_2 occurs, the agent is disappointed, and his expected utility is reduced by a term which is the product of the disappointment aversion β and the expected disappointment. Note that when $\beta = 0$ this expression reduces to the traditional expected utility.

Solving (5) for $V(\alpha)$ yields

$$V(\alpha) = p_1 \frac{1}{1+p_2\beta} u(w_0 + \alpha z_1) + p_2 \frac{1+\beta}{1+p_2\beta} u(w_0 + \alpha z_2). \tag{6}$$

Notice that the DA implies that $V(\alpha)$ depends on the modified value of the original probabilities p_1 and p_2. These probabilities, $p_1 \frac{1}{1+p_2\beta}$ and $p_2 \frac{1+\beta}{1+p_2\beta}$, give more weight to the unfavorable event and less weight to the favorable one. It is convenient to rewrite the expression (6) as

$$V(\alpha) = q_1 u(w_0 + \alpha z_1) + q_2 u(w_0 + \alpha z_2), \tag{7}$$

where

$$q_1 = p_1 \frac{1}{1+p_2\beta} \quad \text{and} \quad q_2 = 1 - q_1. \tag{8}$$

We shall call q_1 and q_2 disappointing probabilities. They coincide with the original probabilities when $\beta = 0$.

The first-order condition for portfolio choice with DA expected utility is

$$V'\left(\alpha_D^*\right) = q_1 z_1 u'\left(w_0 + \alpha_D^* z_1\right) + q_2 z_2 u'\left(w_0 + \alpha_D^* z_2\right) = 0, \tag{9}$$

where α_D^* is the optimal portfolio choice. As before, to get an analytical solution, it is helpful to calculate the first-order condition when $\alpha_D^* = 0$

$$\begin{aligned} V'(0) &= q_1 z_1 u'(w_0) + q_2 z_2 u'(w_0) \\ &= u'(w_0)\, E_D(z) = 0, \end{aligned}$$

where $E_D(z)$ is the expected value of z computed using the disappointing probabilities. Since the marginal utility of wealth is positive, the first-order condition is satisfied if the adjusted expected value $E_D(z)$, is equal to zero. Using the definitions in (8), this implies

$$\begin{aligned} E_D(z) &\equiv q_1 z_1 + q_2 z_2 \\ &= \frac{E(z) + p_2 z_2 \beta}{1 + p_2 \beta} = 0. \end{aligned}$$

Thus, in order to have $\alpha_D^* = 0$, it must be

$$E(z) = -p_2 z_2 \beta \quad \text{or} \quad E_D(z) = 0.$$

In contrast to the traditional portfolio model, the amount of wealth invested in the risky asset is now equal to zero *if and only if* the expected excess return is equal to the expected disappointment $-p_2 z_2 > 0$ times the disappointment aversion β. So, under DA it might be better not to invest in the risky asset even when the expected return of the gamble is positive. As a final point, note that if disappointment aversion is zero, then z reduces to x.

3.1 "Small" Risks and Disappointing Aversion

To find an analytical solution to the portfolio choice under DA preferences, we now proceed as in the previous section. As before, we study the portfolio choice for small risks. But now for "small risks" we mean that when k tends to zero the expected excess return tends to $-p_2 z_2 \beta > 0$, and not to zero as in the traditional portfolio choice.

Thus, write the excess return as

$$z = k\mu + \varepsilon, \qquad \text{where } E_D(\varepsilon) = 0. \tag{10}$$

Note that the distribution of ε depends on β since the probabilities of the two states (q_1 and q_2) now reflect disappointment aversion. When disappointment aversion is equal to zero, however, ε coincides with y so that $z = x$.

As before, expanding the first-order condition (9) around $k = 0$ and using the fact that $z_i(0) = \varepsilon_i$ implies $\alpha_D^*(0) = 0$, we get (see Appendix 2 for the details)

$$\mu u'(w_0) + u''(w_0)\,\alpha_D^{*\prime}(0)\left(q_1\varepsilon_1^2 + q_2\varepsilon_2^2\right) = 0. \tag{11}$$

Finally, using the approximation $\alpha_D^*(k) = \alpha_D^*(0) + k\alpha_D^{*\prime}(0)$, we get the optimal portfolio choice[2]

$$\alpha_D^*(k) = \frac{E_D(z)}{var_D(z)}\frac{1}{A(w_0)}.$$

It is interesting to compare the two solutions of the optimal shares in the two alternative frameworks:

$$\alpha^*(k) = \frac{E(x)}{var(x)}\frac{1}{A(w_0)} \quad \text{and} \quad \alpha_D^*(k) = \frac{E_D(z)}{var_D(z)}\frac{1}{A(w_0)}. \tag{12}$$

Inspecting the two formulas makes clear that they are equal when $\beta = 0$. Thus, the DA preferences are one-parameter generalization of the preferences based on the EU theory. From the previous conditions we can easily compute the share of wealth invested in the risky asset

$$\frac{\alpha^*(k)}{w_0} = \frac{E(x)}{var(x)}\frac{1}{R(w_0)}, \qquad \frac{\alpha_D^*(k)}{w_0} = \frac{E_D(z)}{var_D(z)}\frac{1}{R(w_0)},$$

where $R(w_0)$ is the coefficient of relative risk aversion.

3.2 An Example

To make clear the implication of our results for the portfolio choice, it can be of help an example.

Assume that the utility function is CRRA, $u(w) = \frac{w^{1-\gamma}}{1-\gamma}$ with $0 < \gamma < \infty$, so that the coefficient of relative risk aversion is γ. Since it is reasonable to assume that relative risk aversion is somewhere between 1 and 4, we set $\gamma = 2$. To simplify, we

---

[2] Notice that

$$E_D(z) = k\mu + E_D(\varepsilon) = k\mu \quad \text{and} \quad var_D(z) = q_1\varepsilon_1^2 + q_2\varepsilon_2^2$$

since

$$E_D(\varepsilon) = q_1\varepsilon_1 + q_2\varepsilon_2 = 0$$

by the relationship (10).

also normalize to one the final value of the wealth invested in the riskless asset, so that $w_0 = 1$. Finally, assume that the deterministic component of the excess return is $\mu = 0.08$.

To begin with, suppose that the distribution of the excess return is generated by the following binomial process

$$y_1 = 0.1, \quad y_2 = -0.1, \quad \text{with} \quad p_1 = p_2 = 0.5.$$

Under EU and setting $k = 0$, we know that the optimal portfolio share of the risk asset is zero, $\alpha^* = 0$, because the risk premium is zero.

We now introduce a disappointment aversion degree equal to $\beta = 1$, a value consistent with that of Tversky and Kahneman (1991). Then, under DA the probabilities become

$$q_1 = p_1 \frac{1}{1 + p_2\beta} = \frac{1}{3} \quad \text{and} \quad q_2 = p_2 \frac{1 + \beta}{1 + p_2\beta} = \frac{2}{3}.$$

Clearly, to have an optimal share of the risky security equal to zero under DA, i.e. $\alpha_D^* = 0$, we must change the outcomes in such a way that with these new probabilities we have again a risk premium equal to zero. One way to do this is to add the expected disappointment to the first outcome while leaving unchanged the other:

$$\varepsilon_1 = y_1 - \beta \frac{p_2}{p_1} y_2 = 0.2, \qquad \varepsilon_2 = y_2,$$

where we named ε the new random variable thus created. Note once again that with $\beta = 0$, the two random variables y and ε coincide.

The excess return is now defined using ε and thus z:

$$z = k\mu + \varepsilon.$$

Making use of this definition and setting $k = 0$, we get the following mean returns under the two alternative assumption of EU and DA preferences

$$E(z) = E(\varepsilon) = 0.05, \quad \text{and} \quad E_D(z) = E_D(\varepsilon) = 0,$$

whereas the variances are

$$var(z) = 0.0225 \quad \text{and} \quad var_D(z) = 0.02.$$

Thus, for the two different preferences the optimal shares of the risky asset are

$$\alpha^* = \frac{E(z)}{var(z)} \frac{1}{\gamma} = 1.11 \quad \text{and} \quad \alpha_D^* = \frac{E_D(z)}{var_D(z)} \frac{1}{\gamma} = 0.$$

Thus, if the mean of the excess return is equal to 0.05, the EU theory implies a share greater than 110% of total wealth. On the contrary, we discover that under DA preferences the share reduces to zero.

Assume now that $k = 0.1$. This small change affects expected returns

$$E(z) = k\mu + E(\varepsilon) = 0.058$$

and

$$E_D(z) = k\mu + E_D(\varepsilon) = 0.008$$

but does not affect the variances. In this case the shares are respectively

$$\alpha^* = 1.29, \quad \alpha_D^* = 0.2.$$

In short, the difference remains very large.[3]

4 Continuous Random Variables

Up to now we have assumed that agent faces a two-state random variable. This raises the question whether the optimal decision found above also holds with continuous random returns. We will now verify that this is true.

4.1 The Certainty Equivalent

Let's assume that the uncertain return of the risky asset is a continuous random variable. The disappointment-averse utility function (5) is now defined as

$$V(\alpha) = E[u(w)] - \beta \int_{-\infty}^{z_c} [u(w_c) - u(w)] f(z)\, dz, \tag{13}$$

where for brevity we write $w = w_0 + \alpha z$ and $w_c = w_0 + \alpha z_c$ is the certainty equivalent. There is no substantial difference between this definition and the one we gave in the two state model. Here, as before, the negative events are scaled down by the factor β. What really changes is what we mean with "negative" or disappointing events.

---

[3] If we solve numerically the first-order condition, we have for $k = 0$

$$\alpha^* = 1.21, \quad \alpha_D^* = 0,$$

whereas for $k = 0.1$

$$\alpha^* = 1.45, \quad \alpha_D^* = 0.207.$$

Comparing these results with those obtained in the text, it follows that the approximation works well.

Here disappointing states are those whose realization are below the certainty equivalent. In the two states world there was no problem in defining which is the bad state: it was simply that with the smallest payoff. With continuous random variable disappointment arises when the realization is below the certainty equivalent, which as we will see in a moment is itself an endogenous variable.

By definition, the certainty equivalent w_c is the amount of wealth that would leave the agent indifferent between that payoff and the lottery, that is

$$u(w_c) = V(\alpha).$$

Substituting in (13), we have

$$u(w_c) = E[u(w)] - \beta \int_{-\infty}^{z_c} [u(w_c) - u(w)] f(z)\, dz \tag{14}$$

or

$$u(w_c) = \left\{1 + \beta \int_{-\infty}^{z_c} f(z)\, dz\right\}^{-1} \left\{E[u(w)] + \beta \int_{-\infty}^{z_c} u(w) f(z)\, dz\right\}. \tag{15}$$

This utility function computes the value of the certainty equivalent when preferences are characterized by DA. As before, when $\beta = 0$ the previous definition reduces to the traditional expected utility. When $\beta \neq 0$, the same equation implicitly defines the new probability distribution under DA. Indeed, expression (15) is equivalent to

$$u(w_c) = \frac{\int_{z_c}^{\infty} u(w) f(z)\, dz}{1 + \beta \int_{-\infty}^{z_c} f(z)\, dz} + \frac{(1+\beta) \int_{-\infty}^{z_c} u(w) f(z)\, dz}{1 + \beta \int_{-\infty}^{z_c} f(z)\, dz}. \tag{16}$$

Comparing this last condition to the one of the binary model, i.e. (6), we can easily check that the assumption of continuous random return does not affect the qualitative results obtained earlier. In both models the DA assigns an higher weight β to the unfavorable events, i.e. those with payoffs smaller than the certainty equivalent w_c. The new probability distribution is given by

$$f_D(z) = \begin{cases} \frac{f(z)}{1+\beta \int_{-\infty}^{z_c} f(z) dz} & \text{if } z \geq z_c, \\ \\ \frac{(1+\beta) f(z)}{1+\beta \int_{-\infty}^{z_c} f(z) dz} & \text{if } z < z_c. \end{cases} \tag{17}$$

4.2 The Portfolio Choice

Note that the definition of the certainty equivalent w_c in (15) depends on α. This has two implications. On the one hand, we cannot compute the certainty equivalent w_c without knowing the optimal portfolio choice α. On the other, the optimal α depends on the certainty equivalent w_c, which is unknown at the outset. This means that with continuous distributions the optimal portfolio choice under DA is characterized by the presence of two endogenous variables.

If w_c were known, (13) could be solved for in the same way as in the standard expected utility framework. The only difference would be that for states below w_c, the utility has to be scaled down by β. However, w_c is itself a function of the outcome of optimization (that is, w_c is a function of α). Hence, we must solve simultaneously two equations to compute the optimal values of α and w_c. One is given by (14). The other by the first-order condition.

The DA investor's problem can be solved by maximizing $V(\alpha)$ over α

$$\max_{\alpha} V(\alpha) = \max_{\alpha} u(w_c)$$

which from the definition of w_c (15) implies[4]

$$E\left[zu'(w)\right] + \beta \int_{-\infty}^{z_c} \left[zu'(w)\right] f(z)\, dz = 0 \tag{18}$$

or by (17)

$$E_D\left(zu'(w)\right) = 0.$$

Hence, to get the optimal values for w_c and α we must solve the system of equation (15) and (18).

4.3 DA Preferences with Continuous Distributions and "Small" Risks

As before, to find an analytical solution we assume that the excess return has the form

$$z = k\mu + \varepsilon$$

so that when k tends to zero the optimal portfolio choice is equal to zero as well. From our previous discussion we know that $\alpha_D^* = 0$ implies that $E(\varepsilon) > 0$ in such a way that $E_D(\varepsilon) = 0$.

To determine this critical value in the case of continuous distributions, we focus on the first-order condition for $\alpha_D^* = 0$. Since this condition is a function of the certainty equivalent w_c, we have also to use (14). Setting to zero the portfolio choice, we have

$$u(w_c)|_{\alpha_D=0} + \beta \int_{-\infty}^{z_c} \left[u(w_c)|_{\alpha_D=0}\right] f(z)\, dz = u(w_0) + \beta \int_{-\infty}^{z_c} u(w_0)\, f(z)\, dz$$

[4] The derivation of the first-order condition is not so simple as it might appear because the certainty equivalent w_c also depends on the portfolio choice α. See Appendix 3 for a formal derivation.

which is obviously satisfied when $w_c = w_0$. That is, when $\alpha_D = 0$, the certainty equivalent coincides with w_0, the final value of wealth invested in the risk-free asset.

We now use this result in the first-order condition, (18). Taking into account that $w_c \geq w = w_0 + \alpha z$ with $w_c = w_0$, implies $z_c \leq 0$, we set to 0 the upper limit of the integral.

With this in mind, the first-order condition becomes

$$u'(w_0)\, E(z) + \beta u'(w_0) \int_{-\infty}^{0} z f(z)\, dz = 0$$

or

$$E(z) + \beta \int_{-\infty}^{0} z f(z)\, dz = 0 \tag{19}$$

or also

$$\begin{aligned} E_D(z) &= \int_0^{\infty} z \frac{f(z)}{1 + \beta \int_{-\infty}^{0} f(z)\, dz} dz + (1+\beta) \int_{-\infty}^{0} z \frac{f(z)}{1 + \beta \int_{-\infty}^{0} f(z)\, dz} dz \\ &= \int_{-\infty}^{+\infty} z f_D(z)\, dz = 0, \end{aligned}$$

where the distribution $f_D(z)$ is defined as before with $z_c = 0$.

Assuming as above small risks and expanding the first-order condition around $k = 0$, we get

$$\begin{aligned} 0 = E\Bigg\{ & \varepsilon u'\left(w_0 + \alpha_D^*(0)\,\varepsilon\right) \\ & + \Big[\mu u'\left(w_0 + \alpha_D^*(0)\,\varepsilon\right) + \varepsilon u''\left(w_0 + \alpha_D^*(0)\,\varepsilon\right) \alpha_D^*(0)\,\mu \\ & + \varepsilon^2 u''\left(w_0 + \alpha_D^*(0)\,\varepsilon\right) \alpha_D^{*\prime}(0)\Big] k \\ & + \beta \int_{-\infty}^{0} \varepsilon u'\left(w_0 + \alpha_D^*(0)\,\varepsilon\right) f(\varepsilon)\, d\varepsilon \\ & + \beta \Big[\int_{-\infty}^{0} \mu u'\left(w_0 + \alpha_D^*(0)\,\varepsilon\right) f(\varepsilon)\, d\varepsilon \\ & + \int_{-\infty}^{0} \varepsilon u''\left(w_0 + \alpha_D^*(0)\,\varepsilon\right) \alpha_D^*(0)\,\mu f(\varepsilon)\, d\varepsilon \\ & + \int_{-\infty}^{0} \varepsilon^2 u''\left(w_0 + \alpha_D^*(0)\,\varepsilon\right) \alpha_D^{*\prime}(0)\, f(\varepsilon)\, d\varepsilon \Big] k \Bigg\}, \end{aligned}$$

where use has been made of the fact that $z(0) = \varepsilon$.

Since $\alpha_D^*(0) = 0$, the previous condition can be rewritten as

$$0 = u'(w_0)\left[E(\varepsilon) + \beta\int_{-\infty}^{0}\varepsilon f(\varepsilon)\,d\varepsilon\right] + k\mu u'(w_0)\left[1 + \beta\int_{-\infty}^{0}\varepsilon f(\varepsilon)\,d\varepsilon\right]$$
$$+ku''(w_0)\,\alpha_D^{*\prime}(0)\left[E\left(\varepsilon^2\right) + \beta\int_{-\infty}^{0}\varepsilon^2 f(\varepsilon)\,d\varepsilon\right].$$

Notice that by (19) the first term on the right-hand side is zero. Thus, the first-order condition simplifies to

$$0 = \mu - A(w_0)\,\alpha_D^{*\prime}(0)\,\frac{E\left(\varepsilon^2\right) + \beta\int_{-\infty}^{0}\varepsilon^2 f(\varepsilon)\,d\varepsilon}{1 + \beta\int_{-\infty}^{0}\varepsilon f(\varepsilon)\,d\varepsilon}$$

which, given the distribution (17), becomes

$$0 = \mu - A(w_0)\,\alpha_D^{*\prime}(0)\,var_D(\varepsilon), \tag{20}$$

where as before $var_D(\varepsilon)$ is the variance of ε, given the distribution f_D.

To obtain an analytical solution in this set up, we use the same approximation used above, i.e. $\alpha_D^*(k) = \alpha_D^*(0) + k\alpha_D^{*\prime}(0)$. Hence, the optimal portfolio choice under DA and with continuous probability distributions is

$$\begin{aligned}\alpha_D^* &= \frac{k\mu}{var_D(\varepsilon)}\frac{1}{A(w_0)}\\ &= \frac{E_D(\varepsilon)}{var_D(\varepsilon)}\frac{1}{A(w_0)}.\end{aligned}$$

4.4 An Example

Let's suppose the excess return follows a normal distribution. We will calibrate the volatility of the distribution setting it equal to 15%. If as before $\beta = 1$, the solution of the equation

$$E(\varepsilon) + \beta\int_{-\infty}^{0}\varepsilon f(\varepsilon)\,d\varepsilon = 0$$

is $E(\varepsilon) \approx 0.041$. This means that ε is characterized by the normal distribution $N \sim (0.041, 0.0225)$. Assume now that the utility function is Bernoullian, $u(w) = \ln(w)$. When $k = 0$, we get two different expected values for the two alternative distributions

$$E(z) = E(\varepsilon) = 0.041, \quad E_D(\varepsilon) = 0,$$

whereas the two variances are equal.

Using the same normalization as before for w_0 and the CRRA utility function with a coefficient of relative risk aversion equal to 2, the optimal portfolio shares are respectively

$$\alpha^* \simeq 0.92, \quad \alpha_D^* = 0.$$

Suppose instead that $k = 0.1$ and $\mu = 0.08$, so that the new means are $E(z) = k\mu + E(\varepsilon) = 0.049$ and $E_D(x) = k\mu + E_D(\varepsilon) = 0.008$. The optimal shares in this case are[5]

$$\alpha^* = 1.098, \quad \alpha_D^* = 0.177.$$

5 Conclusions

In this paper we have proposed a behavioral approach to portfolio choice by adopting the theory of disappointment aversion of Gul (1997). We have shown how disappointment aversion affects the optimal portfolio choice when risk is small. The analytical solution we found has a functional form very similar to the traditional result of Samuelson (1969) and Merton (1990): the optimal percentage of wealth invested in the risky asset is equal to the product of the market price of risk and the reciprocal of risk aversion. However, under DA the probabilities are appropriately modified by disappointment aversion. These probabilities capture the effect of disappointment: the outcomes below the certainty equivalent are weighted more heavily than the outcomes above. In this perspective, any risk averse agent will prefer to reduce the amount of risky asset held in the portfolio over the amount predicted by the expected utility theory. This result allows to accommodate the puzzling prediction of the standard portfolio model where the proportion of the risky asset is a rather large proportion of the financial portfolio.

The model presented above may appear narrow. It captures something about the belief of the agent in a static context, but do not provide a prediction in a dynamic framework. Our future aim is to extend this basic model to a dynamic context.

Appendix 1

This appendix contains the mathematical details for the portfolio choice and "small risks" under EU. We first rewrite the first-order condition using the definition of $x(k) = k\mu + y$

$$E\left[x(k)\, u'\left(w_0 + \alpha^*(k)\, x(k)\right)\right] = 0, \tag{21}$$

where we write $x(k)$ to highlight that the excess return depends on k.

[5] Solving numerically the first-order condition, we have for $k = 0$

$$\alpha^* = 0.9, \quad \alpha_D^* = 0,$$

whereas for $k = 0.1$

$$\alpha^* = 1.066, \quad \alpha_D^* = 0.18.$$

Comparing these results with those obtained in the text, it turns out again that the approximation works very well.

Using a Taylor expansion around the point $k = 0$, (21) can be written as

$$\begin{aligned} E\{x(0)\,u'\left(w_0 + \alpha^*(0)\,x(0)\right) + [\mu u'\left(w_0 + \alpha^*(0)\,x(0)\right) \\ +x(0)\,u''\left(w_0 + \alpha^*(0)\,x(0)\right)\alpha^*(0)\,\mu \\ +[x(0)]^2\,u''\left(w_0 + \alpha^*(0)\,x(0)\right)\alpha^{*\prime}(0)]k\} = 0 \end{aligned}$$

Since $x(0) = y$ so that $\alpha^*(0) = 0$ (because $E(y) = 0$), the first-order condition becomes

$$\mu u'(w_0) + Ey^2 u''(w_0)\,\alpha^{*\prime}(0) = 0$$

which is (3) in the text.

Appendix 2

This appendix contains the mathematical details for the portfolio choice and "small risks" under DA. Expanding the first-order condition (9) around $k = 0$, we get

$$\begin{aligned} q_1\{z_1(0)\,u'\left(w_0 + \alpha_D^*(0)\,z_1(0)\right) + [\mu u'\left(w_0 + \alpha_D^*(0)\,z_1(0)\right) \\ +z_1(0)\,u''\left(w_0 + \alpha_D^*(0)\,z_1(0)\right)\alpha_D^*(0)\,\mu + [z_1(0)]^2\,u''\left(w_0 + \alpha_D^*(0)\,z_1(0)\right)\alpha_D^{*\prime}(0)]k\} \\ +q_2\{z_2(0)\,u'\left(w_0 + \alpha_D^*(0)\,z_2(0)\right) + [\mu u'\left(w_0 + \alpha_D^*(0)\,z_2(0)\right) \\ +z_2(0)\,u''\left(w_0 + \alpha_D^*(0)\,z_2(0)\right)\alpha_D^*(0)\,\mu + [z_2(0)]^2\,u''\left(w_0 + \alpha_D^*(0)\,z_2(0)\right)\alpha_D^{*\prime}(0)]k\} = 0. \end{aligned}$$

Since $\alpha_D^*(0) = 0$ and $z_i(0) = \varepsilon_i$, this reduces to

$$\begin{aligned} q_1\left[\mu u'(w_0) + \varepsilon_1^2 u''(w_0)\,\alpha_D^{*\prime}(0)\right] \\ +q_2\left[\mu u'(w_0) + \varepsilon_2^2 u''(w_0)\,\alpha_D^{*\prime}(0)\right] = 0. \end{aligned}$$

Finally, remembering that $q_1 + q_2 = 1$, the first-order condition can be further simplified by letting

$$\mu u'(w_0) + u''(w_0)\,\alpha_D^{*\prime}(0)\left(q_1\varepsilon_1^2 + q_2\varepsilon_2^2\right) = 0$$

which is (11) in the text.

Appendix 3

This appendix contains the formal derivation of the first-order condition with DA. The derivation is as follows. Start from the definition of $V(\alpha)$

$$V(\alpha) = \frac{E[u(w)] + \beta\int_{-\infty}^{z_c}[u(w)]\,f(z)\,dz}{1 + \beta\int_{-\infty}^{z_c} f(z)\,dz}.$$

Taking into account that $w = w_0 + \alpha z$, so that $z_c = \frac{w_c - w_0}{\alpha}$, we substitute in the upper limit of the integrals and rewrite $V(\alpha)$ as

$$V(\alpha) = \frac{E[u(w)] + \beta \int_{-\infty}^{\frac{w_c - w_0}{\alpha}} [u(w)] f(z)\, dz}{1 + \beta \int_{-\infty}^{\frac{w_c - w_0}{\alpha}} f(z)\, dz}.$$

The first-order condition is

$$\frac{dV(\alpha)}{d\alpha} = \frac{E[zu'(w)] - \beta \frac{w_c - w_0}{\alpha^2} u(w_c) f(z_c) + \beta \int_{-\infty}^{\frac{w_c - w_0}{\alpha}} (zu'(w)) f(z)\, dz}{D}$$
$$+ \frac{E[u(w)] + \beta \int_{-\infty}^{\frac{w_c - w_0}{\alpha}} [u(w)] f(z)\, dz}{D^2} \beta \frac{w_c - w_0}{\alpha^2} f(z_c) = 0, \quad (22)$$

where $D = 1 + \beta \int_{-\infty}^{\frac{w_c - w_0}{\alpha}} f(z)\, dz$.

Noting that

$$DV(\alpha) = Du(w_c) = E[u(w)] + \beta \int_{-\infty}^{\frac{w_c - w_0}{\alpha}} [u(w)] f(z)\, dz$$

we can rewrite the expression (22) as follows

$$\frac{dV(\alpha)}{d\alpha} = \frac{E[zu'(w)] - \beta \frac{w_c - w_0}{\alpha^2} u(w_c) f(z_c) + \beta \int_{-\infty}^{\frac{w_c - w_0}{\alpha}} [zu'(w)] f(z)\, dz}{D} -$$
$$+ \frac{u(w_c)}{D} \beta \frac{w_c - w_0}{\alpha^2} f(z_c) = 0,$$

which further simplifies to

$$\frac{dV(\alpha)}{d\alpha} = E\left[zu'(w)\right] + \beta \int_{-\infty}^{\frac{w_c - w_0}{\alpha}} \left[zu'(w)\right] f(z)\, dz = 0,$$

which is the first-order condition (18) in the main text.

References

Ang, A., Bekaert, G., & Liu, J. (2005). Why stocks may disappoint. *Journal of Financial Economics, 76*, 471–508.

Arrow, K. (1965). The theory of risk aversion. In K. Arrow (Ed.), *Aspects of the theory of risk-bearing*. Helsinki: Yriö Jahnssonin säätiö. Reprinted in *Individual choice under certainty and uncertainty, Collected Papers of K. J. Arrow* (Vol. 3), Blackwell, Oxford, 1984.

Bernartzi, S., & Thaler, R. (1995). Myopic loss aversion and the equity premium puzzle. *Quarterly Journal of Economics, 110*, 73–92.

Brandolini, A., Cannari, L., D'Alessio, G., & Faiella, I. (2004). *Household wealth distribution in Italy in the 1990s* (Temi di discussione 530).

Campbell, J. Y. (2003). Consumption-based asset pricing. In G. Constantinides, M. Harris, & R. Stulz (Eds.), *Handbook of the economics of finance* (Chap. 13, pp. 803–887). Amsterdam: Elsevier.

Dimson, E., Marsh, P., & Staunton, M. (2002). *Triumph of the optimists. 101 years of global investment returns*. Princeton: Princeton University Press.

Faiella, I., & Neri, A. (2004). *La ricchezza delle famiglie italiane e americane* (Temi di discusssione 501, pp. 1–59).

Giannetti, M., & Koskinen, Y. (2009). Investor protection, equity returns, and financial globalization. *Journal of Financial and Quantitative Analysis* (forthcoming).

Gollier, C. (2001). *The economics of risk and time*. Cambridge: MIT Press.

Gul, F. (1997). A theory of disappointment aversion. *Econometrica, 59*, 667–686.

Heaton, J., & Lucas, J. F. (1996). Evaluating the effects of incomplete markets on risk sharing and asset pricing. *Journal of Political Economy, 104*, 443–487.

Kanheman, D., & Tversky, A. (1979). Prospect theory: An analysis of decision under risk. *Econometrica, 47*, 263–291.

Mankiw, N. G., & Zeldes, S. P. (1991). The consumption of stockholders and nonstockholders. *Journal of Financial Economy, 29*, 97–112.

Mehra, R., & Prescott, E. (1985). The equity premium: A puzzle. *Journal of Monetary Economics, 15*, 145–161.

Mehra, R., & Prescott, E. (2003). The equity premium in retrospect. In G. Constantinides, M. Harris, & R. Stulz (Eds.), *Handbook of the economics of finance* (Chap. 14). Amsterdam: Elsevier.

Merton, R. C. (1990). *Continuous-time finance*. Cambridge: Blackwell.

Samuelson, P. A. (1969). Lifetime portfolio selection by dynamic stochastic programming. *Review of Economics and Statistics, 51*, 239–246.

Segal, U., & Spivak, A. (1990). First order versus second order risk aversion. *Journal of Economic Theory, 51*, 111–125.

Tversky, A., & Kahneman, D. (1991). Loss aversion in riskless choice: A reference dependent model. *Quarterly Journal of Economics, 107*, 1039–1061.

A Simple Agent-based Financial Market Model: Direct Interactions and Comparisons of Trading Profits

Frank Westerhoff

1 Introduction

In the recent past, a number of interesting agent-based financial market models have been proposed. These models successfully explain some important stylized facts of financial markets, such as bubbles and crashes, fat tails for the distribution of returns and volatility clustering. These models, reviewed, for instance, in Chen, Chang, and Du (in press); Hommes (2006); LeBaron (2006); Lux (in press); Westerhoff (2009), are based on the observation that financial market participants use different heuristic trading rules to determine their speculative investment positions. Note that survey studies by Frankel and Froot (1986); Menkhoff (1997); Menkhoff and Taylor (2007); Taylor and Allen (1992) in fact reveal that market participants use technical and fundamental analysis to assess financial markets. Agent-based financial market models obviously have a strong empirical microfoundation.

Recall that technical analysis is a trading philosophy built on the assumption that prices tend to move in trends (Murphy, 1999). By extrapolating price trends, technical trading rules usually add a positive feedback to the dynamics of financial markets, and thus may be destabilizing. Fundamental analysis is grounded on the belief that asset prices return to their fundamental values in the long run (Graham and Dodd, 1951). Buying undervalued and selling overvalued assets, as suggested by these rules, apparently has a stabilizing impact on market dynamics. In most agent-based financial market models, the relative importance of these trading strategies varies over time. It is not difficult to imagine that changes in the composition of applied trading rules - such as a major shift from fundamental to technical trading rules - may have a marked impact on the dynamics of financial markets.

One goal of our paper is to provide a novel view on how financial market participants may select their trading rules. We do this by recombining a number of building blocks from three prominent agent-based financial market models. Let us briefly recapitulate these models:

F. Westerhoff
University of Bamberg, Department of Economics, Feldkirchenstrasse 21, D-96045 Bamberg, Germany,
e-mail: frank.westerhoff@uni-bamberg.de

G.I. Bischi et al. (eds.), *Nonlinear Dynamics in Economics, Finance and the Social Sciences*, DOI 10.1007/978-3-642-04023-8_17,

- Brock and Hommes (1997, 1998) developed a framework in which (a continuum of) financial market participants endogenously chooses between different trading rules. The agents are boundedly rational in the sense that they tend to pick trading rules which have performed well in the recent past, thereby displaying some kind of learning behavior. The performance of the trading rules may be measured as a weighted average of past realized profits, and the relative importance of the trading rules is derived via a discrete choice model. Contributions developed in this manner are often analytically tractable. Moreover, numerical investigations reveal that complex endogenous dynamics may emerge due to an ongoing evolutionary competition between trading rules. Note that in such a setting, agents interact only indirectly with each other: their orders have an impact on the price formation which, in turn, affects the performance of the trading rules and thus the agents selection of rules. Put differently, an agent is not directly affected by the actions of others.
- In Kirman (1991, 1993), an influential opinion formation model with interactions between a fixed number of agents was introduced. In Kirman's model, agents may hold one of two views. In each time step, two agents may meet at random, and there is a fixed probability that one agent may convince the other agent to follow his opinion. In addition, there is also a small probability that an agent changes his opinion independently. A key finding of this model is that direct interactions between heterogeneous agents may lead to substantial opinion swings. Applied to a financial market setting, one may therefore observe periods where either destabilizing technical traders or stabilizing fundamental traders drive the market dynamics. Note that agents may change rules due to direct interactions with other agents but the switching probabilities are independent of the performance of the rules.
- The models of Lux (1995, 1998) and Lux and Marchesi (1999, 2000) also focus on the case of a limited number of agents. Within this approach, an agent may either be an optimistic or a pessimistic technical trader or a fundamental trader. The probability that agents switch from having an optimistic technical attitude to a pessimistic one (and vice versa) depends on the majority opinion among the technical traders and the current price trend. For instance, if the majority of technical traders are optimistic and if prices are going up, the probability that pessimistic technical traders turn into optimistic technical traders is relatively high. The probability that technical traders (either being optimistic or pessimistic) switch to fundamental trading (and vice versa) depends on the relative profitability of the rules. However, a comparison of the performance of the trading rules is modelled in an asymmetric manner. While the attractiveness of technical analysis depends on realized profits, the popularity of fundamental analysis is given by expected future profit opportunities. This class of models is quite good at replicating several universal features of asset price dynamics.

Each of these approaches has been extended in various interesting directions. There are also alternative strands of research in which the dynamics of financial markets is driven, for instance, by nonlinear trading rules or wealth effects. For related models see Chiarella (1992); Chiarella, Dieci, and Gardini (2002); Day and Huang

(1990); de Grauwe and Grimaldi (2006); de Grauwe, Dewachter, and Embrechts (1993); Farmer and Joshi (2002); Li and Rosser (2001, 2004); Rosser et al. (2003); Westerhoff (2008); Westerhoff and Dieci (2006), among many others.

In this paper, we seek to recombine key ingredients of the three aforementioned approaches to come up with a simple model that is able to match the stylized facts of financial markets and that offers a novel perspective on how agents may be influenced in selecting their trading rules. In our model, we consider direct interactions between a fixed number of agents, as in Kirman approach. However, the switching probabilities are not constant over time but depend on the recent performance of the rules. To avoid asymmetric profit measures, as in the models of Lux and Marchesi, we define a fitness function along the lines of the models of Brock and Hommes, i.e., we approximate the fitness (attractiveness) of a rule by a weighted average of current and past myopic profits. Replication of the dynamics of agent-based models is often a challenging undertaking, which is why these models are sometimes regarded with skepticism. A second goal of our paper is thus to come up with a setting for which replication of our results is rather uncomplicated, even, as we hope, for the (interested) layman.

Our paper is organized as follows. In Sect. 2, we present our approach. In Sect. 3, we show that our model may mimic some stylized facts of financial markets. We also explore how a change in the number of agents and in the frequency of their interactions affects the dynamics. In Sect. 4, we check the robustness of our results. The last section offers some conclusions.

2 A Basic Model

Let us first preview the structure of our model. We assume that prices adjust with respect to the current excess demand. The excess demand, in turn, depends on the orders submitted by technical and fundamental traders. While technical traders base their orders on a trend-extrapolation of past prices, fundamental traders place their bets on mean reversion. The relative impact of these two trader types evolves over time. We assume that agents regularly meet each other and talk about their past trading performance. As a result, traders may change their opinion and switch to a new trading strategy. In particular, the time-varying switching probabilities depend on the relative success of the rules. Numerical simulations will reveal that the fractions of technical and fundamental trading rules evolve over time, which is exactly what gives rise to interesting asset price dynamics. Now we are ready to turn to the details of the model.

As in Farmer and Joshi (2002), the price adjustment is due to a simple log-linear price impact function. Such a function describes the relation between the quantity of an asset bought or sold in a given time interval and the price change caused by these orders. Accordingly, the log of the price of the asset in period $t + 1$ is quoted as

$$P_{t+1} = P_t + a(W_t^C D_t^C + W_t^F D_t^F) + \alpha_t, \tag{1}$$

where a is a positive price adjustment coefficient, D^C and D^F stand for orders generated by technical and fundamental trading rules, and W^C and W^F denote the fractions of agents using these rules. Excess buying (selling) thus drives prices up (down). Since our model only provides a simple representation of real financial markets, we add a random term to (1). We assume that α is an IID normal random variable with mean zero and constant standard deviation σ^α.

The goal of technical analysis is to exploit price trends (see Murphy (1999) for a practical introduction). Since technical analysis typically suggests buying the asset when prices increase, orders triggered by technical trading rules may be written as

$$D_t^C = b(P_t - P_{t-1}) + \beta_t. \tag{2}$$

The first term of the right-hand side of (2) stands for transactions triggered by an extrapolation of the current price trend. The reaction parameter is positive and captures how strongly the agents react to this price signal. The second term reflects additional random orders to account for the large variety of technical trading rules. As in (1) we assume that shocks are normally distributed, i.e., β is an IID normal random variable with mean zero and constant standard deviation σ^β.

Fundamental analysis (see Graham and Dodd (1951) for a classical contribution) presumes that prices may disconnect from fundamental values in the short run. In the long run, however, prices are expected to converge towards their fundamental values. Since fundamental analysis suggests buying (selling) the asset when the price is below (above) its fundamental value, orders generated by fundamental trading rules may be formalized as

$$D_t^F = c(F_t - P_t) + \gamma_t, \tag{3}$$

where c is a positive reaction parameter and F is the log of the fundamental value. Note that we assume that traders are able to compute the true fundamental value of the asset. In order to allow for deviations from the strict application of this rule, we include a random variable γ in (3), where γ is IID normally distributed with mean zero and constant standard deviation σ^γ.

For simplicity, the fundamental value is set constant, i.e.,

$$F_t = 0. \tag{4}$$

Alternatively, the evolution of the fundamental value may be modelled as a random walk and we will do this later on. However, in order to show that the dynamics of a financial market may not depend on fundamental shocks, we abstain from this for the moment.

We furthermore assume that there are N traders in total. Let K be the number of technical traders. We are then able to define the weight of technical traders as

$$W_t^C = \frac{K_t}{N}. \tag{5}$$

Similarly, the weight of fundamental traders is given as

$$W_t^F = \frac{N - K_t}{N}. \tag{6}$$

Obviously, (5) and (6) imply that $W_t^F = 1 - W_t^C$.

The number of technical and fundamental trades is determined as follows. As in Kirman (1991, 1993), we assume that two traders meet at random in each time step, and that the first trader will adopt the opinion of the other trader with a certain probability $(1-\delta)$. In addition, there is a small probability ϵ that a trader will change his attitude independently. Contrary to Kirmans approach, however, the probability that a trader converts another trader is asymmetric and depends on the current and past myopic profitability of the rules (indicated by the fitness variables A^C and A^F, which we define in the sequel). Suppose that technical trading rules have generated higher myopic profits than fundamental trading rules in the recent past. Then it is more likely that a technical trader will convince a fundamental trader than vice versa. Similarly, when fundamental trading rules are regarded as more profitable than technical trading rules, the chances are higher that a fundamental trader will successfully challenge a technical trader. Thus, we express the transition probability of K as

$$K_t = \begin{cases} K_{t-1} + 1 & \text{with probability } p_{t-1}^{+} = \frac{N-K_{t-1}}{N}\left(\varepsilon + (1-\delta)_{t-1}^{F\to C}\frac{K_{t-1}}{N-1}\right) \\ K_{t-1} - 1 & \text{with probability } p_{t-1}^{-} = \frac{K_{t-1}}{N}\left(\varepsilon + (1-\delta)_{t-1}^{C\to F}\frac{N-K_{t-1}}{N-1}\right), \\ K_{t-1} & \text{with probability } 1 - p_{t-1}^{+} - p_{t-1}^{-} \end{cases} \tag{7}$$

where the probability that a fundamental trader is converted into an technical trader is

$$(1-\delta)_{t-1}^{F\to C} = \begin{cases} 0.5 + \lambda & \text{for } A_t^C > A_t^F \\ 0.5 - \lambda & \text{otherwise} \end{cases} \tag{8}$$

and the probability that a technical trader is converted into a fundamental trader is

$$(1-\delta)_{t-1}^{C\to F} = \begin{cases} 0.5 - \lambda & \text{for } A_t^C > A_t^F \\ 0.5 + \lambda & \text{otherwise} \end{cases}, \tag{9}$$

respectively.

Finally, we measure the fitness (attractiveness) of the trading rules as

$$A_t^C = (\exp[P_t] - \exp[P_{t-1}])\, D_{t-2}^C + d A_{t-1}^C, \tag{10}$$

and

$$A_t^F = (\exp[P_t] - \exp[P_{t-1}])\, D_{t-2}^F + d A_{t-1}^F, \tag{11}$$

respectively. Formulations (10) and (11) are as in Westerhoff and Dieci (2006) which, in turn, were inspired by Brock and Hommes (1998). Note that the fitness of a trading rule depends on two components. First, the agents take into account the most recent performance of the rules, indicated by the first terms of the right-hand side. The timing we assume is as follows. Orders submitted in period $t-2$ are executed at the price stated in period $t-1$. Whether or not these orders produce myopic profits then depends on the realized price in period t. Second, the agents have a memory. The memory parameter $0 \leq d \leq 1$ measures how quickly current myopic profits are discounted. For $d = 0$, agents obviously have no memory, while for $d = 1$ they compute the fitness of a rule as the sum of all observed myopic profits.

3 Some Simulation Results

The dynamics of international financial markets display certain stylized facts (Cont, 2001; Lux and Ausloos, 2002; Mantegna and Stanley, 2000). These universal features include (1) a random walk-like behavior of prices, (2) the sporadic appearance of bubbles and crashes, (3) excess volatility, (4) fat tails of the distribution of returns, and (5) volatility clustering. To be able to replicate these properties, we have selected the following parameter setting:[1]

$$a = 1, b = 0.05, c = 0.02, d = 0.95, \epsilon = 0.1, \lambda = 0.45, \sigma^{\alpha} = 0.0025,$$
$$\sigma^{\beta} = 0.025, \sigma^{\gamma} = 0.0025.$$

In the remaining part of the paper, we explore the dynamics of the model for different values of N. In particular, we increase N from 25 to 100 and to 500. In addition, for the case $N = 500$ we consider that there is more than one direct interaction between agents per trading time step.

3.1 Setting 1: $N = 25$

In our first experiment, we assume that there are only $N = 25$ agents. Of course, in real markets we usually observe a much larger number of traders. In the first step, it can be assumed that these agents reflect the trading activities of larger trading institutions or of groups of agents who collectively behave in the same manner (think, for instance, of group pressure). However, in the next subsections we increase the number of agents.

The seven panels of Fig. 1 aim at illustrating what kind of dynamics our model may produce for a limited number of speculators. In the top panel, we see the

[1] Interested readers should note that calibrating agent-based financial market models may be a time-consuming and pain-staking trial and error process. Some initial progress in estimating such models has recently been reported by Alfarano, Lux, and Wagner (2005); Boswijk, Hommes, and Manzan (2007); Manzan and Westerhoff (2007); Westerhoff and Reitz (2003); Winker, Gilli, and Jeleskovic (2007).

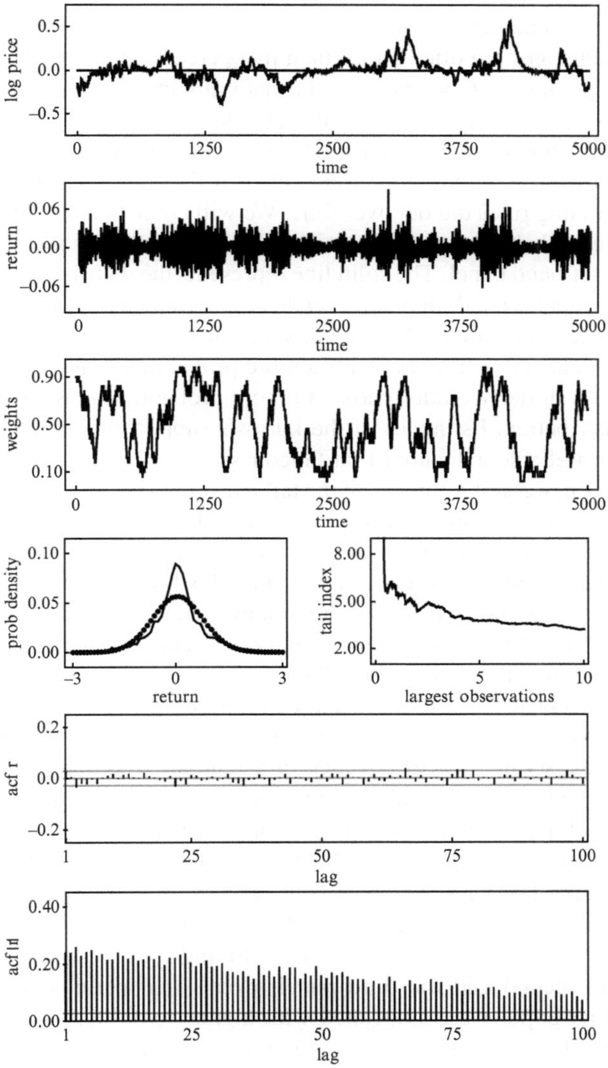

Fig. 1 The panels show the evolution of log prices, the returns, the weights of technical analysis, the distribution of the returns (the *dotted line* gives the corresponding normal distribution), estimates of the tail index, and the autocorrelation coefficients of raw and absolute returns, respectively. The simulation is based on 5,000 time steps and $N = 25$ traders. The remaining parameters are specified in Sect. 3

development of log prices. As can be seen, prices move erratically around their fundamental values. There are periods where prices are close to the fundamental value but occasionally larger bubbles set in. A prominent example is given around time step 4,000, where the distance between log prices and log fundamental values is about 0.5, implying a substantial overvaluation of about 65%.

In the second panel, returns, defined as log price changes, are plotted. Note that extreme price changes are often larger than five percent, although the fundamental value is fixed. A constant fundamental value naturally implies that the entire volatility should be regarded as excess volatility. The third panel depicts the evolution of the weights of technical and fundamental trading strategies. As can be seen, there is a permanent evolutionary competition between the rules. Neither technical nor fundamental trading rules die out over time. We will come back to this soon.

In the two panels below, we characterize the distribution of the returns. Let us start with the left-hand panel. The solid line represents the distribution of the returns of our model, whereas the dotted line visualizes a normal distribution with identical mean and standard deviation. A closer inspection reveals that the distribution of returns of our financial market model has more probability mass in the center, less probability mass in the shoulder parts and more probability mass in the tails than the normal distribution. Estimates of the kurtosis support this view. However, the kurtosis is an unreliable indicator of fat-tailedness.

For this reason, we plot estimates of the tail index in the right-hand panel, varying the number of the largest observations from 0% to 10%. For this particular simulation run we obtain a tail index of about 3.7 (using the largest 5% of the observations). We found for other simulation runs that the tail index hovers around the range from 3.5 to 4.5, which may be slightly too high on average. Most tail indices estimated from real financial data seem to range between 3 and 4, and are almost always captured by the interval 2–5 (e.g., Lux 2009).

In the last two panels, we plot the autocorrelation functions for raw returns and for absolute returns, respectively. Absence of significant autocorrelation between raw returns suggests that prices advance in a random walk-like manner. Despite the sporadic development of bubbles and crashes, it is thus hard to predict prices within our model. However, the autocorrelation coefficients for absolute returns are clearly significant and decay slowly. The autocorrelation coefficients are even positive for more than 100 lags. This is also in agreement with the second panel, and is a clear sign of volatility clustering, as observed in many real financial markets. From Fig. 1 we can also understand what is driving the dynamics of our model. Comparing the second and the third panel reveals that periods where technical analysis is rather popular are associated with higher volatility. Also, bubbles may be triggered in these periods. The trend-extrapolating (and highly noisy) nature of technical analysis has obviously a destabilizing impact on the dynamics. Note that technical analysis is quite profitable during the course of a bubble. As a result, more traders learn about this due to their interactions with other traders. Since technical analysis consequently gains in popularity, bubbles may possess some kind of momentum. A major shift from technical to fundamental analysis may be witnessed when a bubble collapses. A dominance of fundamental analysis then leads to a period where prices are closer towards fundamental values and where volatility is less dramatic.[2]

[2] What causes the everlasting competition between the trading strategies? Since prices fluctuate randomly it is hard for traders to make systematic and consistent long-run profits, i.e., the difference in the fitness of the two competing rules oscillates somehow around zero, which, in turn, causes

3.2 Setting 2: $N = 100$

Now we turn to the case with $N = 100$ traders. Figure 2 may be directly compared with Fig. 1, since it is based on the same simulation design. The only difference is that the number of traders is quadrupled. As indicated by the third panel, the popularity of the trading strategies now varies only very slowly over time. Therefore, there are extremely long periods where one or the other trading strategy dominates the market, which has some obvious consequences for the dynamics. For instance, between time steps 1,500 and 2,700 the majority of traders rely on fundamental analysis, and hence we find a period where prices are more or less in line with fundamental values and where absolute returns are rather low. Afterwards, technical analysis gains in strength and for the next 2,000 time steps volatility is elevated. Since the model is calibrated to daily data, 2,000 time steps correspond to a time span of about 8 years. Although some stylized facts may still be replicated for $N = 100$ agents, the dynamics of our model appears less convincing than before. Apparently, to generate realistic dynamics, the popularity of technical and fundamental trading rules has to vary more quickly, at least from a technical point of view. If there are only 25 traders, it may – in an extreme scenario – only take 25 time steps to accomplish a regime change from pure technical to pure fundamental analysis (or vice versa). An increase in the number of agents naturally increases the duration of such a complete regime switch. As seen in Fig. 2, regime changes may take a very long time if the number of agents is equal to 100 (of course, internal and external factors delay regime changes). In the next section, we try to show that this is not directly a problem of setting the number of agents too high. To achieve a reasonable fit of actual market dynamics with our model, the relation between the number of agents and the number of direct interactions between them per trading time step has to be within a certain range.

3.3 Setting 3: $N = 500$

Let us increase the number of agents up to $N = 500$. In addition, let us assume that there is not only one direct interaction between the agents per trading time step but that there are 20 contacts. Clearly, we now always run the interaction part of the model 20 times before we iterate the trading part of the model. As a result, the whole system may then again complete a full regime turn from pure fundamental to pure technical analysis (or the other way around) within 25 trading time steps.

Figure 3 presents the results. The qualitative similarities between Figs. 1 and 3 are striking. We recover bubbles and crashes, excess volatility, fat tails for the dis-

---

repeated swings in opinion. Put differently, if one of the rules outperformed the other one, it would also dominate the market. In addition, traders may change their opinion independently of market circumstances.

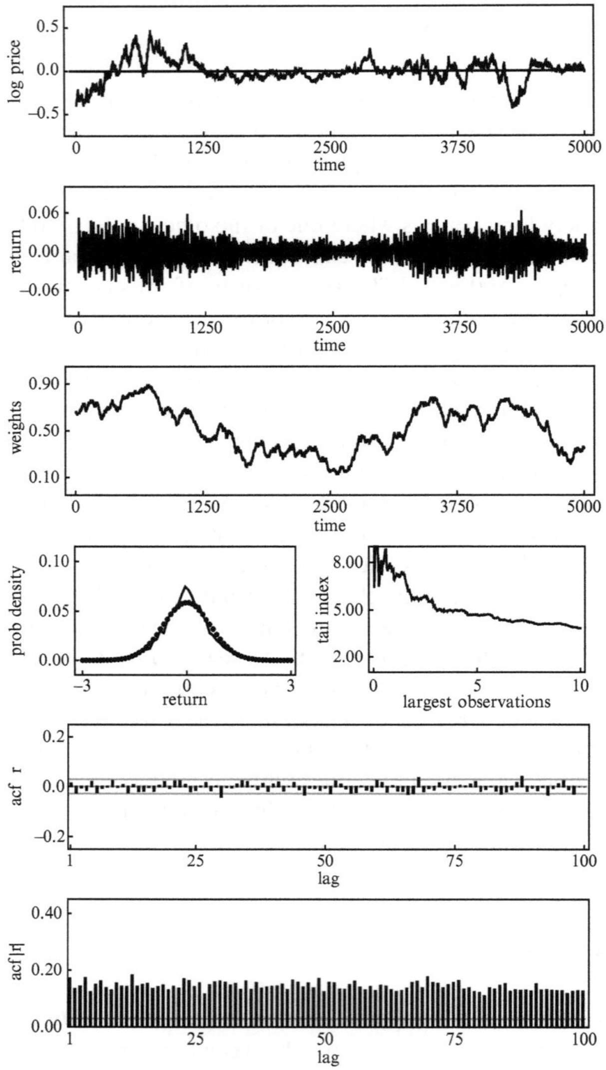

Fig. 2 The same simulation design as in Fig. 1, except that we now consider $N = 100$ agents

tribution of the returns, absence of autocorrelation for raw returns, and volatility clustering, i.e., our model again mimics key stylized facts of financial markets quite well.

Two further comments are required. Note first that periods of high volatility may or may not be associated with bubbles and crashes. It may thus happen that prices fluctuate wildly around fundamental values. We consider it interesting that there is

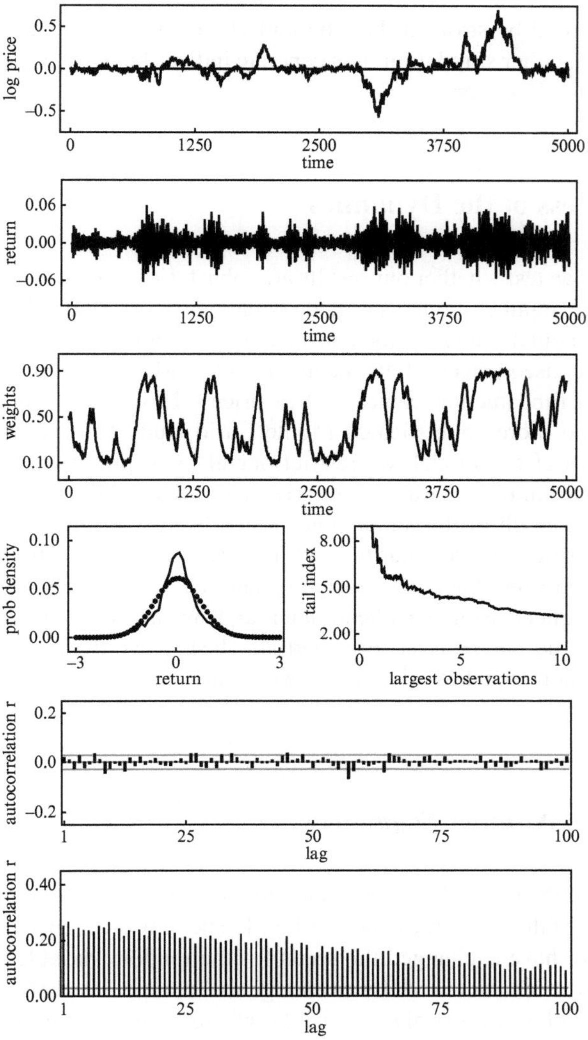

Fig. 3 The same simulation design as in Fig. 1, except that we now consider $N = 500$ agents and 20 direct interactions per trading time step

not a strict one-to-one relation between high volatility and bubble periods.[3] Finally, although the model once again generates a distribution which deviates from the normal distribution, in the sense that there is more probability mass in its tails, the

[3] This implies that technical analysis may also outperform fundamental analysis in a non-bubble period; otherwise its weight – which is mostly driven by the agent (imitative) learning behavior – would not have increased.

fat-tailedness could be stronger. For the underlying simulation run we compute a tail index of 4.3. Other simulation runs generate indices between 3.5 and 4.5, as was the case for $N = 25$ traders.

4 Robustness of the Dynamics

In this section, we test whether our results are robust. First, we explore the impact of different random number sequences on the dynamics of our model. Next, we assume that the fundamental value is not constant but follows a random walk. Finally, we study the consequences of financial market crashes by introducing extreme shocks both in fundamental values and in prices. However, instead of performing a larger and more sophisticated Monte Carlo study to check the robustness of the dynamics of our model, we restrict ourselves to presenting and discussing some additional simulation runs. The reason for doing this is that we strongly believe in the strength of the human eye, which has a remarkable ability to identify both regularities and irregularities in time series. It is also instructive to inspect single simulation runs. Phenomena such as bubbles and crashes or volatility outbursts are infrequent, irregular phenomena, and by measuring them with certain statistics their true nature is at least partially lost. Nevertheless, we ascertained that a more elaborate statistical analysis would also confirm the robustness of the dynamics.[4]

4.1 Random Number Sequences

Let us start with the issue of different random number sequences. Figure 4 displays four repetitions of the first three panels of Fig. 1. The only difference between Figs. 1 and 4 is that we have exchanged the seeds for the random variables. Note that all simulation runs are characterized by an endogenous competition between the trading rules. Volatility clustering is always visible, whereas bubbles and crashes may be absent for longer time periods or may evolve on a smaller scale. However, and this is one of the reasons why we should pay attention to these simulation runs, the panels show us that even after a very long time period without significant mispricing the next bubble may be just about to kick in. This warning may have a philosophical attitude but, given the common sense of policy makers, it seems important to us to note that even a stable period of, say, 10 years does not guarantee that the future will also be stable. A major bull or bear market period may just be days away without much forewarning.

[4] Also modest changes in the parameter setting do not destroy the model ability to mimic actual asset price dynamics reasonably well.

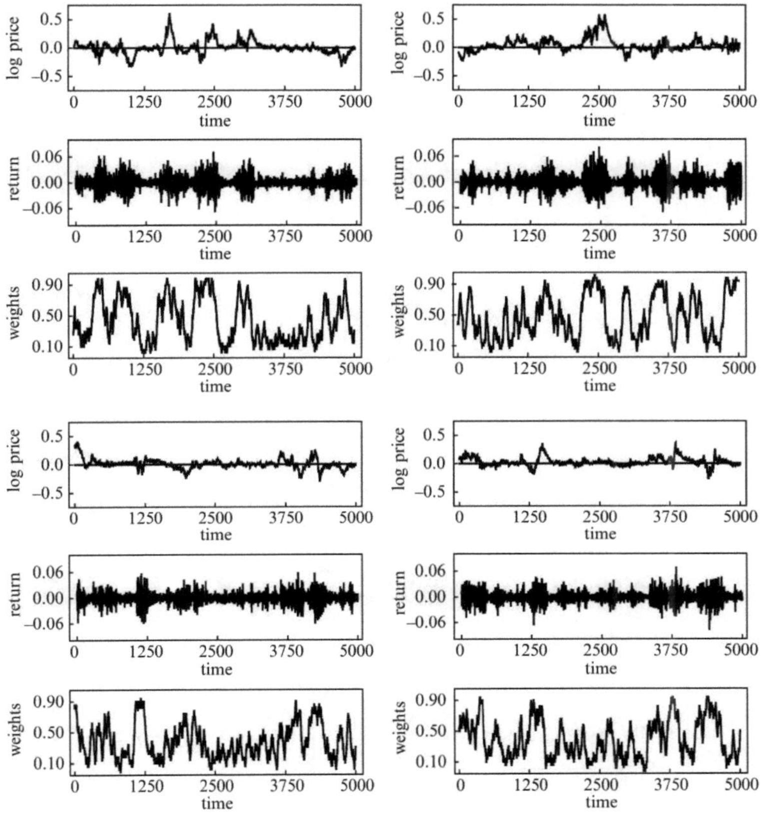

Fig. 4 Four repetitions of Fig. 1 using different random number streams

Figure 5 extends the analysis for $N = 100$ traders. In all simulation runs we see that the degree of volatility clustering is presumably exaggerated. The reason for this is that swings in opinion take too much time. Finally, Fig. 6 demonstrates that our model may generate realistic dynamics for a scenario with $N = 500$ agents and 20 direct interactions per trading time step.

4.2 Evolution of the Fundamental Value

So far, we have assumed that the fundamental value is constant. In the following, we explore the dynamics of our model when the fundamental value follows a random walk. To be precise, the log of the fundamental value now evolves as

$$F_t = F_{t-1} + \eta_t \tag{12}$$

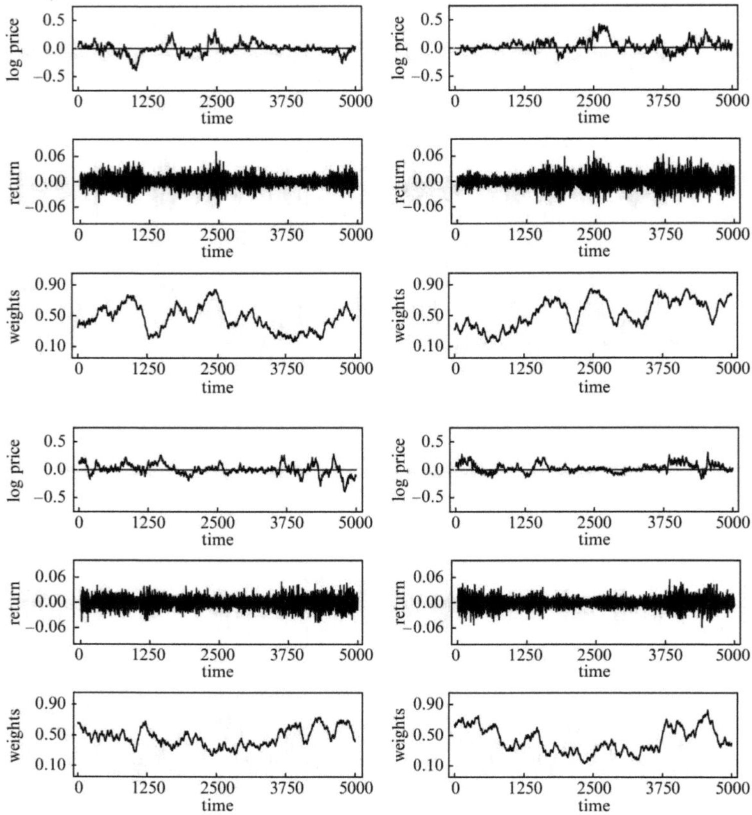

Fig. 5 Four repetitions of Fig. 2 using different random number streams

The fundamental shocks η are normally distributed with mean zero and constant standard deviation σ^{η}.

How do fundamental shocks change the dynamics of the model? It may be surprising to see that the statistical properties of our dynamics are more or less independent of (12), as is visible in the two scenarios depicted in Fig. 7. In the top three panels, the standard deviation of the fundamental shocks is $\sigma^{\eta} = 0.0065$ while in the bottom three panels it is $\sigma^{\eta} = 0.013$. Since the standard deviation of the returns of Fig. 3 (with a constant fundamental value) is about 0.013, we thus assume in the first (second) scenario that the fundamental shocks are half as volatile (as volatile) as these returns. Apart from that, the simulation design remains as it was in Fig. 3, i.e., there are 500 traders and 20 interactions per trading time step. We also rely on the same random number sequences. Of course, due to the evolution of the fundamental value also the course of the price is affected, yet its statistical properties are quite robust. Interestingly, the volatility is not amplified through the fundamental shocks. For both scenarios we obtain volatility estimates which are close to 0.013.

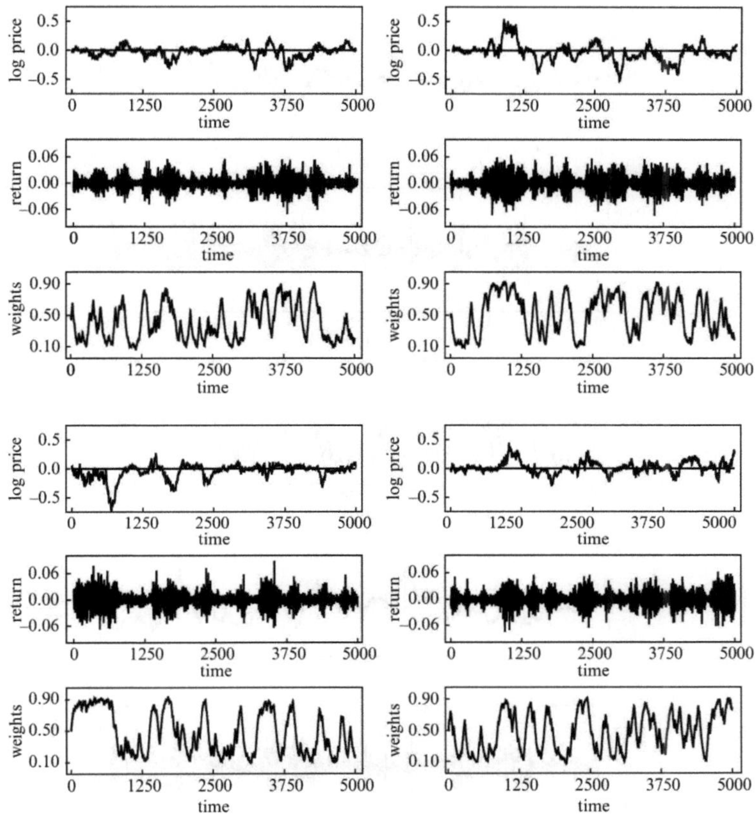

Fig. 6 Four repetitions of Fig. 3 using different random number streams

4.3 *Financial Market Crashes*

Finally, we investigate how the dynamics of our model reacts to extreme market crashes. In the first three panels of Fig. 8, we assume that the log of the fundamental value is 0 until time step 800 but then drops sharply to −0.5. Apparently, this corresponds to an extreme negative fundamental crash. Everything else is again as in Fig. 3. The top panel reveals that also the price crashes, yet not as quickly. It takes about 450 time steps (which corresponds to a time span of almost two years) before prices have reached their fundamental value again. Afterward, prices fluctuate in a similar way as in Fig. 3, expect that the price level is now shifted downwards. In the bottom set of panels, we introduce a price crash, i.e., we hold the fundamental value constant but set the price of period 800 equal to −0.5. After this (exogenous) price crash, the market remains for some time strongly undervalued but then prices

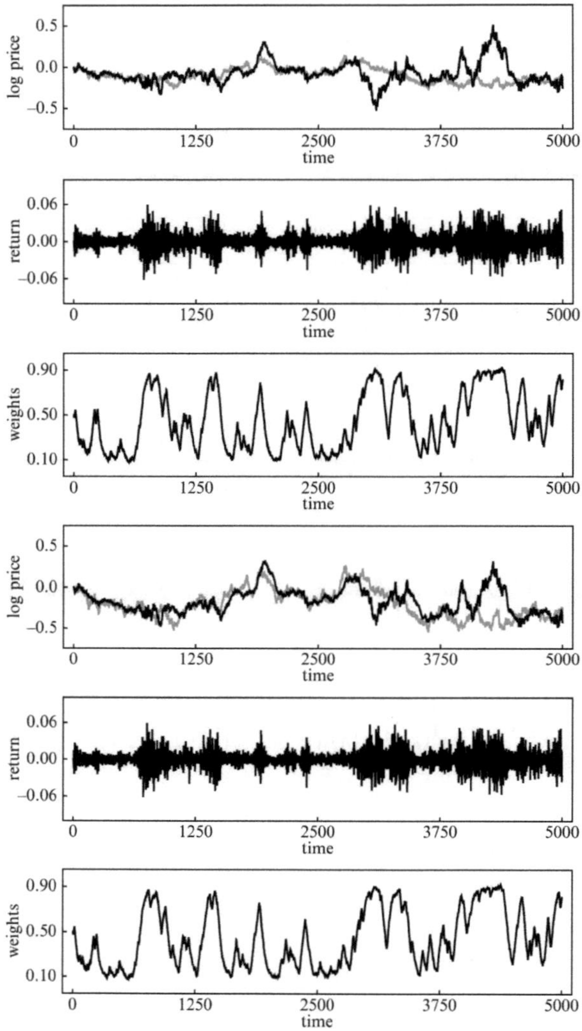

Fig. 7 The same simulation design as in Fig. 3, except that the fundamental value (given by the *gray line*) now follows a random walk with a standard deviation of 0.0065 (*top set of panels*) and 0.013 (*bottom set of panels*)

recover and the dynamics become again comparable to the dynamics represented in Fig. 3. To sum up, both crash scenarios have a temporary impact on the dynamics. In the long run, however, the model dynamics apparently digests such crashes and then behaves as usual.

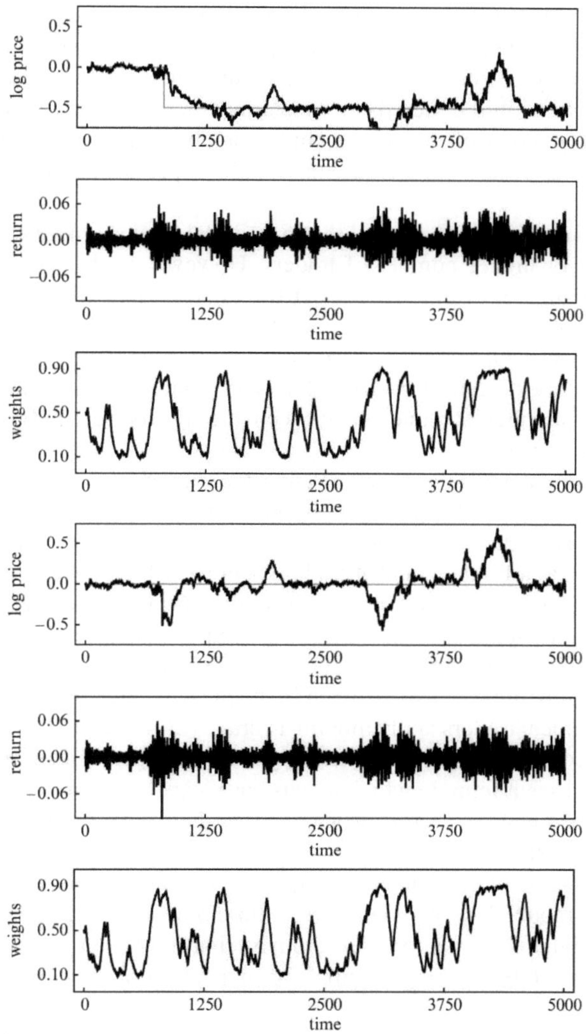

Fig. 8 The same simulation design as in Fig. 3, except that the log of the fundamental value (the log of the price) is set to -0.5 in the top set of panels (bottom set of panels) in period 800

5 Conclusions

The goal of this paper is to develop a simple agent-based financial market model with direct interactions between the market participants. When the traders meet each other within our model, they compare the past success of their trading rules. Should an agent discover that his opponent uses a more profitable strategy, it is quite likely that he/she will change his/her strategy. Simulations reveal that such a setting may incorporate a permanent evolutionary competition between the trading rules. For instance, there may be periods where fundamental analysis dominates the markets. Prices then fluctuate in the vicinity of their fundamental values. However, at some point in time a major shift towards technical analysis may set in and the market becomes unstable. Besides an increase in volatility, spectacular bubbles and crashes may materialize.

Moreover, we have demonstrated that our model may generate realistic dynamics for a lower or higher number of traders. However, in the latter case we have to increase the number of interactions per trading time step. Otherwise the relative importance of the trading rules is not flexible enough due to the assumed tandem recruitment process. Of course, one could also consider increasing the number of agents further, say, to 5,000 traders. Realistic dynamics may still be recovered as long as the number of contacts between the agents per trading time step is appropriately adjusted.

One interesting extension of the current setup may be to consider that (also) the probability that an agent changes his opinion independently from social interactions is state dependent. One could, for instance, assume that the probability to switch from a technical to a fundamental attitude is relatively high if fundamental analysis outperforms technical analysis. In this sense, the agents would then (also) display some kind of individual economic reasoning behavior. Another worthwhile investigation may be to consider different technical trading strategies instead of the simple trend-continuation rule we have assumed in the present paper. How do the dynamics look like if technical traders apply moving average rules with longer time windows? Moreover, we consider only random meetings between agents in our model. It would be interesting to see a setup in which agents have a social network. Interactions and the resulting dynamics may, for instance, be studied for a simple lattice or more complex network structures.

Finally, we would like to point out that, with a bit of experience, it is quite simple to program our model. It should therefore be possible, even for interested laymen, to reproduce the dynamics of our model. From a scientific point of view, replication of results is important. Everything required for such an exercise is given in our paper.

References

Alfarano, S., Lux, T., & Wagner, F. (2005) Estimation of agent-based models: the case of an asymmetric herding model. *Computational Economics, 26*, 19–49.

Boswijk, P., Hommes, C., & Manzan, S. (2007) Behavioral heterogeneity in stock prices. *Journal of Economic Dynamics and Control, 31*, 1938–1970.

Brock, W. & Hommes, C. (1997) A rational route to randomness. *Econometrica, 65*, 1059–1095.

Brock, W. & Hommes, C. (1998) Heterogeneous beliefs and routes to chaos in a simple asset pricing model. *Journal of Economic Dynamics and Control, 22*, 1235–1274.

Chen, S.-H., Chang, C.-L., & Du, Y.-R. Agent-based economic models and econometrics. *Knowledge Engineering Review*, in press.

Chiarella, C. (1992) The dynamics of speculative behavior. *Annals of Operations Research, 37*, 101–123.

Chiarella, C., Dieci, R., & Gardini, L. (2002) Speculative behaviour and complex asset price dynamics: a global analysis. *Journal of Economic Behavior and Organization, 49*, 173–197.

Cont, R. (2001) Empirical properties of asset returns: stylized facts and statistical issues. *Quantitative Finance, 1*, 223–236.

Day, R. & Huang, W. (1990) Bulls, bears and market sheep. *Journal of Economic Behavior and Organization, 14*, 299–329.

De Grauwe, P. & Grimaldi, M. (2006) Heterogeneity of agents, transactions costs and the exchange rate. *Journal of Economic Dynamics and Control, 29*, 691–719.

De Grauwe, P., Dewachter, H., & Embrechts, M. (1993) *Exchange rate theory chaotic models of foreign exchange markets*. Blackwell: Oxford.

Farmer, D. & Joshi, S. (2002) The price dynamics of common trading strategies. *Journal of Economic Behavior and Organization, 49*, 149–171.

Frankel, J. & Froot, K. (1986) Understanding the U.S. dollar in the eighties: the expectations of chartists and fundamentalists. *Economic Record, 62*, 24–38.

Graham, B. & Dodd, D. (1951) *Security analysis*. New York: McGraw-Hill.

Hommes, C. (2006) Heterogeneous agent models in economics and finance. In L. Tesfatsion & K. Judd (Eds.) *Handbook of computational economics, Volume 2: agent-based computational economics*. Amsterdam: North-Holland.

Kirman, A. (1991) Epidemics of opinion and speculative bubbles in financial markets. In M. Taylor (Ed.) *Money and financial markets*. Blackwell: Oxford.

Kirman, A. (1993) Ants, rationality, and recruitment. *Quarterly Journal of Economics 108*, 137–156.

LeBaron, B. (2006) Agent-based computational finance. In L. Tesfatsion & K. Judd (Eds.) *Handbook of computational economics, Volume 2: agent-based computational economics*. Amsterdam: North-Holland.

Li, H. & Rosser, B. (2001) Emergent volatility in asset markets with heterogeneous agents. *Discrete Dynamics in Nature and Society, 6*, 171–180.

Li, H. & Rosser, B. (2004) Market dynamics and stock price volatility. *European Physical Journal B, 39*, 409–413.

Lux, T. (1995) Herd behavior, bubbles and crashes. *Economic Journal, 105*, 881–896.

Lux, T. (1998) The socio-economic dynamics of speculative markets: interacting agents, chaos, and the fat tails of return distributions. *Journal of Economic Behavior and Organization, 33*, 143–165.

Lux, T. (2009) Applications of statistical physics in finance and economics. In B. Rosser Jr (Ed.), *Handbook of research on complexity*. Cheltenham: Edward Elgar.

Lux, T. Financial power laws: empirical evidence, models and mechanisms. In C. Cioffi-Revilla (Ed.), *Power laws in the social sciences: discovering complexity and non-equilibrium dynamics in the social universe*, in press.

Lux, T. & Ausloos, M. (2002) Market fluctuations I: scaling, multiscaling, and their possible origins. In A. Bunde, J. Kropp, & H. Schellnhuber (Eds.) *Science of disaster: climate disruptions, heart attacks, and market crashes*. Berlin: Springer.

Lux, T. & Marchesi, M. (1999) Scaling and criticality in a stochastic multi-agent model of a financial market. *Nature, 397*, 498–500.

Lux, T. & Marchesi, M. (2000) Volatility clustering in financial markets: a micro-simulation of interacting agents. *International Journal of Theoretical and Applied Finance, 3*, 675–702.

Mantegna, R. & Stanley, E. (2000) *An introduction to econophysics*. Cambridge: Cambridge University Press.

Manzan, S. & Westerhoff, F. (2007) Heterogeneous expectations, exchange rate dynamics and predictability. *Journal of Economic Behavior and Organization, 64*, 111–128.

Menkhoff, L. (1997) Examining the use of technical currency analysis. *International Journal of Finance and Economics, 2*, 307–318.

Menkhoff, L. & Taylor, M. (2007) The obstinate passion of foreign exchange professionals: technical analysis. *Journal of Economic Literature, 45*, 936–972.

Murphy, J. (1999) *Technical analysis of financial markets*. New York: New York Institute of Finance.

Rosser, B., Ahmed, E., & Hartmann, G. (2003) Volatility via social flaring. *Journal of Economic Behavior and Organization, 50*, 77–87.

Taylor, M. & Allen, H. (1992) The use of technical analysis in the foreign exchange market. *Journal of International Money and Finance, 11*, 304–314.

Westerhoff, F. (2008) The use of agent-based financial market models to test the effectiveness of regulatory policies. *Jahrb Nat Statist, 228*, 195–227.

Westerhoff, F. (2009) Exchange rate dynamics: a nonlinear survey. In B. Rosser Jr (Ed.) *Handbook of research on complexity*. Cheltenham: Edward Elgar.

Westerhoff, F. & Dieci, R. (2006) The effectiveness of Keynes-Tobin transaction taxes when heterogeneous agents can trade in different markets: a behavioral finance approach. *Journal of Economic Dynamics and Control, 30*, 293–322.

Westerhoff, F. & Reitz, S. (2003) Nonlinearities and cyclical behavior: the role of chartists and fundamentalists. *Studies in Nonlinear Dynamics and Econometrics, 7*(4), 3.

Winker, P., Gilli, M., & Jeleskovic, V. (2007) An objective function for simulation based inference on exchange rate data. *Journal of Economic Interaction and Coordination, 2*, 125–145.

Global Bifurcations in a Three-Dimensional Financial Model of *Bull and Bear* Interactions

Fabio Tramontana, Laura Gardini, Roberto Dieci, and Frank Westerhoff

1 Introduction

In a previous paper Tramontana et al. (2009), we developed a three-dimensional discrete-time dynamic model in which two stock markets of two countries, say H(ome) and A(broad), are linked via and with the foreign exchange market. The latter is modelled in the sense of Day and Huang (1990), i.e. it is characterized by a nonlinear interplay between technical traders (or chartists) and fundamental traders (or fundamentalists). In the absence of connections, the foreign exchange market is driven by the iteration of a one-dimensional cubic map, which has the potential to produce a regime of alternating and unpredictable bubbles and crashes for sufficiently large values of a key parameter, which captures the speculative behavior of chartists. Such a dynamic feature, first observed and explained by Day and Huang (1990) in their stylized model of financial market dynamics, can be understood with the help of bifurcation analysis: an initial situation of bi-stability (two coexisting, attracting *non-fundamental* steady states around an unstable *fundamental* equilibrium) evolves into coexistence of cycles or chaotic intervals within two disjoint *bull* and *bear regions*, which eventually merge via a homoclinic bifurcation. By introducing connections between markets (i.e. by allowing stock market traders to be active abroad), the endogenous fluctuations originating in one of the markets spread throughout the whole three-dimensional system. It therefore becomes interesting to investigate how the coupling of the markets affects the *bull and bear* dynamics of the model. With regard to this, in Tramontana et al. (2009) we already performed a thorough analytical and numerical study of two simplified lower-dimensional cases, where connections are either totally absent (each market evolves according to an independent one-dimensional map) or occur in one direction (a two-dimensional system evolves independently of the third dynamic equation). Also a short analysis of the stability of the equilibria of the three-dimensional model was there started, arguing that the global (homoclinic) bifurcations may still be a characteristic of the

F. Tramontana (✉)
Università di Ancona, Dipartimento di Economia Piazza Martelli, 8 60121 Ancona, Italy,
e-mail: f.tramontana@univpm.it

G.I. Bischi et al. (eds.), *Nonlinear Dynamics in Economics, Finance and the Social Sciences*, DOI 10.1007/978-3-642-04023-8_18,

dynamics. This investigation is precisely the object of the present paper. We shall analyze the dynamic behavior of the complete three-dimensional model, following the approach adopted in Tramontana et al. (2009), based mainly on the numerical and graphical detection of the relevant global bifurcations. Although analytical conditions for such global bifurcation, mainly homoclinic bifurcations, are difficult to be formalized, their existence and occurrence can be numerically detected. As it is standard in the qualitative study of dynamic behaviors, the transverse crossing between stable and unstable sets of unstable cycles, leading to homoclinic trajectories, give numerical tools which may be considered as proofs in a given numerical example.

The structure of the paper is as follows. In Sect. 2 we briefly describe the three-dimensional model of interacting stock and foreign exchange markets. The main results regarding the lower-dimensional subcases explored in Tramontana et al. (2009) are summarized in Sect. 3. Section 4 deals with the dynamics of the complete three-dimensional model by discussing, in particular, the steady state properties and the existence of multiple equilibria (Sect. 4.1), the homoclinic bifurcations of the *non-fundamental* steady states (Sect. 4.2) and of the *fundamental* equilibrium (Sect. 4.3), and the so-called *final bifurcation* (Sect. 4.4). Section 5 concludes the paper.

2 The Dynamic Model

This model describes the joint evolution of two stock markets (denoted as H and A), denominated in different currencies, and the related foreign exchange market. While the two stock prices (P_t^H and P_t^A, respectively) adjust over time depending on the excess demand for stock generated by national and foreign fundamental traders, the exchange rate[1] (S_t) depends on the excess demand of currency H. The latter consists of *(i)* demand for currency by heterogeneous speculators (technical and fundamental traders) who explicitly focus on the foreign exchange market and *(ii)* demand for currency by stock market traders who invest abroad, who obviously buy/sell foreign currency to conduct stock market transactions. In the following, we denote as F^H, F^A and F^S the fundamental values of the two stock prices and the exchange rate, respectively. Assuming, for the sake of simplicity, a linear price impact function, prices in the three markets jointly evolve according to the following laws of motion:

$$P_{t+1}^H = P_t^H + a^H(D_{F,t}^{HH} + D_{F,t}^{HA}), \tag{1}$$

$$P_{t+1}^A = P_t^A + a^A(D_{F,t}^{AA} + D_{F,t}^{AH}), \tag{2}$$

[1] Here we define the exchange rate as the price, expressed in currency A, of one unit of currency H.

$$S_{t+1} = S_t + d(P_t^H D_{F,t}^{HA} - \frac{P_t^A}{S_t} D_{F,t}^{AH} + D_{C,t}^S + D_{F,t}^S), \tag{3}$$

where a^H, a^A and d are positive parameters, and where the demand terms appearing on the right-hand sides of the above equations have the following definitions and meaning:

- $D_{F,t}^{HH} = b^H(F^H - P_t^H)$, $b^H > 0$, is the demand[2] for stock H by the fundamental traders (or fundamentalists) from country H.
- $D_{F,t}^{HA} = c^H[(F^H - P_t^H) + \gamma^H(F^S - S_t)]$, $c^H = 0$, $\gamma^H > 0$, is the demand for stock H by the fundamental traders from country A.
- $D_{F,t}^{AA} = b^A(F^A - P_t^A)$, $b^A > 0$, is the demand for stock A by the fundamental traders from country A.
- $D_{F,t}^{AH} = c^A[(F^A - P_t^A) + \gamma^A(\frac{1}{F^S} - \frac{1}{S_t})]$, $c^A = 0$, $\gamma^A > 0$, is the demand for stock A by the fundamental traders from country H.
- $D_{C,t}^S = e(S_t - F^S)$, $e > 0$, and $D_{F,t}^S = f(F^S - S_t)^3$, $f > 0$, are the demands of currency H by chartists and fundamentalists, respectively, who enter speculative positions in the foreign exchange market. In particular, chartist demand coefficient, e, turns out to be an important bifurcation parameter in our analysis.

The following additional comments about agents' demands are useful:

(1) Fundamentalists seek to profit from mean reversion, so that they submit buying orders (positive demand) when the market is undervalued (the price is below fundamental) and selling orders (negative demand) when the market is overvalued.

(2) In addition, foreign fundamentalists may also benefit from exchange rate movements, and therefore their demand function also includes a term that is dependent on the observed mispricing in the foreign exchange market; in particular, traders from H to A take into account the reciprocal values of the exchange rate and its fundamental.

(3) In the foreign exchange market, chartists believe in the persistence of *bull* markets or *bear* markets and therefore optimistically buy (pessimistically sell) currency H as long as the exchange rate is high (low). Fundamentalists seek to exploit misalignments using a nonlinear trading rule. As long as the exchange rate is close to its fundamental value, fundamentalists are relatively cautious, but the greater the mispricing, the more aggressive they become.

(4) Finally, $P_t^H D_{F,t}^{HA}$ represents the demand for currency H generated by stock market orders from A to H, and symmetrically $P_t^A D_{F,t}^{AH}$ is the demand for currency A generated by stock market orders from H to A: the latter is

[2] The demand for stock is given in real units.

converted into an amount of currency H of opposite sign, via the reciprocal exchange rate $\frac{1}{S_t}$.

By specifying all of the demand terms in (1)–(3), we obtain a three-dimensional dynamical system with the following structure

$$\begin{cases} P_{t+1}^H = G^H(P_t^H, S_t), \\ S_{t+1} = G^S(P_t^H, S_t, P_t^A), \\ P_{t+1}^A = G^A(S_t, P_t^A). \end{cases} \tag{4}$$

In particular, for $c^H = c^A = 0$ the structure of the system (4) simplifies into three independent dynamic equations, $P_{t+1}^H = G^H(P_t^H)$, $S_{t+1} = G^S(S_t)$, $P_{t+1}^A = G^A(P_t^A)$, of which that for exchange rate S is nonlinear (of cubic type), whereas the two stock prices P^H and P^A evolve linearly. More interestingly, for $c^A = 0$ but $c^H > 0$ the system is of the type

$$\begin{cases} P_{t+1}^H = G^H(P_t^H, S_t), \\ S_{t+1} = G^S(P_t^H, S_t), \\ P_{t+1}^A = G^A(P_t^A), \end{cases} \tag{5}$$

that is to say, P^A decouples from the system, whereas P^H and S co-evolve in a two-dimensional nonlinear dynamical system. Both such lower-dimensional subcases were analyzed in detail in Tramontana et al. (2009). The main findings of such an analysis are summarized in the following section.

3 Summaries of the 1D and 2D Models

In this section, we recall the main results regarding the simplified, lower-dimensional subcases analyzed in Tramontana et al. (2009).

3.1 One-Dimensional Case

In the absence of interactions, $c^H = c^A = 0$, each market evolves as a one-dimensional dynamical system. Stock markets are represented by simple linear equations and in each of them the unique fundamental steady state is globally stable, at least for reasonable values of the price and demand adjustment parameters. The law of motion of the foreign exchange market is nonlinear, determined by iteration of a cubic map, with three fixed points: namely, two *non-fundamental* steady states, say P_1 and P_2, relative to each of the two unimodal branches of the cubic map, surrounding an unstable *fundamental* steady state (O). The map is symmetric with respect to the fundamental value, which is why the bifurcations involving P_1 and P_2 are synchronized. The bifurcation analysis with respect to parameter e highlighted the route to chaotic *bull and bear* dynamics of the model. The (synchronized) period-doubling bifurcations of P_1 and P_2, followed by the usual cascade of flip bifurcations and the homoclinic bifurcations of the two steady states, lead to the coexistence of two symmetric intervals (around P_1 and P_2, respectively), each characterized by chaotic dynamics (in the sense of chaos of full measure on an interval). Due to the noninvertibility of the map, within this range of values of parameter e the basins of the two coexisting attractors have a disconnected structure, each being made up of an infinite sequence of intervals which alternate on the real line with that of the competing attractor. For higher values of parameter e, the two attractors and their basins merge together via a homoclinic bifurcation of the fundamental steady state O. After this point, the exchange rate dynamics, previously confined to below or above the fundamental value, depending on the initial condition, wanders within a unique chaotic interval around the fundamental steady state, alternating *bull and bear* market episodes in an unpredictable manner. A *final bifurcation* then occurs when the unique attractor touches the border between its basin and the *basin of infinity*, $\mathcal{B}_\infty$, after which the generic trajectory is divergent.

A crucial tool for the bifurcation analysis, strictly associated with the noninvertibility of the map, is represented by the critical points (local extrema and their iterates), which are at the boundary of chaotic intervals, and their contacts with the unstable steady states.

3.2 Two-Dimensional Case

By introducing a partial connection between stock markets A and H (namely, by allowing investors from country A to trade in country H), the latter turns out to coevolve with the foreign exchange market (whereas market A is still decoupled from the system). As a result, we have a system of two coupled equations, one linear and one nonlinear. In particular, in the nonlinear equation for the exchange rate we also have a feedback from stock market H, which makes the dynamics even more intricate. One difference with the one-dimensional case is that now a unique steady state exists for small values of e. Another difference is that the symmetry property is lost. Apart from this, in the two-dimensional case we still observe the same multiple steady state structure (when e is large enough) and a similar sequence of local and global bifurcations. More precisely, we highlighted the homoclinic bifurcations that involve the saddle equilibria P_1 and P_2 first (albeit now in an asynchronous manner) and then the fundamental equilibrium O. Due to this sequence of bifurcations (also called *interior* and *exterior crises* in Grebogi et al. (1983), the system has a transition across different dynamic scenarios: from coexisting attracting *bull and bear* chaotic regions, to the disappearance of one of them, to the merging of the two regions into a unique wider chaotic area. The resulting dynamic outcome is a coupled *bull and bear* market behavior of stock price H and the exchange rate, which may switch across different regions of the two-dimensional phase space with apparently random behavior. In all cases, the bifurcation mechanisms are basically due to contacts[3] between invariant sets – such as stable manifolds of saddles – and the boundary of chaotic attractors, the latter being made up of portions of *critical curves* of the noninvertible two-dimensional map (see Mira et al., 1996). Finally, also the bifurcation leading to the disappearance of the unique chaotic attractor is similar to that of the one-dimensional case. Such a two-dimensional analysis has been largely carried out with the help of numerical simulation and graphical visualization. In particular, the tool of the *critical curves* has suggested how the basins of attraction may acquire a disconnected structure.

4 Analysis of the 3D Model

In this section we deal with the complete three-dimensional model, mainly with the help of numerical simulation. Our analysis will show that the dynamic phenomena highlighted in Tramontana et al. (2009) also persist in the full model, and can be detected and understood by extending the approach and techniques used in the lower-dimensional cases to a three-dimensional setup. In particular, we are also able in the full model to detect and explain the sequence of local and global bifurcations that determine the transition between different dynamic regimes: namely, from a unique attracting fundamental equilibrium to coexistence of attracting non-fundamental equilibria, to more complex coexisting attractors, up to the homoclinic bifurcations which bring about a regime of *bull and bear* market fluctuations, first established by Day and Huang (1990) in a one-dimensional setup, characterized by apparently random switches of prices across different regions of the phase space.

In the full model, stock market traders from countries A and H are allowed to trade in both markets, i.e. $c^H > 0$ and $c^A > 0$. In this case, the two stock prices and the exchange rate are all interdependent, and the model has the complete structure (4). The system (4), expressed in deviations[4] from fundamental values, $x = (P^H - F^H)$, $y = (S - F^S)$ and $z = (P^A - F^A)$, is represented by a map $T : \mathbb{R}^3 \to \mathbb{R}^3$ that takes the following form:

[3] Following Mira et al. (1996) we call *contact bifurcation* any contact between two closed invariant sets of different kinds. A contact bifurcation may have several different dynamic effects, depending on the nature of the invariant sets.

[4] Although we work with deviations, in all the following numerical experiments we have checked that original prices never become negative.

$$T: \begin{cases} x_{t+1} = x_t - a^H[(b^H + c^H)x_t + c^H\gamma^H y_t], \\ y_{t+1} = y_t - d\Big[c^H(x_t + F^H)(x_t + \gamma^H y_t) \\ \qquad +c^A \dfrac{z_t + F^A}{y_t + F^S}\left(\gamma^A \dfrac{y_t}{F^S(y_t + F^S)} - z_t\right) - ey_t + fy_t^3\Big], \\ z_{t+1} = z_t - a^A\Big[(b^A + c^A)z_t - c^A\gamma^A \dfrac{y_t}{F^S(y_t + F^S)}\Big]. \end{cases} \tag{6}$$

The model is not tractable analytically. Apart from the *fundamental* fixed point, say $O = (0, 0, 0)$, whose existence can be immediately checked, we cannot solve explicitly for the coordinates of further possible non-fundamental equilibria, nor can we obtain easily interpretable analytical conditions for their existence. A brief discussion of the steady states is provided in the following subsection.

4.1 Fixed Points and Multistability

By imposing the fixed point condition to (6), we obtain the following system of equations

$$\left(b^H + c^H\right)x + c^H\gamma^H y = 0, \tag{7}$$

$$c^H\left(x + F^H\right)\left(x + \gamma^H\right) + c^A \frac{z + F^A}{y + F^S}\left(\gamma^A \frac{y}{F^S(y + F^S)} - z\right) - ey + fy^3 = 0, \tag{8}$$

$$\left(b^A + c^A\right)z - c^A\gamma^A \frac{y}{F^S(y + F^S)} = 0. \tag{9}$$

We observe from (7) and (9) that any steady state must belong to both the plane of equation:

$$y = -\frac{x}{q^H} \tag{10}$$

and the surface of equation

$$z = q^A \frac{y}{(y + F^S)}, \tag{11}$$

where

$$q^H := \frac{c^H\gamma^H}{b^H + c^H}; \quad q^A := \frac{c^A\gamma^A}{(b^A + c^A)F^S}.$$

This implies that when the steady state exchange rate is above the fundamental value ($y > 0$), steady state price A is then above the fundamental value ($z > 0$), whereas steady state price H is below the fundamental value ($x < 0$), and vice versa. From now on, we will label the region $y > 0, z > 0, x < 0$ as the *bull region* and region $y < 0, z < 0, x > 0$ as the *bear region*. By substituting (10) and (11) into (8), we are able to express condition (8) in terms of the steady state (deviation of) price H only, as follows:

$$x\left[\frac{f}{(q^H)^3}x^2 + b^H x + \left(b^H F^H - \frac{e}{q^H}\right) + M(x)\right] = 0, \tag{12}$$

where

$$M(x) := b^A q^H q^A \frac{q^H F^S F^A - x(q^A + F^A)}{(q^H F^S - x)^3}.$$

Therefore, besides the fundamental solution $x = 0$, further possible solutions are the roots of the expression in square brackets in (12). Note that for $c^A = 0$, and therefore $q^A = 0$ and $M(x) = 0$, the x-coordinates of further possible steady states are the solutions of a quadratic equation, and their existence was discussed in detail in Tramontana et al. (2009).[5]

In contrast, if $c^A > 0$, it becomes impossible to solve (12) analytically. When c^A is small enough, we may expect a steady state structure qualitatively similar to that of the two-dimensional subcase $c^A = 0$, with two further steady states initially appearing simultaneously in the *bull region*, via saddle-node bifurcation.[6]

However, if c^A is large enough, as is the case of the following numerical experiments, as we shall see, it is difficult to detect the appearance of further equilibria and their initial location with respect to the fundamental. We remark that the analytical investigation of the local stability properties of fundamental fixed point $O = (0, 0, 0)$ is also a difficult task. The Jacobian matrix evaluated at O is given by

$$J(O) : \begin{bmatrix} 1 - a^H(b^H + c^H) & -a^H c^H \gamma^H & 0 \\ -dc^H F^H & 1 - d\left[c^H F^H \gamma^H + \dfrac{c^A F^A \gamma^A}{(F^S)^3} - e\right] & \dfrac{dc^A F^A}{F^S} \\ 0 & \dfrac{a^A c^A \gamma^A}{(F^S)^2} & 1 - a^A(b^A + c^A) \end{bmatrix} \tag{13}$$

and its eigenvalues (which are roots of a third-order polynomial) cannot be solved for explicitly. Nor can we write down tractable analytical conditions for the eigenvalues to be smaller than one in modulus. We shall now study the local and global bifurcations via numerical investigation, supported by our knowledge of the model behavior in the simplified, two-dimensional case. In fact, as we shall see, the analysis performed in Tramontana et al. (2009) provides important guidelines for understanding the dynamic phenomena occurring in this more complex three-dimensional model.

With the parameter setting used in following numerical simulations (as well as with other similar constellations of parameters) we do not observe the appearance of the non-fundamental equilibria via saddle-node bifurcation. Instead, a pitchfork

[5] Moreover, in this case, in which market A decouples from the system, (11) reduces to $z = 0$.

[6] This is also confirmed by numerical simulations.

bifurcation of the fundamental steady state seems to occur, leading to the appearance of two new stable equilibria, one in the *bull region* and one in the *bear region*, at the same parameter value at which the fundamental becomes unstable.[7] The situation resulting from the local bifurcation of the fundamental steady state is in any case qualitatively the same as for the two-dimensional subcase. That is, the phase space is shared amongst the basins of attraction of two non-fundamental equilibria, separated by the stable set of the (saddle) fundamental steady state. From now on, the bifurcation sequences involving the two coexisting equilibria (or, more generally, the two coexisting attractors) follow paths similar to those observed in the two-dimensional model, albeit involving stable and unstable manifolds in higher dimensions. In this paper we confirm and strengthen almost all of the results of the two-dimensional case, albeit via numerical simulations only. We shall describe various kinds of homoclinic bifurcations, following the same scheme of the study carried out in Tramontana et al. (2009).

Our base parameter selection is the following: $a^H = 0.41, b^H = 0.11, c^H = 0.83$, $F^H = 4.279$, $\gamma^H = 0.3$, $d = 0.35$, $f = 0.7$, $F^S = 6.07$ (which are the same parameters used in the simulations in Tramontana et al. (2009), enabling a direct comparison), $a^A = 0.43$, $b^A = 0.21$, $c^A = 0.9$, $\gamma^A = 0.36$ and $F^A = 1.1$. In order to sufficiently distinguish the model from the two-dimensional case studied in Tramontana et al. (2009) (where c^A is zero), we have chosen a value of c^A that is much further away from zero, and even higher than c^H. Bifurcations similar to those described below are observed with several other parameter constellations. The numerical analysis performed in the Appendix shows that O loses stability when one eigenvalue becomes equal to 1 at $e \simeq 0.125$. We argue that this corresponds to a pitchfork bifurcation, because we observe the simultaneous appearance of two further equilibria, which we denote as P_1 (in the *bear region*) and P_2 (in the *bull region*). Figure 1 shows the asymptotic dynamics in the three-dimensional phase space for increasing values of e. We can see that the fundamental fixed point is unstable and that two new stable fixed points exist. The stable fixed points are located on opposite sides with respect to plane $y = 0$ (i.e. $S = F^S$), as shown in Fig. 1a. Since only one eigenvalue of $J(O)$ becomes larger than 1, while two other eigenvalues are real and still smaller than one in absolute value, the fundamental equilibrium is a saddle with a one-dimensional unstable manifold (made up of two branches, connecting O with P_1 on one side and O with P_2 on the opposite side), while the stable set W^s_O of the origin is a two-dimensional manifold, which separates the two basins of attraction of the two coexisting fixed points. In other words, the frontier of the basins of attraction of P_1 and P_2, say $\mathcal{B}_1$ and $\mathcal{B}_2$, respectively, includes surface W^s_O.

Moreover, it is easy to see that divergent behavior is also possible, so that the basin of divergent trajectories, $\mathcal{B}_\infty$, also exists (and will be involved in the final bifurcation, as we shall see below). The structure of the basins after the appearance

[7] We remark that this is just numerical evidence, and we cannot exclude the existence of a sequence similar to that described for the 2D model (i.e. a Saddle-Node bifurcation immediately followed by a Transcritical), occurring in a very narrow range of parameter e.

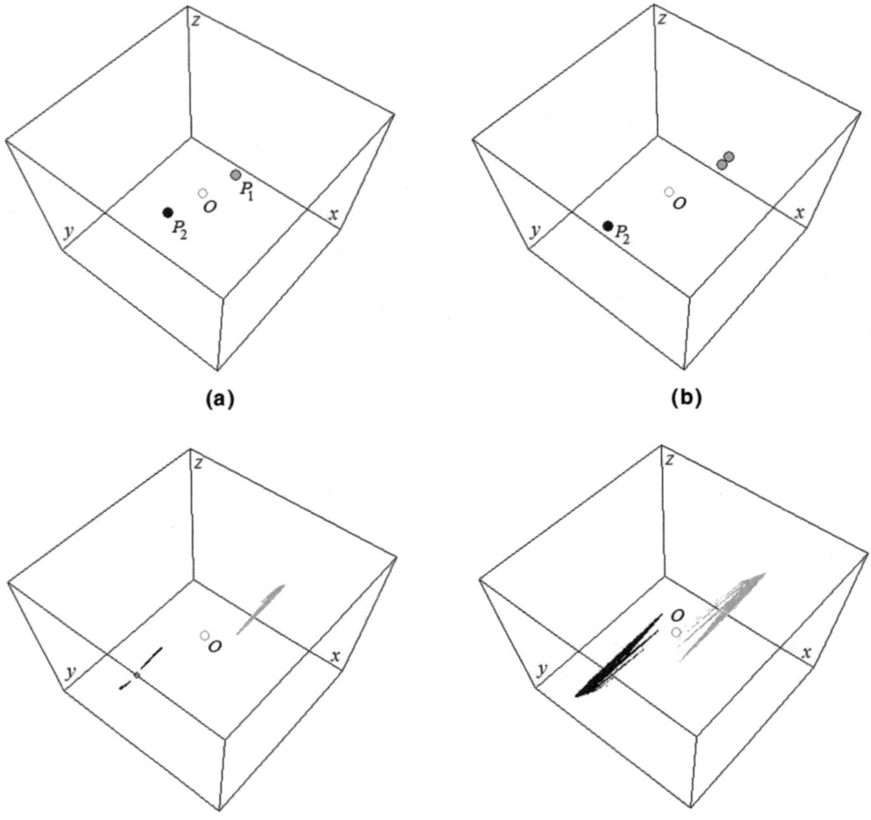

Fig. 1 Coexisting attractors for increasing values of parameter e and other parameters according to our base selection. In (**a**), $e = 0.89$, the attractors are two stable fixed points P_1 and P_2. In (**b**), $e = 2.43$, there is coexistence of the stable fixed point P_2 and a stable 2-cycle. In (**c**), $e = 3.576$, one chaotic attractor (*blue*, in the *bear region*) consists of a unique piece (after the homoclinic bifurcation of the fixed point P_1) while the other chaotic attractor (*red*, in the *bull region*) is made up of two disjoint pieces, on opposite sides with respect to the fixed point P_2. In (**d**), $e = 4.1841$, both have become one-piece attractors, and the *light grey* one approaches the stable set of the fundamental fixed point in the origin. The boxes are centered in O and the range of all axes is $[-3, +3]$

of the two new attractors is shown in Fig. 2, where a cross section of a plane through the fundamental fixed point $O = (0, 0, 0)$ is considered. The value of the parameter is $e = 0.89$, as in Fig. 1a, and two attracting fixed points coexist. In the cross section, the different grey tonalities belong to different basins of attractions. We denote in grey the basin of the fixed point P_1 (in the *bear region*). The basin of fixed point P_2 (in the *bull region*) is light grey, while points generating divergent trajectories, and thus belonging to the basin $\mathcal{B}_\infty$, are shown in dark grey.

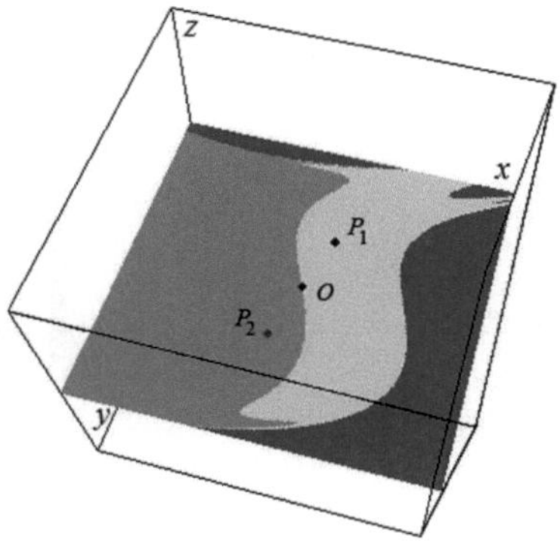

Fig. 2 Cross section along a plane through the fundamental fixed point $O = (0, 0, 0)$. The value of parameter is $e = 0.89$, as in Fig. 1a. The box is centered in O and the range of axes is $[-3, +3]$ for all state variables. In the cross section, different colors denote different basins of attractions. Basin $\mathcal{B}_1$ of the fixed point P_1 is in *grey*, basin $\mathcal{B}_2$ of the fixed point P_2 is in *light grey*, basin $\mathcal{B}_\infty$ is in *dark gray*

As already conducted in Tramontana et al. (2009) of our study, we analyze the sequence of bifurcations occurring when parameter e is increased. We first show a bifurcation diagram of the asymptotic behavior of state variable y as a function of parameter e. The diagram (Fig. 3) highlights how a sequence of bifurcations very similar to those observed in the two-dimensional case also occurs in the three-dimensional case and, as expected, in an asynchronous manner (because in the full model, as well as in the two-dimensional subcase studied in Tramontana et al., 2009, there is no symmetry with respect to the origin). In Fig. 3a, the initial condition is taken close to the fixed point P_1, while in Fig. 3b the starting point is close to the other fixed point P_2. The global bifurcations first involve the attractor associated with equilibrium P_1 (in blue) and then that associated with P_2 (in red).

4.2 Homoclinic Bifurcation of Equilibria P_1 and P_2

As noted above, after their appearance, both locally stable fixed points undergo a cascade of flip bifurcations (in an asynchronous manner), leading to chaos (see Figs. 3 and 1). In particular, in Fig. 1c we can see that the attracting set in the *bull region* (in red) is still made up of two disjoint pieces, located on opposite sides with respect to the unstable fixed point P_2, while the second attractor, located in the *bear region* (in blue), is already a one-piece chaotic attractor. Although we do not have the coordinates of the unstable fixed points P_1 and P_2, we can state that in this case (Fig. 1c) fixed point P_1 is already homoclinic, at least on one side, because it belongs to the invariant chaotic area (and it is probably located on its boundary, as it occurs in the 2D model), while the fixed point P_2 is not yet homoclinic (because it is not jet included in the chaotic area), although it will be involved later

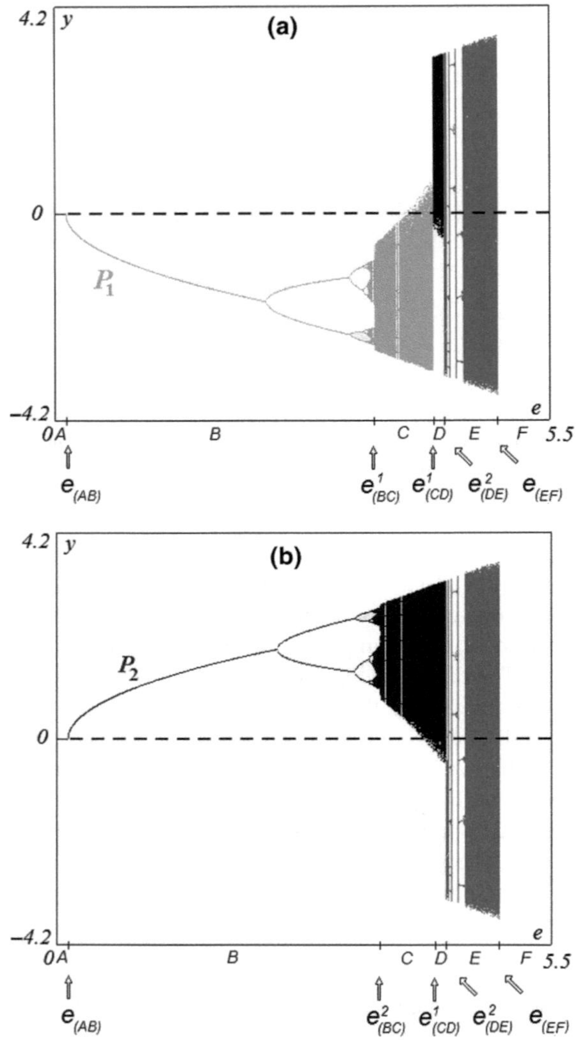

Fig. 3 Bifurcation diagrams of y vs. parameter e, ranging from 0 to 5.5. The range of e is subdivided into different intervals. In interval A the only attractor is the fundamental equilibrium O. Its pitchfork bifurcation occurs at $e \simeq 0.125$, after which two new stable equilibria appear. In (**a**) the initial condition is close to the fixed point P_1, in (**b**) it is close to P_2. In interval B we observe a complete *route to chaos* for each fixed point. The homoclinic bifurcation of P_1 occurs at $e^1_{(BC)} \simeq 3.56$, which results in the one-piece chaotic attractor in *light grey*. The homoclinic bifurcation of P_2 occurs at $e^1_{(BC)} \simeq 3.6$, leading to the one-piece chaotic attractor in *dark grey*. In interval C there is coexistence of two one-piece chaotic attractors. The upper bound of interval C corresponds to a homoclinic bifurcation of the origin. In (**a**) the *light grey* chaotic attractor disappears at the first homoclinic bifurcation of the origin, which occurs at $e^1_{(CD)} \simeq 4.185$, so that for any e in interval D ($e^1_{(CD)} < e < e^2_{(DE)}$) the unique attractor is the *dark grey* one. The second homoclinic bifurcation of the origin occurs at $e^2_{(DE)} \simeq 4.3$ and leads to an *explosion* of the chaotic attractor into a wider region (in *grey*). This unique chaotic attractor exists up to its final bifurcation at $e_f \simeq 5.03$

(i.e. for larger e) in a homoclinic bifurcation, causing the reunion of the two pieces of chaotic attractor around P_2. Figure 1d indeed shows the situation existing after both the first homoclinic bifurcations of equilibria P_1 and P_2 have occurred.

In the bifurcation diagram in Fig. 3 we plot the asymptotic behavior of the state variable y, as e varies in the range $[0, 5.5]$. In the interval of values of e denoted by A, the only attracting set is the fundamental equilibrium O. Its pitchfork bifurcation occurs at $e \simeq 0.125$, after which we have the appearance of two further stable equilibria. In (a) the initial condition is taken close to the fixed point P_1, while in (b) it is taken close to the fixed point P_2. The fixed points are stable up to their flip bifurcation, which occurs for P_1 first, and then for P_2. In the interval denoted by B we observe the typical *route to chaos* for each fixed point, and the parameter values $e^1_{(BC)}$ and $e^2_{(BC)}$ are the homoclinic bifurcation values of P_1 and P_2, respectively, at which the reunion of two pieces into one single chaotic attractor takes place. In the proposed example, we first observe this bifurcation in the *bear region*, at $e^1_{(BC)} \simeq 3.56$ (leading to the one-piece chaotic attractor in blue), and then in the *bull region*, at $e^2_{(BC)} \simeq 3.6$ (leading to the one-piece chaotic attractor in red).

The coexistence of two disjoint attractors in the *bull* and *bear regions* is coupled with an increase of complexity in the structure of the related basins of attraction $\mathcal{B}_1$ and $\mathcal{B}_2$. An example is shown in Fig. 4: in (a) we show the two disjoint attractors and in (b) a cross section through the origin shows the basins in different colors. $\mathcal{B}_1$, in pink, is the locus of initial points converging to the chaotic attractor in blue, and $\mathcal{B}_2$, in orange, is the locus of points converging to the chaotic attractor in red, while the gray points belong to basin $\mathcal{B}_\infty$. Note that the basins are now disconnected: within the region that approximately coincides with basin $\mathcal{B}_1$ of Fig. 2 there are now also points belonging to basin $\mathcal{B}_2$ and to $\mathcal{B}_\infty$; at the same time, points belonging to basin $\mathcal{B}_1$ and to $\mathcal{B}_\infty$ are now located in the region previously belonging to basin $\mathcal{B}_2$ in Fig. 2. This phenomenon is again due to contact bifurcations of the basins of attraction with *critical sets* (*critical surfaces*, in this three-dimensional case). That is, denoting by $J(x, y, z)$ the Jacobian matrix of the map (6) and by SC_{-1} the locus

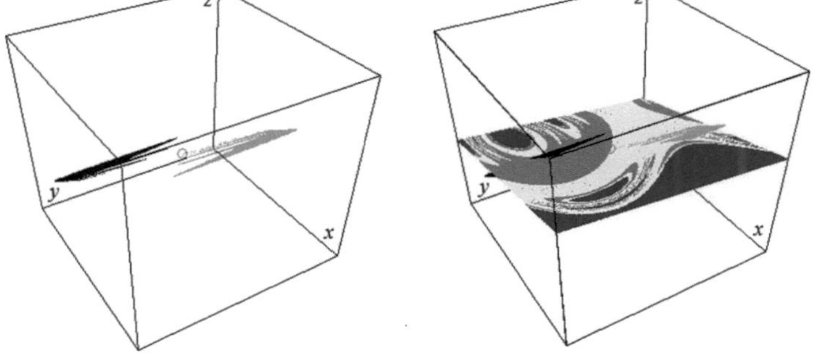

Fig. 4 Coexisting attractors at $e = 4.1841$. The boxes are centered in O and the range of axes is $[-3, +3]$. In (**b**) a plane through the origin O is shown, along which different colors denote different basins of attraction, as in Fig. 2

of points defined by the equation $det\ J(x, y, z) = 0$, this set plays the same role of the *critical points* x^m_{-1} and x^M_{-1} in the one-dimensional map corresponding to the subcase $c^H = c^A = 0$, and to the *critical curve* LC_{-1} in the two-dimensional map corresponding to $c^A = 0$, both analyzed in Tramontana et al. (2009). The image of SC_{-1} under map T gives a surface, $SC := T(SC_{-1})$, which is responsible for the contact bifurcations of the basins of attraction. In the 3D phase space this critical surface SC separates regions of points with a different number of rank-1 preimages. When basin $\mathcal{B}_1$ (or basin $\mathcal{B}_\infty$) touches the critical surface SC and then crosses it, a portion of the basin, say H', enters a region of the phase space whose points have a higher number of preimages, thus leading to the appearance of new portions of the basin. Such portions consist of the region (volume) $T^{-1}(H')$, located around the critical surface SC_{-1}, and of its further preimages.

4.3 Homoclinic Bifurcation of the Fundamental Equilibrium O

The coexistence of two attractors located in two disjoint *bull* and *bear regions* ends at the first homoclinic bifurcation of the origin. The section of the basins of attractions in Fig. 4b shows that the chaotic attractor in the bear region, colored blue, is very close to the boundary of its basin $\mathcal{B}_1$. Moreover, we know that the frontier of the two basins $\mathcal{B}_1$ and $\mathcal{B}_2$ includes the two-dimensional stable set W^s_O of the fundamental fixed point O, which is now a set with a complex structure. Thus, from the closeness of the chaotic area to the origin we can argue that in the parameter situation shown in Fig. 4b we are already very close to this first homoclinic bifurcation of the origin (while the second one is due to the other chaotic attractor, which is still far from the origin).

At the fixed point itself the stable set W^s_O is a surface tangent to the plane generated by the eigenvectors associated with the two stable eigenvalues of the Jacobian matrix $J(O)$. A portion of this plane (tangent at the origin to the surface W^s_O) is shown in Fig. 5, at $e = 4.1841$ (same parameter value as in Fig. 4). At this value the

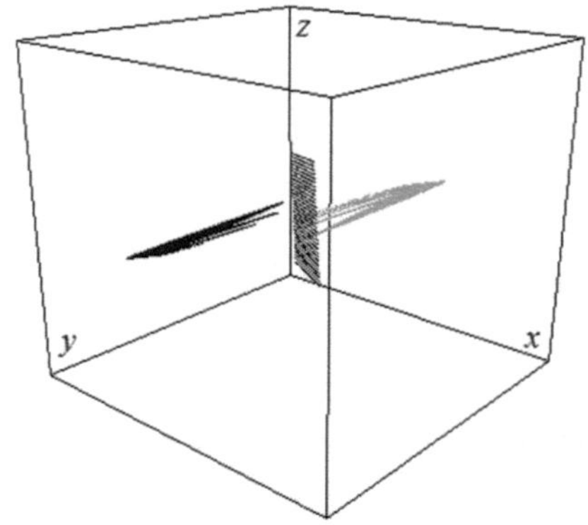

Fig. 5 Coexisting attractors for $e = 4.1841$ (as in Fig. 4) and a portion of the plane through the origin O, tangent to the stable set W^s_O

eigenvalues of the Jacobian matrix $J(O)$ are given by $\xi_1 = 2.1725$, $\xi_2 = -0.5341$ and $\xi_3 = 0.5216$. The eigenvectors associated with eigenvalues ξ_2 and ξ_3 (less than 1 in modulus), say v_2 and v_3, respectively, are given by

$$v_2 = \begin{pmatrix} -0.6066 \\ -0.769 \\ -0.2017 \end{pmatrix}, \quad v_3 = \begin{pmatrix} -0.2582 \\ -0.2836 \\ -0.9235 \end{pmatrix}$$

and the tangent plane is generated by these two vectors. We can see that in Fig. 5 the tangent plane is already crossed by the chaotic attractor in blue. This means that we are not far from the parameter value at which a contact with surface W_O^s occurs. Since one branch of the unstable set W_O^u of the origin tends to the chaotic attractor, the contact of the chaotic attractor with the stable set of the origin is also a contact between the unstable set W_O^u and the stable set W_O^s, leading to the first homoclinic bifurcation of the fixed point O.

After the contact, the stable and unstable sets have infinitely many transverse intersections. However, the chaotic set associated with the origin is not observable in the asymptotic dynamics. In fact, as a result of this bifurcation we have the disappearance of the chaotic attractor in the bear region. That is, the previous light grey chaotic attractor has now turned into a chaotic repeller, which also includes homoclinic trajectories on one side of the origin. We recall that a contact bifurcation causing the disappearance of a chaotic attractor always lives a chaotic repeller in its place in the phase space. This chaotic repeller is formed by all the unstable cycles previously existing in the chaotic set, and the related stable sets, or insets (see Mira et al., 1996 for further details). Thus although in Fig. 6a we observe only one attractor on one side of the stable set of the origin, we know that on the

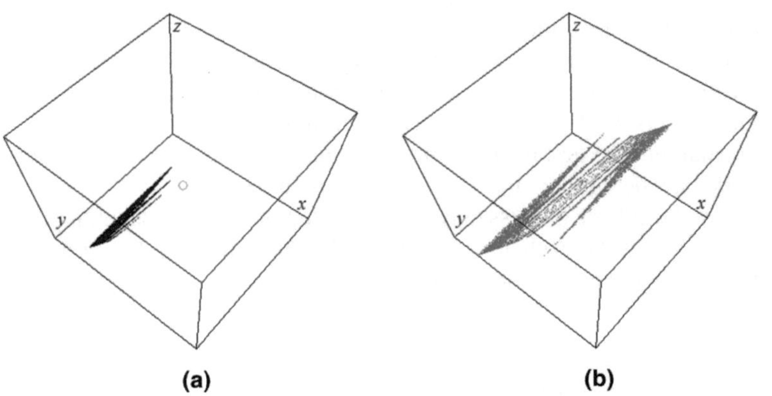

Fig. 6 (**a**) Unique chaotic set in the *bull region*, at $e = 4.208$, after the first homoclinic bifurcation of the fundamental equilibrium O. (**b**) Unique chaotic set covering both the *bull* and *bear regions*, at $e = 4.761$, after the second homoclinic bifurcation of the fundamental equilibrium O. Boxes are centered in O and the range of axes is $[-3, +3]$

other side a chaotic repeller exists, and its existence may be put in evidence when a contact between the existing chaotic attractor and the origin occurs (at another homoclinic bifurcation of the origin). We may *observe* the chaotic repeller via the long chaotic transient of trajectories starting from initial conditions in the old $\mathcal{B}_1$ area: they remain in the old region for several iterations before converging to the chaotic set in the *bull* area.

For this reason, the interval labelled C in Fig. 3a (where two one-piece chaotic attractors coexist) ends with the first homoclinic bifurcation of the origin, which occurs at $e^1_{(CD)} \simeq 4.185$, when the chaotic attractor in the *bear region* disappears and the generic initial condition in that region then converges to the dark grey attractor, in the *bull region*. Similarly to the two-dimensional case, a range of values of the parameter e exists such that the chaotic attractor located (approximately) in the region $S > F^S$ $(y > 0)$, colored dark grey, becomes the only attractor in the phase space (see Fig. 6a). From the asymptotic behavior, shown in Fig. 6a, we cannot observe the chaotic repeller, which we know exists. The chaotic repeller will again be observable after the second homoclinic bifurcation of the origin, which occurs at $e^2_{(DE)} \simeq 4.3$, leading to an *explosion* of the chaotic attractor into a wide region of the phase space, as shown in Fig. 6b.

From Fig. 6a we can see that the tongues of the dark grey chaotic set increasingly approach the fundamental fixed point, and thus we are very close to the second homoclinic bifurcation of O. This bifurcation involves a contact between the branch of W^u_O that converges to the chaotic attractor in the *bull region* and the surface representing the stable set W^s_O. The result of this bifurcation is an explosion of the chaotic set (which includes both the previous chaotic attractor and the previous chaotic repeller), as shown in Fig. 6b.

This brings about a major change of the dynamics. Whatever the initial condition is (from either the *bear* or the *bull region*), the trajectory will wander in both regions, jumping from one to the other after an unpredictable number of iterations. An example of the resulting fluctuations of the state variables is given in Fig. 7. The dynamics we obtain are much more intricate than those observed in Day and Huang (1990). The reason for this is that there is a feedback process from the foreign exchange markets on the stock markets and from the stock markets on the foreign exchange market. The first feedback process generates endogenous dynamics in the stock markets, where otherwise no dynamics would be observable.

The second feedback process may be interpreted as deterministic noise impacting on the evolution of the exchange rate. Also, the three markets demonstrate excess volatility and endogenous bubbles and crashes.

4.4 Final Bifurcation

After the above-described second homoclinic bifurcation of the origin, the region of the phase space covered by the chaotic dynamics becomes wider as parameter e increases. The oscillations of the trajectories increase in amplitude, and we approach

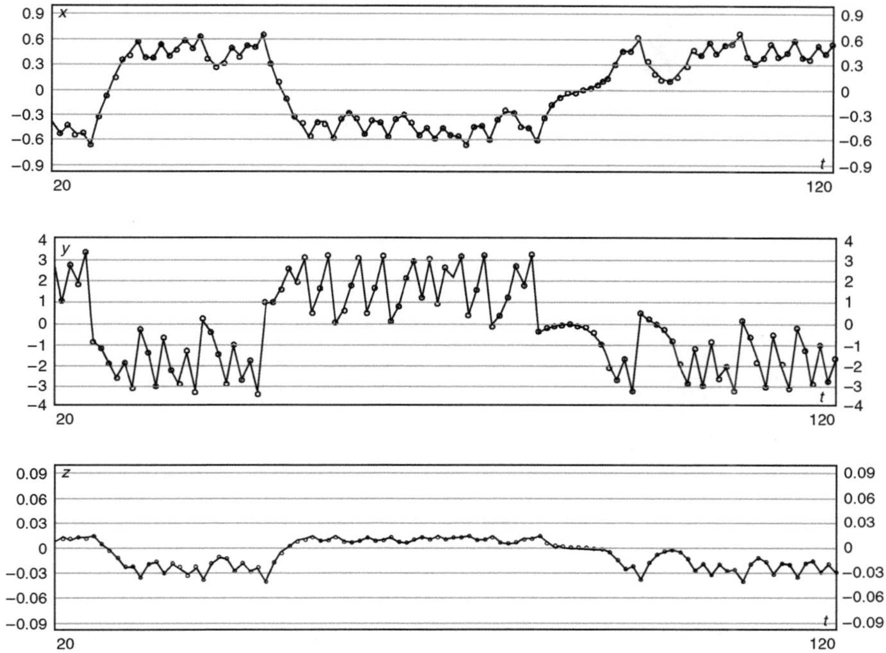

Fig. 7 Trajectories of the state variable x (deviation of stock price P^H from the fundamental price F^H), y (deviation of the exchange rate S from the fundamental exchange rate F^S), z (deviation of stock price P^A from the fundamental price F^A), obtained at $e = 4.75$

a catastrophic situation, after which trajectories will be mainly divergent. In Fig. 3 this bifurcation is revealed by the existence of a unique attractor, colored grey, which covers both chaotic regions and exists up to its final bifurcation at $e_f \simeq 5.03$. The final bifurcation is again given (as in the one- and two-dimensional cases studied in Tramontana et al., 2009) by a contact of the chaotic attractor with the frontier of its basin of attraction. We recall (see Mira et al., 1996) that a contact between an invariant set and the basin of divergent trajectories always leads to a final bifurcation, because the invariant set will no longer exist after the contact, and almost all the points whose trajectory was previously trapped into the invariant set will then have divergent trajectories. In our example this is shown in Fig. 8 where, for a specific value of e close to the final bifurcation, we represent the attractor in black and its basin of attraction in light grey.[8] Dark gray points, as usual, denote points belonging

[8] For a better visualization, the region of the three-dimensional phase space represented in Fig. 8 also includes a set of points that are not economically meaningful (the bottom part of the cube and of the related sections), but the attractor and the contacts that give rise to the *final bifurcation* all belong to the economically relevant zone.

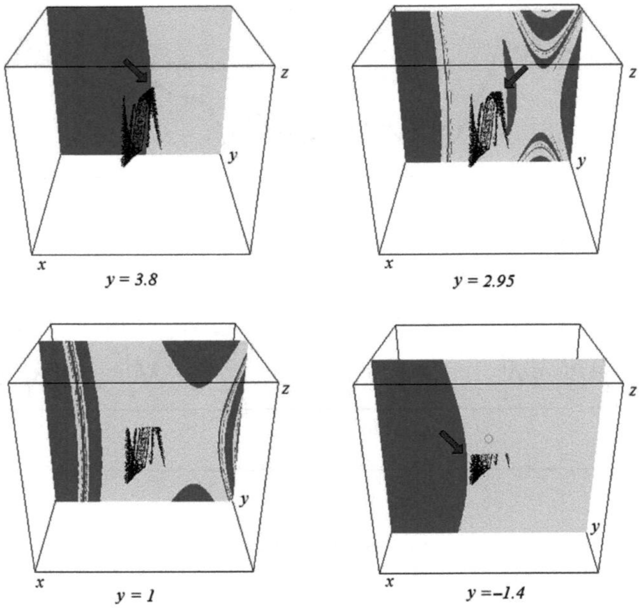

Fig. 8 Chaotic attractor at $e = 5.02$ and four different sections of the three-dimensional phase space through planes of equation $y = k$. The *dark gray points* belong to $\mathcal{B}_\infty$ and thus generate divergent trajectories; the *light grey points* belong to the basin of attraction of the attracting set (in *black*). The *bottom part* of each section corresponds to initial conditions that have no economic relevance ($z < -F^A$), included only for better visualization of the basins. The contacts occur in the meaningful region

to basin $\mathcal{B}_\infty$. Figure 8 shows four different sections with planes of equation $y = k$. In the first hyperplane (at $y = 3.8$), the boundary between the light and dark gray points is a simple line, and this section is still far from the chaotic attractor. In the second cross section (at $y = 2.95$), the boundary has become more complex, and the attractor is crossed: the section of attractor belonging to the plane, still inside the light grey area, is close to the frontier, in the points indicated by an arrow. In the third section (at $y = 1$), the attractor again appears a long distance from the border of the basin. Finally, the last section (at $y = -1.4$) again suggests that a portion of the attractor is close to the frontier, in the points indicated by the arrow.

The contact between two invariant sets of different nature (the chaotic attractor and the frontier of its basin) leads to the final bifurcation, which will leave a chaotic repeller instead of the chaotic attractor. That is, after this final bifurcation the model is no longer meaningful, as the generic trajectory in the phase space is a divergent trajectory (maybe after a long chaotic transient). The chaotic repeller survives in a set of zero measure, almost inaccessible, and includes all of the unstable fixed points and cycles, as well as all of their stable sets.

5 Conclusions

In this paper we have furthered the study of a dynamic model started in Tramontana et al. (2009), where two stock markets are linked with each other due to the trading activity of foreign investors. Connections occur through the foreign exchange market, where demand for currencies, and consequent exchange rate adjustments, are generated partly by international stock market transactions and partly by the trading activity of heterogeneous foreign exchange speculators. The model results in a three-dimensional nonlinear dynamical system, which is able to generate the typical *bull and bear* dynamic behavior already detected and discussed by Day and Huang (1990) in a one-dimensional financial market model with fundamentalists and chartists. The previous study was mainly devoted to the derivation of the model and a thorough analysis of its dynamic behavior in simplified one- and two-dimensional cases, corresponding to situations in which the three markets are at least partially disconnected, due to restrictions to the trading activity of foreign investors.

This present paper is focused on the dynamic behavior of the complete three-dimensional model. We have presented a study of the full model carried out mainly by numerical simulation and graphical visualization, suitable to reveal contact bifurcations between invariant sets of different nature. Following the *road map* provided by the analysis performed in Tramontana et al. (2009), and taking advantage of the techniques employed in the 1D and 2D cases, we have seen that the homoclinic bifurcations also occur in this complete model. Thus, as expected, also in the 3D case it turns out that the typical *bull and bear* dynamics – with seemingly random switches of stock prices and exchange rates across different regions of the phase space – result from a sequence of global bifurcations involving both the non-fundamental steady states and the fundamental equilibrium of the model. Our results thus extend such dynamic mechanisms, which provide a simplified yet intriguing explanation for the emergence of bubbles and crashes in financial markets, to higher-dimensional setups.

Appendix

Given the parameters selection used in this work (i.e. $a^H = 0.41$, $b^H = 0.11$, $c^H = 0.83$, $F^H = 4.279$, $\gamma^H = 0.3$, $d = 0.35$, $f = 0.7$, $F^S = 6.07$, $a^A = 0.43$, $b^A = 0.21$, $c^A = 0.9$, $\gamma^A = 0.36$, $F^A = 1.1$), from (13) the Jacobian matrix of the three-dimensional map evaluated at the fixed point $O = (0, 0, 0)$ becomes

$$J(O) : \begin{bmatrix} 0.6146 & -0.10209 & 0 \\ 1.2430495 & 0.6265274 + 0.35e & 0.057084 \\ 0 & 0.003781256 & 0.05227 \end{bmatrix}$$

so that we look for the necessary and sufficient conditions for O to have all the eigenvalues less than one in modulus, roots of the characteristic polynomial

$$\lambda^3 + A_1\lambda^2 + A_2\lambda + A_3 = 0,$$

where

$$\begin{aligned} A_1 &= -1.7638 - 0.35e, \\ A_2 &= 0.9067 + 0.398055e, \\ A_3 &= -0.1348 - 0.1124e. \end{aligned}$$

Following Farebrother (1973) the eigenvalues of the polynomial given above have to satisfy the following conditions (equivalent conditions can be found in Gandolfo, 1980, Yury's conditions in Elaydi, 1970 and Okuguchi and Irie, 1990):

$$\begin{aligned} &(i) \quad 1 + A_1 + A_2 + A_3 > 0, \\ &(ii) \quad 1 - A_1 + A_2 - A_3 > 0, \\ &(iii)\ 1 - A_2 + A_1 A_3 - (A_3)^2 > 0, \\ &(iv) \quad A_2 < 3. \end{aligned}$$

In our case condition (i) is satisfied for $e < 0.125$ (approximate value). Condition (ii) becomes $3.8053 + 0.8605e > 0$ and is obviously satisfied for positive values of e. Condition (iii) is satisfied for $e < 3.3096$ and $e > 3.54$, while condition (iv) is satisfied for $e < 5.258821$. Starting from values of the parameter e positive and close to 0 and increasing its value, the first condition which is violated is (i), so that $e = 0.125$ is the bifurcation value at which the fixed point loses stability.

References

Day, R. H., & Huang, W. (1990). Bulls, bears and market sheep. *Journal of Economic Behavior and Organization, 14*, 299–329.

Elaydi, S. N. (1970). *An introduction to difference equations*. New York: Springer.

Farebrother, R. W. (1973). Simplified Samuelson conditions for cubit and quartic equations. *The Manchester School of Economic and Social Studies, 41*, 396–400.

Gandolfo, G. (1980). *Economic dynamics: Methods and models*. Amsterdam: North Holland.

Grebogi, C., Ott, E., & Yorke, J. A. (1983). Crises, sudden changes in chaotic attractors, and transient chaos. *Physica D, 7*, 231.

Mira, C., Gardini, L., Barugola, A., & Chatala, J. C. (1996). *Chaotic dynamics in two-dimensional noninvertible maps*. Singapore: World Scientific.

Okuguchi, K., & Irie, K. (1990). The Shur and Samuelson conditions for a cubic equation. *The Manchester School of Economic and Social Studies, 58*, 414–418.

Tramontana, F., Gardini, L., Dieci, R., & Westerhoff, F. (2009). The emergence of "bull and bear" dynamics in a nonlinear model of interacting markets. *Discrete Dynamics in Nature and Society, 2009*, 310471.

A Framework for CAPM with Heterogeneous Beliefs

Carl Chiarella, Roberto Dieci, and Xue-Zhong He

1 Introduction

The Sharpe–Lintner–Mossin (Sharpe 1964; Lintner 1965; Mossin 1966) Capital Asset Pricing Model (CAPM) plays a central role in modern finance theory. It is founded on the paradigm of homogeneous beliefs and a rational representative agent. However, from a theoretical perspective this paradigm has been criticized on a number of grounds, in particular concerning its extreme assumptions about homogeneous beliefs, information about the economic environment, and the computational ability on the part of the rational representative economic agent.

The impact of heterogeneous beliefs among investors on the market equilibrium price has been an important focus in the CAPM literature. A number of models with investors who have heterogeneous beliefs have been previously studied.[1] A common finding in this strand of research is that heterogeneous beliefs can affect aggregate market returns. However, the question remains as to how exactly does heterogeneity affect the market risk of risky assets? In much of this earlier work, the heterogeneous beliefs reflect either differences of opinion among the investors[2] or differences in information upon which investors are trying to learn by using some Bayesian updating rule.[3] Heterogeneity has been investigated in the context of either CAPM-like mean-variance models (for instance, Lintner 1969; Miller 1977; Williams 1977; and Mayshar 1982) or Arrow–Debreu contingent claims models (as in Varian 1985; Abel 1989; 2002; and Calvet et al. 2004).

[1] See, for example, Lintner (1969), Williams (1977), Huang and Litzenberger (1988), Abel (1989), Detemple and Murthy (1994), Zapatero (1998) and Basak (2000).
[2] See, for example, Lintner (1969), Miller (1977), Mayshar (1982), Varian (1985), Abel (1989, 2002) and Cecchetti et al. (2000).
[3] Typical studies include Williams (1977), Detemple and Murthy (1994) and Zapatero (1998).

X.-Z. He (✉)
School of Finance and Economics, University of Technology Sydney, P.O. Box 123, Broadway, Sydney, NSW 2007, Australia,
e-mail: Tony.He1@uts.edu.au

G.I. Bischi et al. (eds.), *Nonlinear Dynamics in Economics, Finance and the Social Sciences*, DOI 10.1007/978-3-642-04023-8_19,

In most of the cited literature, the impact of heterogeneous beliefs is studied for the case of a portfolio of one risky asset and one risk-free asset (for example, Abel 1989; Detemple and Murthy 1994; Zapatero 1998; Basak 2000; and Johnson 2004). In those papers that consider a portfolio of many risky assets and one risk-free asset, investors are assumed to be heterogeneous in their risk preferences and expected payoffs or returns of the risky assets (such as Williams 1977 and Varian 1985), but not in their estimates of variances and covariances. The only exception seems to have been the early contribution of Lintner (1969) in which heterogeneity in both means and variances/covariances is investigated in a mean-variance portfolio context. As suggested by the empirical study of Chan et al. (1999), while future variances and covariances are more easily predictable than expected future returns, the difficulties in doing so should not be understated. These authors argue that "while optimization (based on historical estimates of variances and covariances) leads to a reduction in volatility, the problem of forecasting covariance poses a challenge". Therefore, a theoretical understanding of the impact of heterogeneous beliefs in variances and covariances on equilibrium prices, volatility and asset betas is very important for a proper development of asset pricing theory.

Different from the above literature, various heterogeneous agent models (HAMs) have been developed to characterize the dynamics of financial asset prices resulting from the interaction of heterogeneous agents with different attitudes towards risk and different expectations about the future evolution of asset prices. One of the key elements of this literature is the expectations feedback mechanism, see Brock and Hommes (1997, 1998). We refer the reader to Hommes (2006), LeBaron (2006) and Chiarella et al. (2009) for surveys of recent literature on HAMs. This framework has successfully explained various types of market behaviour, such as the long-term swing of market prices from the fundamental price, asset bubbles and market crashes. It also shows a potential to characterize and explain the stylized facts (for example, Chiarella et al. 2006b; Gaunersdorfer and Hommes 2007; and Farmer et al. 2004) and various kinds of power law behaviour (for instance, Lux 2004; Alfarano et al. 2005; and He and Li 2007) observed in financial markets. However, most of the HAMs analyzed in the literature involve a financial market with only one risky asset[4] and are not in the context of the CAPM. In markets with many risky assets and heterogeneous investors, the impact of heterogeneity on the market equilibrium and standard portfolio theory remains a largely unexplored issue.

This paper is largely motivated by a re-reading of Lintner's early work and recent development in the HAMs literature, in particular, our recent work Chiarella et al. (2006a). Although Lintner's earlier contribution discusses how to aggregate heterogeneous beliefs, the impact of heterogeneity on the market equilibrium price, risk premia and CAPM within the mean-variance framework has not been fully

[4] Except for some recent contributions by Westerhoff (2004), Chiarella et al. (2005, 2007) and Westerhoff and Dieci (2006) showing that complex price dynamics may also result within a multi-asset market framework.

explored. The main obstacle in dealing with heterogeneity is the complexity and heavy notation involved when the number of assets and the dimension of the heterogeneity increase. It might be this notational obstacle that makes the paper of Lintner hard to follow, and renders rather complicated the analysis of the impact of heterogeneity on market equilibrium prices. In this paper, we reconsider the derivation of the traditional CAPM in a discrete time setting for a portfolio of one risk-free asset and many risky assets and provide a simple framework that incorporates heterogeneous beliefs. In contrast to the standard setting we consider heterogeneous agents whose expectations of asset returns are based on statistical properties of past returns and so induce expectations feedback. Different from Chiarella et al. (2006a) where beliefs are formed in terms of the payoff, we assume that agents form their demands based upon heterogeneous beliefs about conditional means and covariances of the risky asset returns. The market clearing prices are determined under a Walrasian auctioneer scenario. In this framework we first construct a "consensus" belief (with respect to the means and covariances of the risky asset returns) to represent the aggregate market belief and derive a heterogeneous CAPM which relates aggregate excess return on risky assets with aggregate excess return on the aggregate market wealth via an aggregate beta coefficient for risky assets. We then extend the analysis to a repeated one-period set up and establish a framework for a dynamic CAPM using a "market fraction" model in which agents are grouped according to their beliefs. We obtain an exact relation between heterogeneous beliefs and the market equilibrium returns and the ex-ante beta-coefficients. The framework developed here could be used for further study of the complicated impact of heterogeneity on the market equilibrium.

The paper is organized as follows. Section 2 derives equilibrium CAPM-like relationships for asset returns in the case of heterogeneous beliefs and relates a "consensus" belief about the expected excess return on each risky asset to a "consensus" belief about expected market return, via aggregate beta coefficients. There follows a discussion about the wealth dynamics and the beta coefficients, and how they relate to the heterogeneous beliefs about the returns on the risky assets. Finally this section also considers explicitly the supply of the risky securities, and derives equilibrium prices, and relates the aggregate beta coefficients to the market equilibrium prices. Section 3 extends the analysis to a repeated one-period set up and obtains a dynamic, "market fraction" multi-asset framework with heterogeneous groups of agents, which generalizes earlier contributions by Brock and Hommes (1998) and Chiarella and He (2001, 2002), and highlights how the aggregate ex-ante beta coefficients may vary over time once agents' beliefs are assumed to be updated dynamically at each time step as a function of past realized returns. The framework is different from that of Chiarella et al. (2007), which uses a market maker mechanism to arrive at the market price, as here we use the Walrasian auctioneer scenario. Section 4 concludes and suggests some directions for future research.

2 The CAPM with Heterogeneous Beliefs

The present section generalizes the derivation of the CAPM relationships, as carried out for instance by Huang and Litzenberger (1988, Sect. 4.15), to the case of investors with heterogeneous beliefs about asset returns. Some of the ideas contained in the present section are adapted from Lintner (1969), where aggregation of individual assessments about future payoffs is performed in a mean-variance framework. However, different from Lintner (1969), the aggregation is explicitly given by constructing a consensus belief, which greatly facilitates the establishment of the CAPM with heterogeneous beliefs.

Consider an economy with many agents who invest in portfolios consisting of a riskless asset and N risky assets with $N \geq 1$. Let r_f be the risk free rate of the riskless asset and $\tilde{r}_j$ be the rate of return of risky asset j, $j = 1, 2, \ldots, N$. Following the standard CAPM setup, we assume that the returns of the risky assets are multivariate (conditionally) normally distributed and the utility function $u_i(x)$ of agent i is twice differentiable, concave and strictly increasing, $i = 1, 2, \ldots, I$. Let W_0^i be the initial wealth of agent i and w_{ij} be the wealth proportion of agent i invested in asset j. Then the end-of-period wealth, $\widetilde{W}_i$, of agent i is given by

$$\widetilde{W}_i = W_0^i \left(1 + r_f + \sum_{j=1}^{N} w_{ij}(\tilde{r}_j - r_f) \right). \tag{1}$$

Following Huang–Litzenberger (Sect. 4.15), the maximization of the expected utility of the portfolio wealth of agent i is characterized by the first order condition:

$$E_i\left[u_i'(\widetilde{W}_i)\right] E_i\left[\tilde{r}_j - r_f\right] = -E_i\left[u_i''(\widetilde{W}_i)\right] Cov_i(\widetilde{W}_i, \tilde{r}_j) \tag{2}$$

for any $j = 1, 2, \ldots, N$, where $E_i(\cdot)$ is the conditional mean and $Cov_i(\cdot, \cdot)$ is the conditional covariance of agent i, characterizing the heterogeneity of the agents in their beliefs. By defining the *global absolute risk aversion* of agent i

$$\theta_i := \frac{-E_i\left[u_i''(\widetilde{W}_i)\right]}{E_i\left[u_i'(\widetilde{W}_i)\right]},$$

condition (2) becomes

$$\theta_i^{-1} E_i\left[\tilde{r}_j - r_f\right] = Cov_i(\widetilde{W}_i, \tilde{r}_j), \qquad j = 1, 2, \ldots, N. \tag{3}$$

Note that by its definition in (1)

$$Cov_i(\widetilde{W}_i, \tilde{r}_j) = W_0^i \sum_{k=1}^{N} w_{ik} Cov_i\left(\tilde{r}_k, \tilde{r}_j\right).$$

It follows that the conditions (3) can be rewritten with vector notation as

$$\theta_i^{-1}\left(E_i[\tilde{\mathbf{r}}] - r_f\mathbf{1}\right) = W_0^i \boldsymbol{\Omega}_i \mathbf{w}_i, \tag{4}$$

where $\tilde{\mathbf{r}} = [\tilde{r}_1, \tilde{r}_2, \ldots, \tilde{r}_N]^\top$, $\mathbf{1} = [1, 1, \ldots, 1]^\top \in \mathbb{R}^N$, $\mathbf{w}_i = [w_{i1}, w_{i2}, \ldots, w_{iN}]^\top$, $\boldsymbol{\Omega}_i = [\sigma_{i,jk}]_{N\times N}$, $j, k = 1, 2, \ldots, N$, and $\sigma_{i,jk} := Cov_i(\tilde{r}_j, \tilde{r}_k)$, $i = 1, \ldots, I$. We assume that the $\boldsymbol{\Omega}_i$ $(i = 1, 2, \ldots, I)$ are positive definite and thus invertible. It follows from (4) that the optimal portfolio $\mathbf{w}_i$ of agent i is given by[5]

$$\mathbf{w}_i = \frac{1}{W_0^i}\theta_i^{-1}\boldsymbol{\Omega}_i^{-1}\left(E_i[\tilde{\mathbf{r}}] - r_f\mathbf{1}\right). \tag{5}$$

Let $W_{m0} = \sum_{i=1}^I W_0^i$ be the total wealth in the economy and $\mathbf{w}_a$ be the proportions of the total wealth in the economy invested in the risky assets. The *market is in equilibrium* when the condition

$$W_{m0}\mathbf{w}_a = \sum_{i=1}^{I} W_0^i \mathbf{w}_i \tag{6}$$

is satisfied.[6] Let $\widetilde{W_m}$ represent the random end-of-period wealth in the economy. Similarly to Huang and Litzenberger (1988, Sect. 4.15), we define the rate of return $\tilde{r}_m$ on the aggregate market wealth as the one which satisfies

$$\widetilde{W_m} = \sum_{i=1}^{I} \widetilde{W_i} = W_{m0}(1 + \tilde{r}_m). \tag{7}$$

Substituting (1) into the right hand side of the first equality of (7) and performing some algebraic manipulations we find that $\tilde{r}_m$ can also be rewritten in terms of aggregate wealth proportions as

$$\tilde{r}_m = r_f + \mathbf{w}_a^\top\left(\tilde{\mathbf{r}} - r_f\mathbf{1}\right). \tag{8}$$

Then we have the following result when the market is in equilibrium.

Proposition 1 *Let* $\Theta = (\sum_{i=1}^I \theta_i^{-1})^{-1}$. *Define a consensus belief in the covariance matrix and the expected return vector, respectively, as*

---

[5] The optimal portfolio $\mathbf{w}_i$ of agent i is only implicitly defined by (5), because in general $\theta_i = \theta_i(\mathbf{w}_i)$ will depend on $\mathbf{w}_i$. Nevertheless, at this stage we are interested in equilibrium relationships involving aggregate beliefs, which do not require $\mathbf{w}_i$ to be made explicit.

[6] The condition (6) is in monetary units, it can also be expressed as aggregate demand (in quantity terms) for risky assets equals aggregate supply (also quantity terms) on dividing throughout by the equilibrium price.

$$\boldsymbol{\Omega}_a = \left(\Theta \sum_{i=1}^{I} \theta_i^{-1} \boldsymbol{\Omega}_i^{-1}\right)^{-1}, \tag{9}$$

$$E_a[\tilde{\mathbf{r}}] = \Theta \boldsymbol{\Omega}_a \sum_{i=1}^{I} \theta_i^{-1} \boldsymbol{\Omega}_i^{-1} E_i[\tilde{\mathbf{r}}]. \tag{10}$$

Then, when the market is in equilibrium:

(1) The vector of proportions $\mathbf{w}_a$ *of the total wealth in the economy invested in the risky assets is given by*

$$\mathbf{w}_a = \frac{1}{W_{m0}} \Theta^{-1} \boldsymbol{\Omega}_a^{-1} \left(E_a[\tilde{\mathbf{r}}] - r_f \mathbf{1}\right). \tag{11}$$

(2) The expected return on the aggregate market wealth

$$E_a(\tilde{r}_m) = r_f + \Theta W_{m0} \sigma_{a,m}^2, \tag{12}$$

where

$$\sigma_{a,m}^2 = \mathbf{w}_a^T \boldsymbol{\Omega}_a \mathbf{w}_a \tag{13}$$

is the variance of the aggregate market wealth return.

(3) The expected returns of the risky assets satisfy

$$E_a[\tilde{\mathbf{r}}] - r_f \mathbf{1} = \boldsymbol{\beta}_a (E_a(\widetilde{r}_m) - r_f), \tag{14}$$

where

$$\boldsymbol{\beta}_a = (\beta_{a,1}, \beta_{a,2}, \ldots, \beta_{a,N})^\top = \frac{1}{\sigma_{a,m}^2} \boldsymbol{\Omega}_a \mathbf{w}_a, \qquad \beta_{a,j} = \sigma_{a,jm} / \sigma_{a,m}^2. \tag{15}$$

Proof. See the appendix

Note that the existence of the consensus covariance matrix $\boldsymbol{\Omega}_a$ follows from the fact that, in (9), $\boldsymbol{\Omega}_a^{-1}$ is a convex combination of positive definite matrices $\boldsymbol{\Omega}_i^{-1}$, which implies that $\boldsymbol{\Omega}_a^{-1}$ is also positive definite, and therefore nonsingular. Note also that when the consensus belief is replaced with the objective and homogeneous belief, Proposition 1 results in the standard CAPM. When agents have heterogeneous beliefs, the consensus beliefs defined in Proposition 1 provides an explicit way to aggregate the heterogeneous beliefs, under which the standard CAPM-like relation (14) holds under the heterogeneous beliefs. Note that ΘW_{m0} can be interpreted as the *aggregate relative risk aversion* of the economy in equilibrium. In particular, when $\theta_i = \theta_0$ for $i = 1, 2, \ldots, I$, we have $\Theta = \theta_0 / I$ and $\Theta W_{m0} = \theta_0 (W_{m0}/I)$, measuring the relative risk aversion of an agent at the average wealth level. The market equilibrium condition (6) allows a non-zero supply of the riskless asset in the economy. If a zero net supply of the riskless asset is assumed when the market is in equilibrium, we then obtain the following corollary.

Corollary 1 *In market equilibrium, if the riskless asset is in zero net supply in the economy, then the risk-free rate r_f is given by*

$$r_f = \frac{\mathbf{1}^\top \boldsymbol{\Omega}_a^{-1} E_a[\tilde{r}] - \Theta W_{m0}}{\mathbf{1}^\top \boldsymbol{\Omega}_a^{-1} \mathbf{1}}. \tag{16}$$

In this case the return $\widetilde{r}_m$ on the aggregate market wealth becomes the return on the market portfolio of the risky assets, and the variance $\sigma_{a,m}^2$ becomes the variance of the market portfolio of the risky assets.

Proof. It follows from $\mathbf{1}^\top \mathbf{w}_a = 1$ and (11) in Proposition 1 that

$$\Theta W_{m0} = \mathbf{1}^T \boldsymbol{\Omega}_a^{-1}(E_a[\tilde{\mathbf{r}}] - r_f \mathbf{1}). \tag{17}$$

Solving for r_f leads to the result.

Corollary 1 shows that the equilibrium risk-free rate r_f is determined endogenously when the riskless asset is in zero net supply in the economy. It in fact depends on the aggregate relative risk aversion coefficient ΘW_{m0} and the consensus beliefs in the expected return and the variance-covariance matrix of the risky assets.

In order to understand the impact of the market wealth on the risk premia and beta coefficients of the risky assets, we provide the following result.

Corollary 2 *In market equilibrium, the expected return of the economy is given by*

$$E_a(\tilde{r}_m) = r_f + \frac{1}{\Theta W_{m0}}(E_a[\tilde{\mathbf{r}}] - r_f \mathbf{1})^\top \boldsymbol{\Omega}_a^{-1}(E_a[\tilde{\mathbf{r}}] - r_f \mathbf{1}), \tag{18}$$

the variance is given by

$$\sigma_{a,m}^2 = \frac{1}{(\Theta W_{m0})^2}(E_a[\tilde{\mathbf{r}}] - r_f \mathbf{1})^\top \boldsymbol{\Omega}_a^{-1}(E_a[\tilde{\mathbf{r}}] - r_f \mathbf{1}), \tag{19}$$

and the beta coefficients can be rewritten as

$$\boldsymbol{\beta}_a = \frac{\Theta W_{m0}}{(E_a[\tilde{\mathbf{r}}] - r_f \mathbf{1})^\top \boldsymbol{\Omega}_a^{-1}(E_a[\tilde{\mathbf{r}}] - r_f \mathbf{1})}(E_a[\tilde{\mathbf{r}}] - r_f \mathbf{1}). \tag{20}$$

Proof. In market equilibrium, (18) follows from (11) and the result $E_a(\widetilde{r}_m) = r_f + \mathbf{w}_a^\top(E_a[\tilde{\mathbf{r}}] - r_f \mathbf{1})$ [see (37) of the appendix]; (19) follows from and (11) and (13); and (20) follows from (11), (15) and (19).

Corollary 2 expresses the equilibrium relationships where the riskless asset is not necessarily in zero net supply. If the riskless asset is in zero net supply the equilibrium relationships turn out to be explicitly independent of the total wealth in the economy.

Corollary 3 *If the riskless asset is in zero net supply in the economy, then the expected return of the market portfolio of the risky assets is given by*

$$E_a(\tilde{r}_m) = r_f + \frac{(E_a[\tilde{\mathbf{r}}] - r_f\mathbf{1})^\top \boldsymbol{\Omega}_a^{-1}(E_a[\tilde{\mathbf{r}}] - r_f\mathbf{1})}{\mathbf{1}^\top \boldsymbol{\Omega}_a^{-1}(E_a[\tilde{\mathbf{r}}] - r_f\mathbf{1})}, \tag{21}$$

the variance of the market portfolio of the risky assets is given by

$$\sigma_{a,m}^2 = \frac{(E_a[\tilde{\mathbf{r}}] - r_f\mathbf{1})^\top \boldsymbol{\Omega}_a^{-1}(E_a[\tilde{\mathbf{r}}] - r_f\mathbf{1})}{(\mathbf{1}^\top \boldsymbol{\Omega}_a^{-1}(E_a[\tilde{\mathbf{r}}] - r_f\mathbf{1}))^2}, \tag{22}$$

and the beta coefficients can be rewritten as

$$\boldsymbol{\beta}_a = \frac{\mathbf{1}^\top \boldsymbol{\Omega}_a^{-1}(E_a[\tilde{\mathbf{r}}] - r_f\mathbf{1})}{(E_a[\tilde{\mathbf{r}}] - r_f\mathbf{1})^\top \boldsymbol{\Omega}_a^{-1}(E_a[\tilde{\mathbf{r}}] - r_f\mathbf{1})}(E_a[\tilde{\mathbf{r}}] - r_f\mathbf{1}). \tag{23}$$

Proof. When the riskless asset in the economy is in zero net supply, we have that (17) holds. Using this to replace ΘW_{m0} with $\mathbf{1}^\top \boldsymbol{\Omega}_a^{-1}(E_a[\tilde{\mathbf{r}}] - r_f\mathbf{1})$ in (18), (19) and (20), we obtain (21), (22) and (23), respectively.

It is interesting to note that, when the risk-free rate r_f is given exogenously and the riskless asset is not in zero net supply, the expected return and variance of the economy and the beta coefficients of the risky assets with the economy[7] depend on the total wealth in the economy. However, when the riskless asset in the economy is in zero net supply, the return of the economy is given by the return of the market portfolio of the risky assets. Consequently, the expected return and variance of the market portfolio and the beta coefficients of the risky assets with the market portfolio do not depend explicitly on the wealth. This difference, generated from the restriction that the riskless asset in the economy be in zero net supply, has the potential to explain the impact of heterogeneous beliefs on the risk-free rate and risk premium puzzles. We refer the reader to He and Shi (2009) for further discussion of this issue. To obtain the equilibrium prices of the risky assets, we assume that agents have CARA exponential utility of wealth functions, so that the global absolute risk aversion of agent i, $\theta_i = -E_i\left[u_i''(\widetilde{W}_i)\right]/E_i\left[u_i'(\widetilde{W}_i)\right]$, and hence the aggregate risk aversion Θ, are constants. Let $\mathbf{z} := [z_1, z_2, \ldots, z_N]^\top$ be the positive supply vector (number of shares) of the risky assets in the economy and denote by $\mathbf{Z} := diag[z_1, z_2, \ldots, z_N]$ the $(N \times N)$ diagonal matrix whose entries are the elements of $\mathbf{z}$. Then the market equilibrium prices of the risky assets can be determined according to the following corollary.

[7] Note we distinguish between beta of the economy when the riskfree asset is not in zero net supply and the beta of the market (obtained when the riskfree asset is in zero net supply).

Corollary 4 *Let* $\mathbf{p}_0 = [p_{01}, p_{02}, \ldots, p_{0N}]^\top$ *be the vector of the prices of the risky assets when the market is in equilibrium. Then*

$$\mathbf{p}_0 = \mathbf{Z}^{-1}\Theta^{-1}\boldsymbol{\Omega}_a^{-1}\left(E_a[\tilde{\mathbf{r}}] - r_f\mathbf{1}\right) = \mathbf{Z}^{-1}\sum_{i=1}^{I}\theta_i^{-1}\boldsymbol{\Omega}_i^{-1}\left(E_i[\tilde{\mathbf{r}}] - r_f\mathbf{1}\right) \tag{24}$$

and the beta coefficients can be written as

$$\boldsymbol{\beta}_a = \frac{W_{m0}}{\mathbf{p}_0^\top \mathbf{Z}\boldsymbol{\Omega}_a\mathbf{Z}\mathbf{p}_0}\boldsymbol{\Omega}_a\mathbf{Z}\mathbf{p}_0. \tag{25}$$

In particular, when the riskless asset is in zero net supply in the economy,

$$\boldsymbol{\beta}_a = \frac{\mathbf{p}_0^\top \mathbf{z}}{\mathbf{p}_0^\top \mathbf{Z}\boldsymbol{\Omega}_a\mathbf{Z}\mathbf{p}_0}\boldsymbol{\Omega}_a\mathbf{Z}\mathbf{p}_0. \tag{26}$$

Proof. Given the positive supply of the risky assets in the economy, the prices of the risky assets when the market is in equilibrium satisfy the condition $W_{m0}\boldsymbol{w}_a = \boldsymbol{Z}\,\boldsymbol{p}_0$. Substituting $\boldsymbol{w}_a$ from (11) into the last condition, we obtain the first equality in (24), the second follows by use of (10). Replacing $E_a[\tilde{\mathbf{r}}] - r_f\mathbf{1}$ with $\Theta\boldsymbol{\Omega}_a\boldsymbol{Z}\,\boldsymbol{p}_0$ in (20) and (23) we then obtain the expressions (25) and (26) for $\boldsymbol{\beta}_a$, respectively.

One of the advantages of the expressions for the beta coefficients in Corollary 4 is that we can use the market information about the observed beta coefficients and market prices to estimate the market consensus covariance matrix $\boldsymbol{\Omega}_a$, which may not be observed or difficult to estimate in a heterogeneous beliefs market. The implications of this observation for empirical studies is left for future research.

3 A Dynamic Market Fraction CAPM

The present section first sets up a framework for a market fraction model with heterogeneous beliefs, which extends contributions developed by Brock and Hommes (1998), Chiarell and He (2001, 2002) and He and Li (2008) in the simple case of a single risky security to a multi-asset framework. We then extend the framework to a repeated one period dynamic CAPM model. Related, but different, studies of the CAPM with heterogeneous beliefs can be found in Böhm and Chiarella (2005) and Böhm and Wenzelburger (2005).

Assume that the I investors can be grouped into a finite number of agent-types, indexed by $h \in H$, where the agents within the same group are homogeneous in their beliefs $E_h[\tilde{\mathbf{r}}]$ and $\boldsymbol{\Omega}_h$, as well as risk aversion coefficient θ_h. Denote I_h, $h \in H$, the number of investors in group h and $n_h := I_h/I$ the market fraction of agents of type h. We then denote by $\mathbf{s} := (1/I)\mathbf{z}$ the supply of shares per investor. Note that, instead of using the aggregate risk aversion coefficient $\Theta := \left(\sum_{i=1}^{I} \theta_i^{-1}\right)^{-1}$ it is convenient to define the "average" risk aversion θ_a as

$$\theta_a := \left(\sum_{h\in H} n_h \theta_h^{-1}\right)^{-1},$$

where obviously $\theta_a = I\Theta$. It follows from Proposition 1 that the aggregate beliefs about variances/covariances and expected returns can be rewritten, respectively, as

$$\boldsymbol{\Omega}_a = \theta_a^{-1}\left(\sum_{h\in H} n_h \theta_h^{-1} \boldsymbol{\Omega}_h^{-1}\right)^{-1}, \qquad E_a[\tilde{\mathbf{r}}] = \theta_a \boldsymbol{\Omega}_a \sum_{h\in H} n_h \theta_h^{-1} \boldsymbol{\Omega}_h^{-1} E_h[\tilde{\mathbf{r}}].$$

Finally, by defining $\mathbf{S} = diag(s_1, s_2, \ldots, s_N)$, the equilibrium prices in (21) can be rewritten as

$$\mathbf{p}_0 = \mathbf{S}^{-1}\theta_a^{-1}\boldsymbol{\Omega}_a^{-1}\left(E_a[\tilde{\mathbf{r}}] - r_f\mathbf{1}\right) = \mathbf{S}^{-1}\sum_{h\in H} n_h\theta_h^{-1}\boldsymbol{\Omega}_h^{-1}(E_h[\tilde{\mathbf{r}}] - r_f\mathbf{1}).$$

We now turn to the process of formation of heterogeneous beliefs and equilibrium prices in a dynamic setting, from time t to time $t+1$. In doing so, we take the view that agents' beliefs about the returns $\tilde{\mathbf{r}}_{t+1}$ in the time interval $(t, t+1)$, which are formed before dividends at time t are realized and prices at time t are revealed by the market, determine the aggregate demand for each risky asset at time t, which in turns produces the equilibrium prices at time t, $\mathbf{p}_t$, via the market clearing conditions. Of course, once prices and dividends at time t are realized, the returns $\mathbf{r}_t$ become known. More precisely, we assume that agents' assessments of the end-of-period joint distribution of the returns $\tilde{\mathbf{r}}_{t+1}$ are formed at time t before the equilibrium prices at time t are determined. These beliefs remain fixed while the market finds its equilibrium vector of current prices, $\mathbf{p}_t$. In particular, the heterogeneous beliefs (or assessments) of agents about the mean and covariance structure of $\tilde{\mathbf{r}}_{t+1}$ are functions of the information up to time $t-1$, which can be expressed as functions of the realized returns $\mathbf{r}_{t-1}$, $\mathbf{r}_{t-2}$, ..., for any group, or belief-type $h \in H$,[8]

$$\boldsymbol{\Omega}_{h,t} := [Cov_{h,t}(\tilde{r}_{j,t+1}, \tilde{r}_{k,t+1})] = \boldsymbol{\Omega}_h(\mathbf{r}_{t-1}, \mathbf{r}_{t-2}, \ldots), \tag{27}$$

$$E_{h,t}[\tilde{\mathbf{r}}_{t+1}] = \mathbf{f}_h(\mathbf{r}_{t-1}, \mathbf{r}_{t-2}, \ldots), \tag{28}$$

where obviously similar representations hold also for the aggregate beliefs $\boldsymbol{\Omega}_{a,t} := [Cov_{a,t}(\tilde{r}_{j,t+1}, \tilde{r}_{k,t+1})]$ and $E_{a,t}[\tilde{\mathbf{r}}_{t+1}]$. The market clearing prices at time t

[8] We use $E_{h,t}(\tilde{r}_{t+1})$ to denote the expectation of $\tilde{r}_{t+1}$ formed at time t by the agents of group h. Similarly for the notation $Cov_{h,t}(\tilde{r}_{j,t+1}, \tilde{r}_{k,t+1})$.

become

$$\mathbf{p}_t = \mathbf{S}^{-1} \sum_{h \in H} n_h \theta_h^{-1} \boldsymbol{\Omega}_{h,t}^{-1} (E_{h,t}[\tilde{\mathbf{r}}_{t+1}] - r_{f,t}\mathbf{1}), \tag{29}$$

or in terms of the consensus beliefs,

$$\mathbf{p}_t = \mathbf{S}^{-1} \theta_a^{-1} \boldsymbol{\Omega}_{a,t}^{-1} (E_{a,t}[\tilde{\mathbf{r}}_{t+1}] - r_{f,t}\mathbf{1}), \tag{30}$$

where $r_{f,t}$ is the riskfree rate over the time period from t to $t+1$.

Next, note that the return $r_{j,t}$ on asset j, realized over the time interval $(t-1,t)$ is given by

$$r_{j,t} = \frac{p_{j,t} + d_{j,t}}{p_{j,t-1}} - 1,$$

where $d_{j,t}$ denotes the realized dividend per share of asset j, $j = 1, 2, \ldots, N$. We can rewrite realized returns in vector notation as

$$\mathbf{r}_t = \mathbf{P}_{t-1}^{-1}(\mathbf{p}_t + \mathbf{d}_t) - \mathbf{1}, \tag{31}$$

where $\mathbf{d}_t := [d_{1,t}, d_{2,t}, \ldots, d_{N,t}]^\top$, and $\mathbf{P}_t := diag(p_{1,t}, p_{2,t}, \ldots, p_{N,t})$. Equation (31), via the market equilibrium prices (29) and the beliefs updating (27) and (28), gives the return $\mathbf{r}_t$ as a function of $\mathbf{r}_{t-1}, \mathbf{r}_{t-2}, \ldots$ and of the realized dividends $\mathbf{d}_t$, which are assumed to follow an exogenous random process in general. Thus the dynamics of prices and returns are determined by both the endogenous dependence of returns on past returns in (31) and the exogenous stochastic dividend process.

We summarize below the dynamical system that describes the market fraction multi-asset model in terms of returns, where the market clearing prices are used as auxiliary variables

Proposition 2 *For the market fraction model, the equilibrium return vector of the risky assets is given by*

$$\mathbf{r}_t = \mathbf{F}(\mathbf{r}_{t-1}, \mathbf{r}_{t-2}, \ldots; \tilde{\mathbf{d}}_t) = \mathbf{P}_{t-1}^{-1}(\mathbf{p}_t + \tilde{\mathbf{d}}_t) - \mathbf{1},$$

where

$$\mathbf{p}_t = \mathbf{S}^{-1} \sum_{h \in H} n_h \theta_h^{-1} \boldsymbol{\Omega}_{h,t}^{-1} (E_{h,t}[\tilde{\mathbf{r}}_{t+1}] - r_{f,t}\mathbf{1}),$$

$\mathbf{P}_t := diag(p_{1,t}, p_{2,t}, \ldots, p_{N,t})$, $\boldsymbol{\Omega}_{h,t} = \boldsymbol{\Omega}_h(\mathbf{r}_{t-1}, \mathbf{r}_{t-2}, \ldots)$, *and* $E_{h,t}(\tilde{\mathbf{r}}_{t+1}) = \mathbf{f}_h(\mathbf{r}_{t-1}, \mathbf{r}_{t-2}, \ldots)$. *Moreover, at the beginning of each time interval* $(t, t+1)$ *the expected returns under the aggregate beliefs (based on information up to time* $t-1$*) satisfy a CAPM-like equation of the type*

$$E_{a,t}[\tilde{\mathbf{r}}_{t+1}] - r_{f,t}\mathbf{1} = \boldsymbol{\beta}_{a,t}(E_{a,t}(\tilde{r}_{m,t+1}) - r_{f,t}),$$

where $\tilde{r}_{m,t+1}$ *is the rate of the return of the aggregate market wealth over the time period from* t *to* $t+1$ *defined by* $\widetilde{W}_{m,t+1} = \widetilde{W}_{m,t}(1 + \tilde{r}_{m,t+1})$, *and* $\widetilde{W}_{m,t}$ *is the aggregate wealth in the economy at time t. Under the dynamical consensus belief, the rate of return on the aggregate market wealth is given by*

$$\tilde{r}_{m,t+1} = r_{f,t} + \frac{1}{\Theta W_{m,t}}(E_{a,t}[\tilde{\mathbf{r}}_{t+1}] - r_{f,t}\mathbf{1})^\top \boldsymbol{\Omega}_{a,t}^{-1}(\tilde{\mathbf{r}}_{t+1} - r_{f,t}\mathbf{1})$$

and the "aggregate" beta coefficients are given by

$$\boldsymbol{\beta}_{a,t} = \frac{\Theta W_{m,t}}{(E_{a,t}[\tilde{\mathbf{r}}_{t+1}] - r_{f,t}\mathbf{1})^\top \boldsymbol{\Omega}_{a,t}^{-1}(E_{a,t}[\tilde{\mathbf{r}}_{t+1}] - r_{f,t}\mathbf{1})}(E_{a,t}[\tilde{\mathbf{r}}_{t+1}] - r_{f,t}\mathbf{1}).$$

As in the discussion of the static framework in Sect. 2, in the case of zero net supply of the riskless asset the relationships in Proposition 2 do not depend explicitly on the wealth in the economy. Thus we can state

Proposition 3 *If the riskless asset is in zero net supply over the time period in the economy, then the equilibrium risk-free rate is given by*

$$r_{f,t} = \frac{\mathbf{1}^\top \boldsymbol{\Omega}_{a,t}^{-1} E_{a,t}[\tilde{\mathbf{r}}_{t+1}] - \Theta W_{m,t}}{\mathbf{1}^\top \boldsymbol{\Omega}_{a,t}^{-1}\mathbf{1}}.$$

Consequently,

$$\tilde{r}_{m,t+1} = \frac{(E_{a,t}[\tilde{\mathbf{r}}_{t+1}] - r_{f,t}\mathbf{1})^\top \boldsymbol{\Omega}_{a,t}^{-1}\tilde{\mathbf{r}}_{t+1}}{(E_{a,t}[\tilde{\mathbf{r}}_{t+1}] - r_{f,t}\mathbf{1})^\top \boldsymbol{\Omega}_{a,t}^{-1}\mathbf{1}}$$

and

$$\boldsymbol{\beta}_{a,t} = \frac{(E_{a,t}[\tilde{\mathbf{r}}_{t+1}] - r_{f,t}\mathbf{1})^\top \boldsymbol{\Omega}_{a,t}^{-1}\mathbf{1}}{(E_{a,t}[\tilde{\mathbf{r}}_{t+1}] - r_{f,t}\mathbf{1})^\top \boldsymbol{\Omega}_{a,t}^{-1}(E_{a,t}[\tilde{\mathbf{r}}_{t+1}] - r_{f,t}\mathbf{1})}(E_{a,t}[\tilde{\mathbf{r}}_{t+1}] - r_{f,t}\mathbf{1}).$$

Note that the "aggregate" betas are time varying due to time varying beliefs about both the first and second moments of the returns distribution.

4 Discussion

Unlike the traditional paradigm of the representative agent and rational expectations, recent literature has directed a great deal of attention to a new paradigm of heterogeneity and bounded rationality. The new paradigm provides a platform for analysing the complicated market behaviour that comes from the interaction of heterogeneous, boundedly rational and adaptive agents and for explaining empirical anomalies which are a challenge for the traditional paradigm. It becomes clear that heterogeneity and bounded rationality play very important roles in our understanding of economic behaviour, in particular, their impact on the financial market. It is widely recognized that heterogeneity can have a significant impact on asset pricing. As one of the fundamental asset pricing equilibrium models, the CAPM plays a very important role in modern finance and economics. However, the framework of the traditional paradigm makes it difficult to examine the impact of heterogeneity and bounded rationality on asset pricing. This paper provides a framework for the analysis of CAPM within the new paradigm.

The main obstacle in dealing with heterogeneity is the complexity and heavy notation involved when the number of assets and the dimension of the heterogeneity increase. Within the mean-variance framework with heterogeneous beliefs, this paper overcomes this obstacle by constructing a consensus belief explicitly in order to characterize the market aggregation of the heterogeneous beliefs. Based on the consensus belief, we are able to set up a general framework for the CAPM to incorporate heterogeneous beliefs. We also extend the framework to a repeated one-period dynamical market fraction model. Within this framework, we are able to characterize exactly the relationships between market belief in equilibrium and heterogeneous beliefs, between the market risk premium of each risky asset and its beta coefficient, and derive the dynamics of beta coefficients and market equilibrium prices.

The framework provided in this paper can be used to examine the impact of various types of heterogeneity and bounded rationality on market prices and risk. For example, we may use the framework to explore the following questions: how do the optimistic or pessimistic views of agents and their confidence about their views influence the risk-free rate, equity premium and market price of the risk? which belief or investment strategy will have significant impact on the market equilibrium price? Recent HAMs literature that considers portfolios of one riskless asset and one risky asset demonstrates that bounded rational behaviour of heterogeneous agents can cause the market to be more complicated and less efficient than the standard paradigm allows for, generating many of the stylized facts and observed market anomalies. Within the framework of the dynamic CAPM with multiple risky assets, we can examine if the traditional diversification effect still holds. We can also study how learning and adaptive behaviour of heterogeneous agents contribute to the survivability of agents and market volatility. In particular, it would be interesting to know if the framework for the dynamic CAPM can be used to explain empirical evidence on the time variation of beta, which measures the time varying risk of risky assets. We believe that the framework established in this paper can be used to tackle such questions and issues, all of which we leave to future research.

Appendix

Proof of Proposition 1. With the optimal portfolio $\mathbf{w}_i$ defined by (5), we sum (5) across i and obtain

$$\sum_{i=1}^{I} W_0^i \mathbf{w}_i = \sum_{i=1}^{I} \theta_i^{-1} \boldsymbol{\Omega}_i^{-1} \left(E_i[\tilde{\mathbf{r}}] - r_f \mathbf{1} \right). \tag{32}$$

In market equilibrium, it follows from (32) that the proportions of the market wealth invested in the risky assets are given by

$$\mathbf{w}_a = \frac{1}{W_{m0}} \sum_{i=1}^{I} W_0^i \mathbf{w}_i = \frac{1}{W_{m0}} \sum_{i=1}^{I} \theta_i^{-1} \boldsymbol{\Omega}_i^{-1} \left(E_i[\tilde{\mathbf{r}}] - r_f \mathbf{1} \right). \tag{33}$$

Using the "consensus" belief about the variance and covariance matrix of returns, $\boldsymbol{\Omega}_a$, defined in (9) of Proposition 1, we have

$$\boldsymbol{\Omega}_a^{-1} = \Theta \sum_{i=1}^{I} \theta_i^{-1} \boldsymbol{\Omega}_i^{-1}, \tag{34}$$

where we recall that $\Theta := \left(\sum_{i=1}^{I} \theta_i^{-1} \right)^{-1}$. Then it follows from (33), (34) and the "consensus" belief about the market aggregate return, $E_a[\tilde{\mathbf{r}}]$, defined in (10) of Proposition 1 that

$$\begin{aligned}
\mathbf{w}_a &= \frac{1}{W_{m0}} \left(\sum_{i=1}^{I} \theta_i^{-1} \boldsymbol{\Omega}_i^{-1} E_i[\tilde{\mathbf{r}}] - \Theta^{-1} \boldsymbol{\Omega}_a^{-1} r_f \mathbf{1} \right) \\
&= \frac{1}{\Theta W_{m0}} \boldsymbol{\Omega}_a^{-1} \left(\Theta \boldsymbol{\Omega}_a \sum_{i=1}^{I} \theta_i^{-1} \boldsymbol{\Omega}_i^{-1} E_i[\tilde{\mathbf{r}}] - r_f \mathbf{1} \right) \\
&= \frac{1}{\Theta W_{m0}} \boldsymbol{\Omega}_a^{-1} \left(E_a[\tilde{\mathbf{r}}] - r_f \mathbf{1} \right),
\end{aligned}$$

from which

$$\boldsymbol{\Omega}_a \mathbf{w}_a = \frac{1}{\Theta W_{m0}} \left(E_a[\tilde{\mathbf{r}}] - r_f \mathbf{1} \right). \tag{35}$$

Then, with the consensus belief, the variance of the market return $\sigma_{a,m}^2 = \mathbf{w}_a^\top \boldsymbol{\Omega}_a \mathbf{w}_a$ is given by

$$\sigma_{a,m}^2 = \frac{1}{\Theta W_{m0}} \mathbf{w}_a^\top \left(E_a[\tilde{\mathbf{r}}] - r_f \mathbf{1} \right), \tag{36}$$

and from (8) the expected market return is given by

$$E_a(\tilde{r}_m) = r_f + \mathbf{w}_a^\top \left(E_a[\tilde{\mathbf{r}}] - r_f \mathbf{1} \right). \tag{37}$$

Both (37) and (36) imply that

$$E_a(\tilde{r}_m) - r_f = \Theta W_{m0} \sigma_{a,m}^2 > 0, \tag{38}$$

that is, the aggregate expected market risk premium is proportional to the aggregate relative risk aversion of the economy and the market risk.

It follows from (35) and (38) that

$$E_a[\tilde{\mathbf{r}}] - r_f\mathbf{1} = \frac{E_a(\tilde{r}_m) - r_f}{\sigma^2_{a,m}}\boldsymbol{\Omega}_a\mathbf{w}_a. \tag{39}$$

The entries of $\boldsymbol{\Omega}_a\mathbf{w}_a$ represent the aggregate covariances between the return on each risky asset and the return on the aggregate market wealth,

$$\boldsymbol{\Omega}_a\mathbf{w}_a = [\sigma_{a,jm}], \qquad \sigma_{a,jm} := Cov_a(\tilde{r}_j, \tilde{r}_m), \qquad j = 1, 2, \ldots, N$$

so that (39) can be rewritten componentwise as

$$E_a(\tilde{r}_j) - r_f = \frac{\sigma_{a,jm}}{\sigma^2_{a,m}}[E_a(\tilde{r}_m) - r_f], \qquad j = 1, 2, \ldots, N, \tag{40}$$

where $\sigma_{a,jm}/\sigma^2_{a,m} = \beta_{a,j}$ represents the aggregate beta coefficient of the j-th risky asset. Equation (40) is the traditional CAPM relation generalized to the case of heterogeneous beliefs. The vector $\boldsymbol{\beta}_a := [\beta_{a,1}, \beta_{a,2}, \ldots, \beta_{a,N}]^\top$ of the aggregate beta coefficients in (39) is thus given by

$$\boldsymbol{\beta}_a = \frac{1}{\sigma^2_{a,m}}\boldsymbol{\Omega}_a\mathbf{w}_a. \tag{41}$$

Acknowledgements We are grateful to the participants at the WEHIA 2006 (Bologna), the COMPLEXITY 2006 (Aix-en-Provence), and MDEF08 (Urbino) for helpful comments and suggestions. Financial support for Chiarella and He from the Australian Research Council (ARC) under Discovery Grant (DP0773776) is gratefully acknowledged. Dieci acknowledges support from MIUR under the project PRIN-2004137559.

References

Abel, A. (1989). *Asset prices under heterogeneous beliefs: Implications for the equity premium* (working paper 09-89). Rodney L. White Center for Financial Research.

Abel, A. (2002). An exploration of the effects of pessimism and doubt on asset returns. *Journal of Economic Dynamics and Control, 26*, 1075–1092.

Alfarano, S., Lux, T., & Wagner, F. (2005). Estimation of agent-based models: The case of an asymmetric herding model. *Computational Economics, 26*, 19–49.

Basak, S. (2000). A model of dynamic equilibrium asset pricing with heterogeneous beliefs and extraneous beliefs. *Journal of Economic Dynamics and Control, 24*, 63–95.

Böhm, V.,& Chiarella, C. (2005) . Mean variance preferences, expectations formation, and the dynamics of random asset prices. *Mathematical Finance, 15*, 61–97.

Böhm,V., & Wenzelburger, J. (2005). On the performance of efficient portfolios. *Journal of Economic Dynamics and Control, 29*, 721–740.

Brock, H., & Hommes, C. (1997). A rational route to randomness. *Econometrica, 65*, 1059–1095.

Brock, H., & Hommes, C. (1998) . Heterogeneous beliefs and routes to chaos in a simple asset pricing model. *Journal of Economic Dynamics and Control, 22*, 1235–1274.

Calvet, L., Grandmont, J.-M., & Lemaire, I. (2004). *Aggregation of heterogeneous beliefs and asset pricing in complete financial markets* (working paper 2004-12). CREST.
Cecchetti, S., Lam, P., & Mark, N. (2000). Asset pricing with distorted beliefs: Are equity returns too good to be true? *American Economic Review, 90*, 787–805.
Chan, L., Karceski, J., & Lakonishok, J. (1999). On portfolio optimization: Forecasting covariance and choosing the risk model. *The Review of Financial Studies, 12*, 937–974.
Chiarella, C., & He, X. (2001). Asset price and wealth dynamics under heterogeneous expectations. *Quantitative Finance, 1*, 509–526.
Chiarella, C., & He, X. (2002). Heterogeneous beliefs, risk and learning in a simple asset pricing model. *Computational Economics, 19*, 95–132.
Chiarella, C., Dieci, R., & Gardini, L. (2005). The dynamic interaction of speculation and diversification. *Applied Mathematical Finance, 12(1)*, 17–52.
Chiarella, C., Dieci, R., & He, X. (2006a). *Aggregation of heterogeneous beliefs and asset pricing theory: A mean-variance analysis* (Quantitative Finance Research Centre Working paper No. 186). University of Technology, Sydney.
Chiarella, C., He, X., & Hommes, C. (2006b). A dynamic analysis of moving average rules. *Journal of Economic Dynamics and Control, 30*, 1729–1753.
Chiarella, C., Dieci, R., & He, X. (2007). Heterogeneous expectations and speculative behaviour in a dynamic multi-asset framework. *Journal of Economic Behavior and Organization, 62*, 402–427.
Chiarella, C., Dieci, R., & He, X. (2009). Heterogeneity, market mechanisms and asset price dynamics. In T. Hens & K. R. Schenk-Hoppe (Eds.), *Handbook of financial markets: Dynamics and evolution* (pp. 277–344). Amsterdam: Elsevier.
Detemple, J., & Murthy, S. (1994). Intertemporal asset pricing with heterogeneous beliefs. *Journal of Economic Theory, 62*, 294–320.
Farmer, J., Gillemot, L., Lillo, F., Mike, S., & Sen, A. (2004). What really causes large price changes. *Quantitative Finance, 4*, 383–397.
Gaunersdorfer, A., & Hommes, C. (2007). A nonlinear structural model for volatility clustering. In G. Teyssiere & A. Kirman (Eds.), *Long memory in economics* (pp. 265–288). Berlin: Springer.
He, X., & Li, Y. (2007). Power law behaviour, heterogeneity, and trend chasing. *Journal of Economic Dynamics and Control, 31*, 3396–3426.
He, X., & Li, Y. (2008). Heterogeneity, convergence and autocorrelations. *Quantitative Finance, 8*, 58–79.
He, X., & Shi, L. (2009). *Portfolio analysis and Zero-Beta CAPM with Heterogeneous beliefs*. (Quantitative Finance Research Centre Working paper No. 244). University of Technology, Sydney.
Hommes, C. (2006). Heterogeneous agent models in economics and finance. In L. Tesfatsion & K. L. Judd (Eds.), *Agent-based computational economics, Handbook of computational economics* (Vol. 2, pp. 1109–1186). Amsterdam: North-Holland.
Huang, C.-F., & Litzenberger, R. (1988). *Foundations for financial economics*. Amsterdam: Elsevier.
Johnson, T. (2004). Forecast dispersion and the cross section of expected returns. *Journal of Finance, 59*, 1957–1978.
LeBaron, B. (2006). Agent-based computational finance. In L. Tesfatsion & K. L. Judd (Eds.), *Agent-based computational economics, Handbook of computational economics* (Vol. 2, pp. 1187–1233). Amsterdam: North-Holland.
Lintner, J. (1965). The valuation of risk assets and the selection of risky investments in stock portfolios and capital budgets. *Review of Economic Studies, 47*, 13–37.
Lintner, J. (1969). The aggregation of investor's diverse judgements and preferences in purely competitive security markets. *Journal of Financial and Quantitative Analysis, 4*, 347–400.
Lux, T. (2004). Financial power laws: Empirical evidence, models and mechanisms. In C. Cioffi (Ed.), *Power laws in the social sciences: Discovering complexity and non-equilibrium in the social universe*. Cambridge: Cambridge University Press.

Mayshar, J. (1982). On divergence of opinion and imperfections in capital markets. *American Economic Review, 73*, 114–128.

Miller, E. (1977). Risk, uncertainty, and divergence of opinion. *Journal of Finance, 32*, 1151–1168.

Mossin, J. (1966). Equilibrium in a capital asset market. *Econometrica, 35*, 768–783.

Sharpe, W. (1964). Capital asset prices: A theory of market equilibrium under conditions of risk. *Journal of Finance, 19*, 425–442.

Varian, H. (1985). Divergence of opinion in complete markets. *Journal of Finance, 40*, 309–317.

Westerhoff, F. (2004). Multiasset market dynamics. *Macroeconomic Dynamics, 8*, 591–616.

Westerhoff, F., & Dieci, R. (2006). The effectiveness of Keynes–Tobin transaction taxes when heterogeneous agents can trade in different markets: A behavioral finance approach. *Journal of Economic Dynamics and Control, 30*, 293–322.

Williams, J. (1977). Capital asset prices with heterogeneous beliefs. *Journal of Financial Economics, 5*, 219–239.

Zapatero, F. (1998). Effects of financial innovations on market volatility when beliefs are heterogeneous. *Journal of Economic Dynamics and Control, 22*, 597–626.

Optimal Monetary Policy for Commercial Banks Involving Lending Rate Settings and Default Rates*

Simone Casellina and Mariacristina Uberti

1 Introduction

In the modern industrial economies, the interest rates dynamics are influenced by the decisions of the Money Authority. With these decisions the Central Bank of a country wields a direct control on the trend of short-term interest rates and, through this way, it is in a position to influence indirectly the long-term interest rates. Typically the Money Authority resorts to this possibility with the aim to limit the fluctuations of the main economic variables. So that an immediate connection is established between the trend of the interest rates and the macro-economic variables with respect to the Central Bank, that is institutionally to have an influence.

Within this framework, for example, the monetary policy rule introduced by Taylor (1993) provides the short-term interest rate as a function of the inflation rate and a measure of the business cycle. This approach is supported by the assumptions that the main objectives of the Central Bank are the control of the raise in prices as well as the real economy fluctuations. The efficaciousness of this relation – that explains in a simple way the behaviour of the short-term rates – has fostered empirical as well as theoretical studies.

One of the more interesting aspects is the analysis of the links between the short-term and the long-term rates. In fact, the Central Bank can influence the behaviour of the short-term rate but the decisions involving investments or financing in durable goods (for example, the purchase of a house) depend on the long-term rates. Therefore, it is important to study how the monetary policy decisions influence the whole term structure of interest rates.

* An earlier version of this paper was presented also at the Bachelier Finance Society 2008 – Fifth World Congress, London, July 2008. The views expressed in the article are those of the authors and do not involve the responsibility of the Bank.

M. Uberti (✉)
Università degli Studi di Torino, Dipartimento di Statistica e Matematica Applicata, Corso Unione Sovietica 218/bis, 10134 Torino, Italy,
e-mail: uberti@econ.unito.it

G.I. Bischi et al. (eds.), *Nonlinear Dynamics in Economics, Finance and the Social Sciences*, DOI 10.1007/978-3-642-04023-8_20,

Recently, Casellina and Uberti (2008) have taken into account a model involved by Svensson (1997) to show that the Taylor rule is the optimal policy rule of a dynamic programming problem where the preferences of the Central Bank are represented by a suitable quadratic intertemporal loss function. In particular, they extend Svensson's model (Svensson, 1997) including the long-term rates through a relation introduced by Campbell and Shiller (1991). From an economic viewpoint, thanks to this extension, it is possible to describe the case where the Central Bank's control variable does not influence straight on the real economy. In Casellina and Uberti's model (Casellina and Uberti, 2008), the observed output gap[1] y_t is assumed to depend on the dynamics of long-term interest rate I_t and this last variable depends on the observed short-term interest rate r_t as well as the expectations of the future short-term interest rate $E_t r_{t+1}$. For the proposed dynamic optimization program, the optimal policy function for the Central Bank turns out to be

$$\begin{aligned} r_t &= r + \pi + \phi_1(\pi_{t-1} - \pi) + \phi_2 y_{t-1} + \phi_3 e_{t-1} + \phi_4 (I_{t-1} - \pi_{t-1}) \\ &\quad + \varpi_1 \varepsilon_{1,t} + \varpi_2 \varepsilon_{2,t} + \varpi_3 \varepsilon_{3,t} + \varpi_4 \varepsilon_{4,t}, \end{aligned} \tag{1}$$

that is a generalization of Taylor's rule where π is the inflation target that for the Central Bank guarantees the best development of the economic system; r is the equilibrium real interest rate; π_t and y_t denote inflation and output gap at date t, respectively; I_t denotes the long-term nominal interest rate and e_t is the exchange rate;[2] $\varepsilon_{j,t}$, $j = 1, \ldots, 4$, are the deviations of the corresponding variables from the equilibrium values where, since

$$\varepsilon_{4,t} = I_t - \frac{1}{2}(r_t + E_t r_{t+1}).$$

ϖ_4 represents the optimal response with respect to a shock acting on long-term interest rate.

Through this relation (1), it is possible to point out how the Central Bank can wield an effect on the economic system. Moreover, this model suitably allows us to study the role of expectations on the inflation rate and the spread between short-term rate and long-term rate.

On the other hand, from an empirical viewpoint, there are many studies where the effects of the long-term interest rates on the policy rule are investigated.

In particular, Christensen and Nielsen (2005) estimate a VECM with one cointegrating relation including Federal funds rate, long-term interest rate, unemployment and inflation and they find a positive, highly significative, relation between short and long-term interest rates. Gerlach-Kristen (2003) points out that the traditional Taylor Rule is unstable when estimated on euro area data and forecast poorly out of sample. They present an alternative reaction function finding a significant role for

[1] The output gap is the difference between the observed output and the potential output at date t.

[2] It is the amount of domestic currency for one unit of foreign currency.

the long rate and they argue this can be explained by seeing long-term interest rate as a proxy for the public's perception of long run inflation.

Casellina and Uberti (2008) stress that the proposed model gives a suitable description of the links among short-term interest rate – the control variable of the Central Bank – and all, except one, of the other variables of the economic system considered. In fact, the long-term interest rate enters in the policy function with a negative sign and this result contrasts with commonly used assumptions, as in McCallum (2005), or with empirical findings, as in Gerlach-Kristen (2003).

In this paper, starting from the above results, the linkages between the dynamics of short-term and long-term interest rates are analyzed from a new perspective: through the determination of the lending interest rates, the Commercial Bank reactions to the business cycle perturbation are examined closely, according to the monetary policy decisions of Central Bank where the short-term rates are fixed through Taylor type rules.

Concerning this, it is significant to look at the observed data series relating to Italy. In particular, as a reference rate we consider the EurIRS (Euro Interest Rate Swap) rate, daily estimated by the European Banking Federation as a mean rate to which the main European banking-houses draw up swap as coverage of interest risk; moreover, the EurIRS rate is frequently used as basic rate to calculate fixed-rates, for example, those of loans: for the validity period of the loan, a fixed-rate can be offered as the EurIRS added to a variable spread depending on the bank, from 0.5% to 3%. In Fig. 1, the charged mean rate for the mortgage loans is compared with the EurIRS up to 25 years. It can generally be noted that the dynamic of the banking lending interest rate follows that of the reference EurIRS rate; the correlation of the changes between the two data series is equal to 0.75. Nevertheless, the observed spread is not a constant (see Fig. 2).

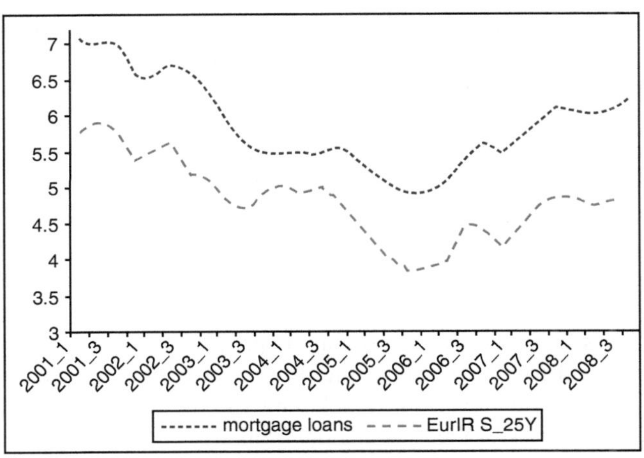

Fig. 1 Comparison between the mean lending rate and the EurIRS rate

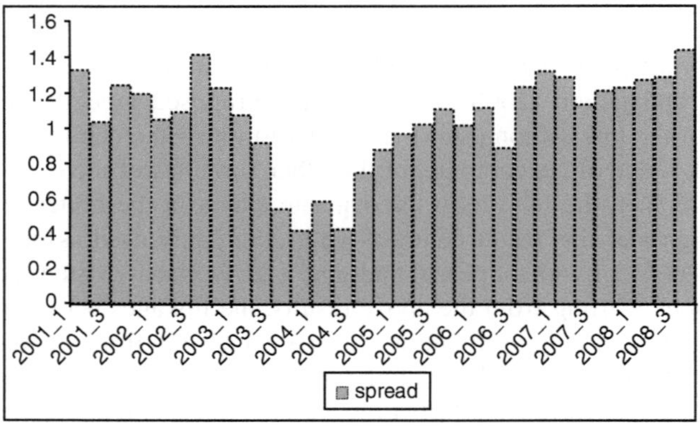

Fig. 2 The observed spread between the lending rate and the EurIRS rate

As Stiglitz and Greenwald (2003) notice, a monetary policy theory that doesn't take into account the rule of the banking-houses is destined to fail. In particular,

> a central function of banks is to determine who is likely to default and, in doing so, banks determine the supply of loans. In recent years, the increasing interest in financial stability as an autonomous policy target has encouraged the analyses of the linkages between the macroeconomic conditions and the banks behaviour. The goal is to assess to what extent macroeconomy affects banks' performance (cyclicality) and whether, in turn, banks actions affects the macroeconomy, reinforcing cyclical fluctuations (procyclicality).

However, the change in the behaviour of the banks, through the business cycle, is not explicitly modelled. Few studies assess the role that macroeconomy uncertainty plays in determining the behaviour of the banks. Therefore, as Marcucci and Quagliariello (2008) notice, the current state of the art is unsatisfactory for two main reasons: (1) no attempt is made in order to model how the banks management varies in changing macroeconomic environments; (2) the effect of the uncertainty regarding future macroeconomic conditions is typically neglected.

Moreover, as regards the areas of recent research in chaotic economic dynamics, we refer to Rosser (1999, 2000) for a wide review of different approaches also involving Policy and Institutions where the motivation of his researches "is the hope that not only deeper understanding of the nature of dynamic processes will be achieved, but that improvements in forecasting in actual markets and economies will be achieved".

In the model proposed in this paper, we refer to a Commercial Bank that fixes the level of rates on investments applying a spread on short-term rate where this last rate is determined by the policy rule of Central Bank. Besides, the long-term interest rate dynamics are determined including also the business cycle effect on the investment portfolio risk of the Commercial Bank.

The advantage of this approach is to include into the monetary policy model the default rate[3] that is not usually involved in the studies traceable in literature, though it has a remarkable significance for the Commercial Banks (as a reference, see Stiglitz and Greenwald, 2003 and the references quoted therein).

The aim is to give a theoretical support to the empirically observed relation between the control variable of the Commercial Bank and the long-term interest rates given by the market (Christensen and Nielsen, 2005; Gerlach-Kristen, 2003). In fact the optimal behaviour of the Commercial Bank is derived as the optimal solution of an original proposed quadratic stochastic dynamic optimization program which represents the minimization of an intertemporal loss function of the Commercial Bank. Moreover, this optimal solution can be looked at as an extension of Taylor type rules.

The proposed model is calibrated by the VAR approach with respect to the Italian quarterly data series from 1990 to 2007. This data base is remarkable because the Bank of Italy collected the data in a systematic way and a very long period of time is covered. The equilibrium as well as the optimal paths of the control variable are numerically studied (Dennis, 2004). The analysis of these dynamics with respect to temporary shocks on the variables of the proposed program gives significant results, in particular, as regards an explanation of the observed changes in the yield curve slope during the cyclical phase of the economy. The variations of the spread between long-term and short-term interest rates are explained through the behaviour of an institution (the Commercial Bank) that, in taking its decisions, takes into account the level of the default rate.

The paper is organized as follows. The perspective of the Commercial Bank is outlined in Sect. 2: the notation and the basic assumptions are introduced in Sect. 2.1; the demand of firms and the decisions of the Commercial Bank are developed in Sects. 2.2 and 2.3, respectively; the proposed intertemporal model is outlined in Sect. 3 and the optimal solution is attained in Sect. 3.1; in Sect. 4, the data used in the implementation of the model are described as well as the simulations based on the model and the comparative results are gathered; Sect. 5 presents the conclusions.

2 The Perspective of the Commercial Bank

2.1 Notation and Assumptions

Let L_t be the total amount of the investments of the bank towards one client at time t (loans, financing, ...). We assume that the Commercial Bank[4] re-finances at any date to pay the short-term interest rate r_t that is directly controlled by the

[3] In this paper the default rate is the ratio between doubtful loan and total loan.

[4] From now on, we refer to Bank for Commercial Bank.

Central Bank, where r is the equilibrium value, and the Bank lends funds at the lending interest rate (i.e. long-term interest rate) $i_t = r_t + d_t$.

In the absence of default, the total profit $P_t = (i_t - r_t)L_t$ of the Bank depends on the applied spread and the total amount of the investments. We can assume that the Bank wishes to maintain the applied spread d_t at a fixed level d and this assumption is motivated by two reasons. First, with the underlying hypothesis of imperfect competition or monopoly, a variation in the spread generates two conflicting effects: on the mean yield of every investment and on the numbers of clients of the Bank. Then the level d can be considered as the level that maximizes the total yield before doubtful loans. On the other hand, if the default rate is introduced in the analysis and it is assumed that it also depends on the interest rates, we note that the Bank cannot indefinitely increase the spread because this rises the default risk (see Stiglitz and Weiss, 1981 for a well known analysis on this adverse selection process). As we will see in the following, if a constant target is fixed for the applied spread, the equilibrium level of the default risk can be determined by suitable links among doubtful loans, investments and interest rate. Then, the target d for the spread can be considered as the value that maximizes the preferences of the Bank with respect to the yield and the risk.

As regards the default rate, if D_t is the financing total amount that at any time goes to default (i.e. the doubtful loan) it expresses a loss for the Bank[5] and therefore the default rate can be so defined

$$DR_t = \frac{D_t}{L_t}. \tag{2}$$

The relation (2) can be regarded as a risk measure of the Bank's credit portfolio and moreover, since the relation among the default rate, the doubtful loans and the investments is non-linear, also the quantity

$$\log DR_t = \log D_t - \log L_t \tag{3}$$

can be considered as an alternative risk measure.

In this paper, we assume that the investment and the financing depend on some macroeconomic variables at any date t, in particular, on the interest rate i_t, the inflation π_t and the output gap y_t as a measure of the business cycle.

2.2 *Firms: The Demand Side*

The investments depend on the decisions of firms and such decisions take into account the evolution of the macroeconomic conditions. Let $Ex_{t+1} = E\left(x_{t+1}|\Omega_t\right)$

[5] We assume a Loss Given Default of 100%.

be the expectation of x_{t+1} that a rational decision maker estimates, given the information Ω_t available at date t.

We assume that the investments depend on the expected values of the output gap Ey_{t+1} as well as of the real interest rate $Ei_{t+1} - E\pi_{t+1}$

$$\begin{aligned} \ln L_t &= \widetilde{L_t} = b_0 t + b_1 E y_{t+1} + b_2 \left(E i_{t+1} - E\pi_{t+1}\right), \\ b_0, b_1 &> 0; \qquad b_2 < 0, \end{aligned}$$

where b_0 is the equilibrium growth rate of the investments.

For the doubtful loans, we assume that they are functions of the output gap[6] y_t and of the real interest rate $i_t - \pi_t$ so that the following relation holds

$$\begin{aligned} \ln D_t &= \widetilde{D_t} = a_0 t + a_1 y_{t-1} + a_2 \left(i_{t-1} - \pi_{t-1}\right), \\ a_0, a_2 &> 0; \qquad a_1 < 0. \end{aligned}$$

The quantity $a_1 y_{t-1} + a_2 \left(i_{t-1} - \pi_{t-1}\right)$ describes the deviation from the trend $a_0 t$. A cyclical phase of slowdown ($y_t < 0$) will induce an increase in doubtful loans. There will be the same effect as a consequence of an increase in the interest rate.

In equilibrium we have $y_t = 0$ and $i_t - \pi_t = r + d - \pi$ and then $\widetilde{D_t} = a_0 t + a_2 (r + d - \pi)$. As regards the growth rate we get that, in equilibrium conditions, $\frac{\Delta D_t}{D_{t-1}} \simeq a_0$ and since $\frac{\partial \widetilde{D_t}}{\partial y_t} = \frac{\partial D_t}{\partial y_t} \frac{1}{D_t} = a_0$ this last parameter is the percentage variation of D_t induced by a variation of y_t.

With regard to the default rate, we assume that the Bank is not in a position to distinguish the clients that will pay off the debts from the ones that will not pay them (if it would be the case, the bank never should work at a loss and the default rate would be null); moreover, the Bank knows that a share of the given credits will not be refunded and that this share is influenced by the behaviour of the real economy and by the level of interest rates therefore, according to (3), the following relation holds

$$\ln DR_t = \widetilde{DR_t} = a_1 y_{t-1} + a_2 \left(i_{t-1} - \pi_{t-1}\right) - b_1 E y_{t+1} - b_2 \left(E i_{t+1} - E\pi_{t+1}\right). \tag{4}$$

In order that the default rate doesn't burst (or it turns negative), we have to assume $a_0 = b_0$. The equilibrium level is $\widetilde{DR} = (a_2 - b_2)(r + d)$.

Relation (4) points out the trade-off between a bigger profit that the Bank can obtain increasing the spread and the higher risk that this implies. In fact, if the Bank increases the interest rate (by means of an increase of the spread) then also the default rate raises

$$\frac{\partial \widetilde{DR_t}}{\partial d_t} = a_2 - b_2 > 0;$$

[6] The output gap is the deviation of the real economy from a situation of full employment of the production resources.

for example, if $a_2 = 0.001$ and $b_2 = -0.001$ a rise of 10 basis points in the spread implies an increase of the default rate of 2%.

Therefore, we can assume that an adverse risk institution wishes to maintain the default rate at a fixed level that will be equal to the equilibrium value ($D = \widetilde{DR}$).

3 Commercial Bank

In the following we assume the Bank's preferences depend on the deviations from the spread target d and from the equilibrium value D of the default rate so that the objective function of the Bank can be described as

$$U_t = \lambda\,(d_t - d)^2 + (1-\lambda)\,(\log DR_t - D)^2\,, \tag{5}$$

where λ, $0 \leq \lambda \leq 1$, is a weight that expresses the relative importance of the two components that characterize the Bank's utility function. Referring to the above assumptions, the first component of (5) can be guessed as a relative to profit while the second one relative to risk. So, for increasing values of λ the Bank is less risk adverse.

The control variable d_t of the Bank must act along two directions: (1) the spread of the Bank must be maintained around to a fixed level d such that the net yield is preserved to a desired level and (2) the risk (i.e. the default rate) must be closed to the given value D consistent with the target on yield.

4 The Model

Following the above outlines, the model for the Bank can be formalized into this quadratic dynamic optimization program

$$\min_{d_t} \sum_{t=0}^{+\infty} \beta^t \left[\lambda\,(d_t - d)^2 + (1-\lambda)\,(\log DR_t - D)^2\right], \tag{6}$$

$$\begin{aligned}
\widetilde{DR}_t &= a_1 y_{t-1} + a_2\,(i_{t-1} - \pi_{t-1}) - b_1 E y_{t+1} - b_2\,(E i_{t+1} - E\pi_{t+1})\,,\\
y_t &= a_{21} y_{t-1} + a_{22}\,(i_{t-1} - \pi_{t-1} - r - d)\,,\\
\pi_t &= \pi + a_{31} y_{t-1} + a_{32}\,(E\pi_{t+1} - \pi)\,,\\
r_t &= r + \pi + a_{41} E y_{t+1} + a_{42}\,(E\pi_{t+1} - \pi) + a_{43}\,(E i_{t+1} - E\pi_{t+1} - r - d),\\
i_t &= r_t + d_t\,,\\
a_{22} &< 0; \qquad a_{21}, a_{31}, a_{32}, a_{41}, a_{42}, a_{43} > 0,
\end{aligned}$$

where $0 < \beta < 1$ is the intertemporal discount factor.

The constraints for the output gap y_t and π_t depict a closed economy (IS-LM curve and augmented Philips curve). The fourth constraint involves the dynamic of the short-term interest rate r_t as the optimal monetary policy rule (1) that minimize the Central Bank loss function of a dynamic optimization intertemporal program on infinite horizon (Casellina and Uberti, 2008); it has a similar structure to Taylor rule types, where the spreads from the equilibrium values are pointed out.

The deviation of the business cycle from the equilibrium level depend on the variations, in real terms, of the interest rate on investments. The rate of inflation strays from equilibrium π as a response of the trend of the real economy and of the expected inflation.

The level of the short rates r_t is determined from the Central Bank on the basis of a monetary policy rule: the short rates are increased as a response to an overheating of the economy or to the expectations of an increase of the inflation rate. In a full employment situation and in the absence of deviation from the inflation rate with respect to the target π, fixed to the Central Bank, the equilibrium level of the short rate is $r + \pi$, i.e. the inflation rate plus a spread. When $r_t > r + \pi$ this indicates that the Central Bank is reacting to an overheating of the real economy ($y_t > 0$) or to an inflationary shock ($\pi_t > \pi$). As suggested by Gerlach-Kristen (2003), the response function of the Central Bank depends also on the changes of the long-term rates.

4.1 *Optimal Solution*

The proposed model (6) is a quadratic intertemporal dynamic program and it is well known that this type of program is relatively tractable and that the corresponding value function is a quadratic form while the optimal policy function is linear (see, e.g. Montrucchio and Uberti, 2001 for a theoretical perspective). As regards the aim of this paper, the numerical approaches for the program (6) is suitable because it enables us to study the optimal paths for the control variable also with respect to temporary shocks on the variables of the economic system. Therefore, Dennis's algorithms (Dennis, 2004) are adapted to solve the program since they allow the constraints to be written in a structural form rather than in a state-space form (see also Casellina and Uberti, 2008).

The optimal behaviour for the Bank with the quadratic intertemporal utility function (5) as in the proposed program (6) turns out to be

$$d_t = d + \gamma_1 y_{t-1} + \gamma_2 \left(i_{t-1} - \pi_{t-1} - r - d\right),$$
$$\gamma_1 > 0; \qquad \gamma_2 < 0.$$

It is noteworthy that also for the Bank and not only for the Central Bank, the optimal solution can be looked at as an extension of Taylor type rules.

5 Data and Comparative Results

The proposed model is calibrated on the basis of the vector auto-regressive (VAR) approach with respect to the Italian quarterly data series from 1990 to 2007. Data regarding the macroeconomic variables comes from the national accounting system, data on loans and doubtful loans comes from the Base Informativa Pubblica (BIP) of the Bank of Italy. The Bank of Italy keeps a data base containing data on the bank loans over 75,000 euro and on all the doubtful loans.

The results we attain from the comparison between the dynamics obtained with the proposed model and the ones empirically observed are particularly interesting thanks to the peculiarity and the wealth of the data base of the Bank of Italy, since the data are collected in a systematic way and covering a very long period of time.

As regards the equilibrium, this is achieved with both $a_{43} < 0$ or $a_{43} > 0$ and there is an univocal effect on the optimal behaviour of the Bank: if $a_{43} < 0$ the parameters have absolute values less than the ones of the case $a_{43} > 0$.

Throughout a recession phase (negative output gap) the demand of funds decreases while it increases the quote of loans that go into the state of default (Fig. 3). Consequently, the default rate rises. Following the policy rule, the Monetary Authority reduces the short-term interest rate with the aim to whet the economy. Also the lending interest rate reduces but in a more considerable way (the spread decreases) trying to limit the default rate (Fig. 4).

It is also significant to analyze how the optimal paths of the spread (the control variable) can vary with respect to a temporary shock on the economy.

In Fig. 5 the Central Bank, following the policy rule, rises the short-term interest rate after a positive shock on real economy. Since the default rate decreases, the Bank increases the spread so the lending interest rate increases more than the short-term interest rate. This implies an increase of the slope of the yield curve. Figure 6

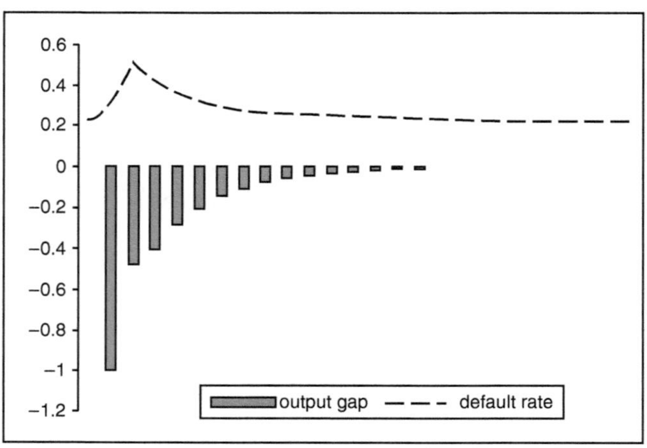

Fig. 3 Reactions of default rate and output gap

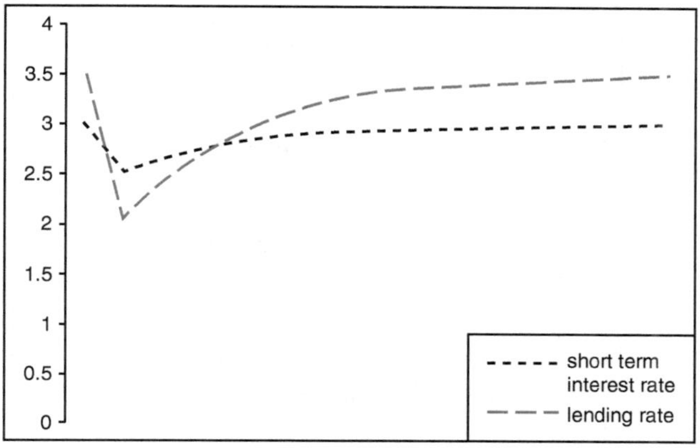

Fig. 4 Reactions of real short and lending interest rates

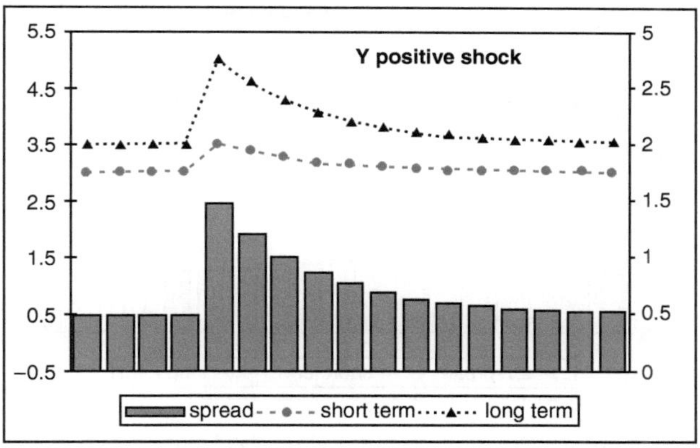

Fig. 5 Estimated optimal impulse response to a positive shock on y_t

shows the opposite situation: a negative shock on real economy leads up to become flat the yield curve, this appends because the Bank reduces the spread trying to reduce the default rate.

Figures 7 and 8 show the dynamic effect of a shock on the inflation rate. This kind of shock influences real economy through two ways: the first one modifies the real interest rate while the second one induces a variation of the short-term nominal interest rate through the policy rule of the Central Bank. A positive shock on the inflation rate reduces the real interest rate and this produces an incentive on real then, because the default rate decreases, the Bank increases the spread. In the opposite situation, a negative shock on the inflation rate induces a reduction of the spread through the increase in the default rate.

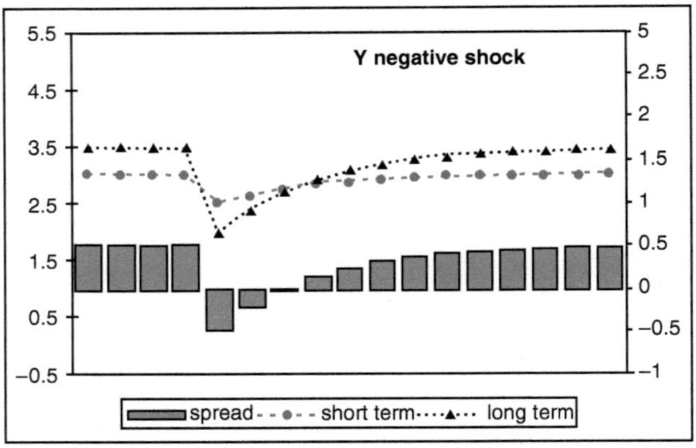

Fig. 6 Estimated optimal impulse response to a negative shock on y_t

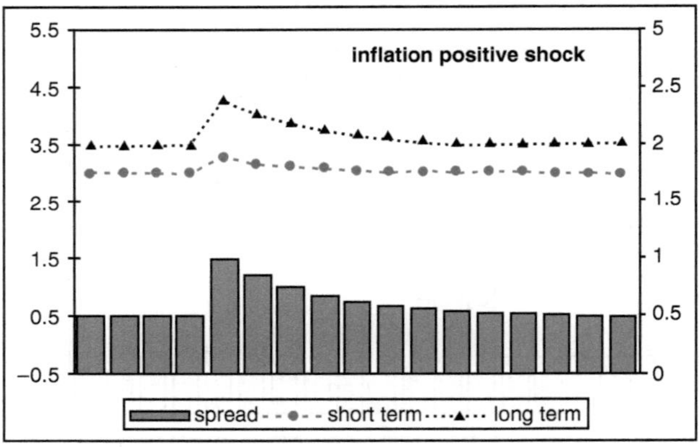

Fig. 7 Estimated optimal impulse response to a positive shock on π_t

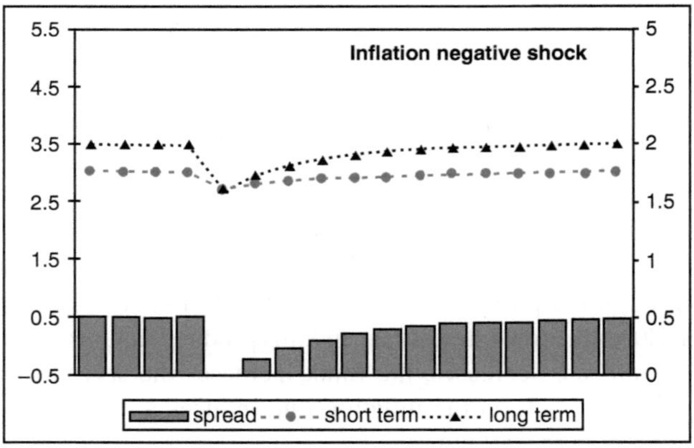

Fig. 8 Estimated optimal impulse response to a negative shock on π_t

6 Conclusions

In this paper the behaviour of the Commercial Bank is analyzed regarding the granting of a credit with respect to the changes of macroeconomic conditions.

Since the Central Bank does not directly control the credit market but can only try to influence it indirectly by controlling the short-term interest rates, this research focus attention on the transmission mechanism between the control variable of the Central Bank, i.e. the short-term interest rate, and the interest rate charges from Commercial Banks.

If it is assumed that at any time the Commercial Bank re-finances itself paying the short-term interest rate – the one controlled by the Central Bank – and it lends money at a higher interest rate, then the control variable of the Commercial Bank is the spread between the lending interest rate and the re-financing interest rate.

The aim of the Commercial Bank is to hold this spread close to a fixed target d that guarantees the maximization of Bank preferences with respect to the yield and the risk. An increase of the rates implies also an increase of the default probability, and for the Commercial Bank this produces a trade-off between the yield and the insolvency risk.

A shock in the real economy (slowing of economic growth) directly influences the variables involved into the objective function of the Commercial Bank: the demand of money and the default probability. A price shock has an indirect influence given by the variation of the re-financing rate.

In the model developed and analyzed in this paper, the Commercial Bank – which knows the adjustment mechanism of the system – must react to these shocks by changing the spread reconciling the objectives to maintain the profit at a given level and to control the risk. To study the dynamics of the involved variables, the problem is modelled as one of intertemporal dynamic programming whose optimal path solution is the optimal behaviour rule of the Commercial Bank.

The main result of this paper is that the proposed model offers an explanation of the observed changes in the yield curve slope during the cyclical phase of the economy. The variations of the spread between long-term and short-term interest rates are explained through the behaviour of an institution (the Commercial Bank) that, in taking its decisions, takes into account the level of the default rate.

Acknowledgements This research was partially supported by MIUR (Ministero dell'Istruzione, dell'Università e della Ricerca scientifica), Italy. The authors are grateful to an anonymous referee for his profitable comments and suggestions.

References

Campbell, J., & Shiller, R. (1991). Yield spreads and interest rate movements: A bird's eye view. *Review of Economic Studies, 58*, 495–514.

Casellina, S., & Uberti, M. (2008). Optimal monetary policy and long-term interest rate dynamics: Taylor rule extensions. *Computational Economics, 32*, 183–198.

Christensen, A. M., & Nielsen, H. B. (2005). *US monetary policy 1988–2004: An empirical analysis* (FRU Working Papers No. 2005/01). Department of Economics (formerly Institute of Economics), University of Copenhagen.

Dennis, R. (2004). Solving for optimal simple rules in rational expectations models. *Journal of Economic Dynamics and Control, 28*, 1635–1660.

Gerlach-Kristen, P. (2003). *Interest rate reaction functions and the TR in the Euro area* (ECB Working Paper series No. 258).

Marcucci, J., & Quagliariello, M. (2008). *Credit risk and business cycle over different regimes*, Banca d'Italia, Temi di discussione (Working Papers), 670.

McCallum, B. (2005). Monetary policy and the term structure of interest rates. *Economic Quarterly, 91*, 1–21.

Montrucchio, L., & Uberti, M. (2001). Quadratic dynamic programming. In *Proceedings of the workshop MDEF2000 (Dynamical Models for Economics and Finance)*, Urbino. Retrieved from http://www.econ.uniurb.it/bischi/MDEF2000.htm.

Rosser, J. B., Jr. (1999). On the complexities of complex economic dynamics. *Journal of Economic Perspectives, 13*, 169–192.

Rosser, J. B., Jr. (2000). *From catastrophe to chaos: A general theory of economic discontinuities*. Boston: Kluwer.

Stiglitz, J., & Greenwald, B. (2003). *Towards a new paradigm in monetary economics*. Cambridge: Cambridge University Press.

Stiglitz, J., & Weiss, A. (1981). Credit rationing in markets with imperfect information. *American Economic Review, 71*, 393–410.

Svensson, L. (1997). Inflation forecast targeting: Implementing and monitoring inflation targets. *European Economic Review, 41*, 1111–1146.

Taylor, J. B. (1993). Discretion versus policy rules in practice. *Carnegie-Rochester Conference Series on Public Policy, 39*, 195–214.

MIX
Papier aus verantwortungsvollen Quellen
Paper from responsible sources
FSC® C105338

If you have any concerns about our products,
you can contact us on
ProductSafety@springernature.com

In case Publisher is established outside the EU,
the EU authorized representative is:
Springer Nature Customer Service Center GmbH
Europaplatz 3, 69115 Heidelberg, Germany

Printed by Libri Plureos GmbH
in Hamburg, Germany